Lecture Notes in Control and Information Sciences

Edited by A. V. Balakrishnan and M. Thoma

Vol. 1: Distributed Parameter Systems: Modelling and Identification
Proceedings of the IFIP Working Conference, Rome, Italy, June 21–26, 1976
Edited by A. Ruberti
V, 458 pages. 1978

Vol. 2: New Trends in Systems Analysis
International Symposium, Versailles, December 13–17, 1976
Edited by A. Bensoussan and J. L. Lions
VII, 759 pages. 1977

Vol. 3: Differential Games and Applications
Proceedings of a Workshop, Enschede, Netherlands, March 16–25, 1977
Edited by P. Hagedorn, H. W. Knobloch, and G. J. Olsder
XII, 236 pages. 1977

Vol. 4: M. A. Crane, A. J. Lemoine
An Introduction to the Regenerative Method for Simulation Analysis
VII, 111 pages. 1977

Vol. 5: David J. Clements, Brian D. O. Anderson
Singular Optimal Control: The Linear Quadratic Problem
V, 93 pages. 1978

Vol. 6: Optimization Techniques
Proceedings of the 8th IFIP Conference on Optimization Techniques, Würzburg, September 5–9, 1977
Part 1
Edited by J. Stoer
XIII, 528 pages. 1978

Vol. 7: Optimization Techniques
Proceedings of the 8th IFIP Conference on Optimization Techniques, Würzburg, September 5–9, 1977
Part 2
Edited by J. Stoer
XIII, 512 pages. 1978

Vol. 8: R. F. Curtain, A. J. Pritchard
Infinite Dimensional Linear Systems Theory
VII, 298 pages. 1978

Vol. 9: Y. M. El-Fattah, C. Foulard
Learning Systems:
Decision, Simulation, and Control
VII, 119 pages. 1978

Vol. 10: J. M. Maciejowski
The Modelling of Systems with Small Observation Sets
VII, 241 pages. 1978

Vol. 11: Y. Sewaragi, T. Soeda, S. Omatu
Modelling, Estimation, and Their Applications for Distributed Parameter Systems
VI, 269 pages. 1978

Vol. 12: I. Postlethwaite, A. G. J. McFarlane
A Complex Variable Approach to the Analysis of Linear Multivariable Feedback Systems
IV, 177 pages. 1979

Vol. 13: E. D. Sontag
Polynomial Response Maps
VIII, 168 pages. 1979

Vol. 14: International Symposium on Systems Optimization and Analysis
Rocquentcourt, December 11–13, 1978;
IRIA LABORIA
Edited by A. Bensoussan and J. Lions
VIII, 332 pages. 1979

Vol. 15: Semi-Infinite Programming
Proceedings of a Workshop, Bad Honnef, August 30 – September 1, 1978
V, 180 pages. 1979

Vol. 16: Stochastic Control Theory and Stochastic Differential Systems
Proceedings of a Workshop of the „Sonderforschungsbereich 72 der Deutschen Forschungsgemeinschaft an der Universität Bonn"
which took place in January 1979 at Bad Honnef
VIII, 615 pages. 1979

Vol. 17: O. I. Franksen, P. Falster, F. J. Evans
Qualitative Aspects of Large Scale Systems
Developing Design Rules Using APL
XII, 119 pages. 1979

Vol. 18: Modelling and Optimization of Complex Systems
Proceedings of the IFIP-TC 7 Working Conference
Novosibirsk, USSR, 3–9 July, 1978
Edited by G. I. Marchuk
VI, 293 pages. 1979

Vol. 19: Global and Large Scale System Models
Proceedings of the Center for Advanced Studies (CAS)
International Summer Seminar
Dubrovnik, Yugoslavia, August 21–26, 1978
Edited by B. Lazarević
VIII, 232 pages. 1979

Vol. 20: B. Egardt
Stability of Adaptive Controllers
V, 158 pages. 1979

Vol. 21: Martin B. Zarrop
Optimal Experiment Design for Dynamic System Identification
X, 197 pages. 1979

For further listing of published volumes please turn over to inside of back cover.

Lecture Notes in
Control and
Information Sciences

Edited by A.V. Balakrishnan and M. Thoma

45

M. Arató

Linear Stochastic Systems with Constant Coefficients

A Statistical Approach

Springer-Verlag
Berlin Heidelberg GmbH 1982

Author
M. Arató
Budapest 1119
Fehérvári u 129
Hungary

ISBN 978-3-540-12090-2 ISBN 978-3-540-39529-4 (eBook)
DOI 10.1007/978-3-540-39529-4

Library of Congress Cataloging in Publication Data
Arató, M. (Mátyás), 1931-
Linear stochastic systems with constant coefficients.
(Lecture notes in control and information sciences ; 45)
Bibliography: p.
Includes index.
1. Stochastic differential equations. 2. Stochastic systems.
I. Title. II. Series
QA274.23.A72 1982 519.2 82-19490

Originally published by Springer-Verlag Berlin Heidelberg New York in 1982

2061/3020-543210

PREFACE

Exactly twenty years ago I finished my "candidate" dissertation at the Moscow State University on statistical problems of multidimensional Gaussian Markovian stationary processes. The problem to find the exact distributions of estimators was posed me by A. N. Kolmogorov in 1959. In those years we also examined the Chandler wobble of the Earth's rotation. The calculations which we made on an M-4 computer now can be carried out without troubles on a calculator or a personal computer.

The current two decades have witnessed an exponential growth of literature on statistics of stochastic processes. A large number of theoretical models appeared, but is seems that there appears to be an ever-widening gap between theory and applications in the area of statistical inference of stochastic processes. The aim of this small book is to attempt to reduce this gap by directing the interest of future researchers to the application aspects of stochastic processes on one side, and to prove that there doesnot exist separately time series analysis (classical statistical treatment) of discrete processes and dynamical treatment of continuous time processes, on the other hand.

Many of the results presented here will be appearing in book form for the first time. This is a research book written for specialists in the common area of applications of statistics and mathematical statistics. The graduate level students should find the book useful.

The topics in the book have been divided into three parts. The first part (Chapter 1) discusses some applications which can convince the reader in the useful conception of Ito's integral in that form which was developed in the fifties, using the Wiener process instead of "white noise" process. As a "surprising" novelty we discuss the exact solution of the estimation problem with additive noise. It turns out that in the constant coefficient case the Riccati equation has an explicite solution. The second part (Chapter 2) discusses the so called elementary Gaussian, i.e., stationary and Markov processes in the discrete and continuous time case as the solutions of difference and differential equations, respectively. The connection between spectral theory and stochastic equations is also shown. The many new developments, with their tremendous range from fundamental theory to specific applications made it difficult to confine ourselves to an elementary treatment every where. The third part (Chapters 3 and 4) contain the statistical investigations of linear stochastic systems, on the basis of continuous time processes as it was proposed by A. N. Kolmogorov in the late forties.

The needed mathematical background is given in a very short Appendix. Those who require a more complete treatment and further generalizatons from the mathematical point of view may be referred to the book of Liptser, Shiryaev or Basawa, Pr. Rao. There I agree with the famous intention of M. Bartlett: "It would, however, be a pity if applied mathematicians or statisticians were put off from using some of the mathematical and statistical techniques available because they did not feel able to absorb all the more pure mathematical theory. As a statistician I find it at times rather exasperating when the mathematics of stochastic processes tends to become so abstract; time spent in wrestling with it can hardly be spared unless, as of course mathematics is best fitted to do, it deepens one's perception of the overall theoretical picture in the probabilistic and statistical sense."

In this book our main purpose is to investigate the most simple dynamical stochastic models, the linear stochastic differential (difference) systems with constant coefficients. At the first moment it seems that such processes are not more interesting, we know everything about them and there are many more sophisticated models which have been studied. From the statistical point of view this is not the case as we have many unsolved problems till now. The elementary processes, if all their components are observable, have the advantage that a set of sufficient statistics exists, which is not the case when we have an elementary process with additive noise and this latter case will be studied here in detail.

I believe that the basic premise in model building is that complicated systems, and all real systems are, as a rule, complicated, do not always need complicated models. It is advisable to fit relatively simple models to the given data and to increase the complexity of the model only if the simpler model is not satisfactory. Of course models with a degree of complexity beyond a certain level often perform poorly and only in this case we shall use more complicated models.

In constructing models for the given data our goal is to understand the process and summarize the entire available set of observations. We say that it is not enough that a model be consistent with the numerical observations, we want the model to be the "simplest." A model that has too many parameters and variables is considered unsatisfactory.

The initial basis of this work is my dissertation written in Moscow and the lectures that I gave at the Budapest University L. Eötvös in the statistics of stochastic processes. In 1974 with A. Benczúr, A. Krámli, and J. Pergel we wrote a preprint in this matter and planned to write it up in book form, but then we learned that the book of Liptser and Shiryaev came out and the idea was dropped. In 1981 A. Kolmogorov encouraged me to summarize some results, which are not well known and widespread, in this formal manner. I tried to give everywhere exact results, as they are available only in a few cases in statistics, and not asymptotic results, but the connection between them is discussed. Later in the second volume I plan to return to the computer programs, data analysis and exercises and to more computer applications in this field.

The "decimal system" of numbering the chapters, sections and subsections has been used. Equations have been numbered separately for each section. The bibliography is cited in [ ] if it is a book and with the author and reference year in case of a paper (e.g., Bartlett (1951)).

I am convinced that even in the investigation of linear stochastic systems there are many gaps and even more in this book, a number of methods and results are not included, which may have been successful in practice, and this is my own responsibility.

ACKNOWLEDGMENTS

I would like to thank first of all my teacher Andrej Nikolaevic Kolmogorov from whom I learned stochastic processes and the statistical approach. I thank him not only for the explanation of an entirely new field of research but for all the continuous help and encouragement he has provided.

I am very grateful to my Hungarian friends, A. Benczúr, A. Krámli, and J. Pergel with whom I discussed many of the problems treated in this book for a couple of years and with whom we wrote a first version in 1974.

I would like to thank my friends in the USSR, Ju. Rozanov, A. Shiryaev, A. Novikov and many others for their help, when I was at the

Moscow State University and Steklov Mathematical Institute,and for encouraging me to undertake this book.

I am pleased to acknowledge the help of Ju. Prokhorov and A. Balakrishnan for giving me the possibility to write this book at the Probability Department of Steklov Mathematical Institute and the System Science Department, University of California, Los Angeles, respectively, and A. Bagchi for many discussions.

I would like to thank Mrs.Loetitia Loberman and Ms. Ginger Nystrom for their careful typing from a marginally legible manuscript, and Miss Andrea Bajusz for typing the final version of this book.

CONTENTS

CHAPTER 1

CASE STUDIES, PROBLEMS AND THEIR STATISTICAL INVESTIGATION

1.1 Introductionary Remarks

The statistical theory of stochastic processes may be regarded as the main tool to find the connection between mathematical investigations of stochastic processes, on one side, and such applications as stochastic control, optimization, filtering, information processes and communication networks, on the other side. The statistical examination of linear dynamical systems with constant coefficients has its beginning in the forties. Both theoretical results and concrete practical applications have shown an accelerated progress in the last twenty years.

Our book may seem in some sense old fashioned, as we are using mostly the classical terminology of mathematical statistics and try to get the results in the framework of this theory. Although we do not deny the influence of communication theory, nonlinear filtering or information processing, a systematic discussion will be given only in estimation theory of these processes, what is called identification in engineering practice, and most of the examples discussed will be of scientific type than technological. The classical theory of mathematical statistics in discrete time stochastic processes (time series) under this pressure of applications became closely connected with the investigations of continuous time processes. And it gives considerable success as the concrete examples illustrate this statement. The first order autoregressive process $\xi(n)$, which fulfils the stochastic difference equation

$$\xi(n) = \varrho\,\xi(n-1) + \varepsilon(n), \tag{1.1.1}$$

where $\varepsilon(n)$ is a Gaussian white noise, having two unknown parameters ϱ, $\sigma_\xi^2 = D^2\xi(n)$, has been investigated for a long time. There were many attempts to find the distribution of the estimators of the unknown parameter ϱ. It turned out (see Ch. 3. Section 3, Theorem 1) that exact distribution we get only in the continuous time case, when $\xi(t)$ is the solution of the stochastic differential equation

$$d\xi(t) = -\lambda\xi(t)dt + dw(t), \tag{1.1.1'}$$

where $w(t)$ is a Wiener process, $\varrho = e^{-\lambda\cdot\Delta t}$

Dealing with applications of general stochastic processes, where unknown parameters may be included, the main part of our research is the accuracy and reliability of our statements. Any estimate of a set of system parameters obtained by processing a sequence of observations will be in uncertainty. The exact values seeked by means of estimators are not known and as one of the measures of their goodness the confidence regions can be used. The size of the confidence region depends not only on the number of measurements but on the "behavior" of the parameter. In many cases if the confidence level is too high we may get infinite confidence intervals even for the mean value of a process (see § 3.4). In almost all cases we shall give the number of observations, n, (or the time period T) to get the desired confidence interval size with given probability.

We say that the random vector $\underline{\xi}(t)^* = (\xi^1(t), \ldots \xi^k(t))$ describes a linear stochastic differential system with constant coefficients if it satisfies the stochastic differential equation

$$d\underline{\xi}(t) = A\underline{\xi}(t)dt + B_w^{1/2}\, d\underline{w}(t) \qquad (1.1.2)$$

where A is an k x k matrix and $\underline{w}(t)$ is the standard Wiener (Brownian motion) process, B_w is a positive semidefinite matrix. We shall prove that the Gaussian processes defined by (1.1.2) are those which were introduced by Doob as "elementary Gaussian processes" (see Doob [1]). It is very important that the elementary Gaussian processes are the same as processes components of which have rational spectral density. And $\underline{\xi}(t)$ is elementary if and only if it is stationary and Markov. The rapid progress of stochastic differential equations in the last 20 years overshadowed the solution of statistical problems of the systems. Here the reason is two fold as in almost all applications of mathematics. Those who have the results are far from the theoretical investigations, and the other, most of the theoretical mathematicians cannot solve the real problems which are arising in examination of real data.

To illustrate advantages of the treatment with continuous time let us take a realization of an AR process (1.1.1) x(1),... , x(100) with $\sigma_\xi^2 = 1$, $E\xi(n) = 0$ and let the maximum likelihood estimate of ϱ equal to 0.5, i.e.

$$\hat{\varrho} = 0.5 \quad , \quad \sigma_\varepsilon^2 = (1-\hat{\varrho}^2)\sigma_\xi^2 = 0.75 \ ,$$

(see this example e.g. in Kashyap, Rao [1], p. 129). Using the normal approximation the $\beta = 0.95$ level confidence limits are the following (see Section 3.5 equations (3.5.8) and Theorem 3), N = 100, is the size of the sample,

$$(1.1.3) \qquad 0.329 = 0.5 - 1.96\sqrt{\frac{1-\hat{\varrho}^2}{99}} < \varrho < 0.671$$

The same result was gotten with the χ^2 approximation by Kashyap, Rao [1]. But using Table 1 in Chapter 3, the limits are on the same $\beta = 0.95$ level:

$$0.39 < \varrho < 0.62$$

Note, that in this case by the last method the confidence bounds in (1.1.3) could be attained by sample size N = 40.

Basawa and Pr. Rao in their book [1] give an extremely good insight into the recent situation of statistics in stochastic processes. Unfortunately, in some special and practically important cases, e.g. for the complex stationary, Markov Gaussian process their discussion seems to be not adequate (see in their book Ch. 9.5 Example 5.1). Equation (1.1.2) in this case has the form

$$(1.1.4) \qquad \begin{aligned} d\xi^1(t) &= -\lambda\xi^1(t)dt - \omega\xi^2(t)dt + a^{1/2}dw^2(t), \\ d\xi^2(t) &= -\lambda\xi^2(t)dt + \omega\xi^1(t)dt + a^{1/2}dw^2(t), \end{aligned}$$

where $w^1(t)$ and $w^2(t)$ are two independent Wiener processes. λ, ω are unknown and a is known. This system of equations describes the rotation of the instantaneous axis of rotation of the earth with respect to the minor axis of the terrestrial ellipsoid, after the elimination of the one year periodic component (see later Section 1.4). Basawa and Pr. Rao in [1], following Taraskin's results, use the asymptotic theory to obtain confidence limits for λ and ω, using the maximum likelihood estimators. It did not become aware of the fact that

$$(1.1.5) \qquad (\hat{\omega} - \omega)\left[\int_0^T((\xi^1(t))^2 + (\xi^2(t))^2)dt\right]^{1/2}$$

is exactly normally distributed (see Theorem 3 in Section 4.3). This fact has no great practical importance in the given example, but for the given time period of observations (T is equal to 80 years) the lower confidence bounds of λ, when the confidence level is greater than

0.97, are negative, where λ is a fortiori positive! The exact distribution of the damping parameter's estimation is given in Chapter 4.

This remark illustrates that using asymptotic results we must carefully check the validity of our assumptions. In the given example $\lambda \cdot T$ must be large enough, which is not the case in the Chandler wobble.

The construction of stochastic dynamic models from empirical time series is practiced in a variety of disciplines, including engineering, economics, and physics. However, there have been few systematic expositions of the major problems facing model builders: the methods of model validation, the determination of confidence of plausible classes of models for the given realization of stochastic process. The central problem in model building is, in our view, the choice of the appropriate class of models and the validation or checking for adequacy of the best fitting models. Detailed validation tests bring out the limitations of the selected class even in those cases when we are convinced that the validation problem, discussed in the framework of terms of the classical theory of parameter estimation, hypothesis testing and desicion theoretic methods will explain that in the statistical problem of stochastic processes we need much more exact results and the difficulties in finding the probability distributions of the estimators, test statistics must be coped.

The validity of our methodolgy developed in the text is demonstrated in this first Chapter by presenting detailed case studies of univariate and multivariate stochastic processes. In this case studies all the important details of parameter estimation, validation, in the framework of confidence limits are included. We believe, in particular, it will be demonstrated that the stochastic models are superior to deterministic models even though the latter are more popular.

We assume in this book that spectral analysis is routinely used in the preliminary analysis of empirical data. However, accurate estimation of the spectral density is difficult, especially when the sample size are small. Sufficient care must be exercised in obtaining inferences based only on spectrum. We are using only rational spectral densities and the validation on spectral estimates will be done by parameter estimation of these densities.

The correlation function B(n) (or B(t)) associated with a stationary, Markov, Gaussian process is an exponentially decaying function of n(or t). It is known that there are also processes e.g. such as atmospheric turbulence, we need to consider models in which the correlation function decays at a rate slower than the exponential rate.

Dealing with mechanical or electromechanical, computer or physical systems the variables whose time histories are available to us are divided into two groups, the so-called inputs and outputs. Usually we assume in such a division that the output variables are those whose behaviour is of particular interest to us. However, they cannot be manipulated directly. They can be influenced by means of the input variables or the independent variables and there may be hidden feedbacks in the system. In economic systems the causal relationship between the variables may not be immediately apparent and in such cases the division into input and output variables may not be quite useful. Generally the complex processes in the linear case are considered in the following way

$$(1.16) \qquad \underline{\xi}(t) = Q\underline{\xi}(t-1) + \sum_{i=1}^{n} G_i \underline{\eta}(t-i) + F\,\underline{f}(t) + \underline{\varepsilon}(t),$$

where $\underline{\xi}(t)$ is the output vector, $\underline{\eta}(t)$ is the input vector, $\underline{\varepsilon}(t)$ is the disturbance vector which is independent of the past history of $\underline{\xi}$ and $\underline{\eta}$, $\underline{f}(t)$ is a deterministic trend vector function. This family is large enough to handle a variety of stochastic sequences. The parameter matrices Q, G, F are assumed to be unknown and have to be estimated from the history. The same equations can be given in the continuous time case.

To choose the appropriate stucture and the primary parameters for the given data, we narrow down our choice to a finite number of different classes of models, so we get the different order autoregressive (AR) models, the autoregressive moving average (ARMA) models, multi-variate autoregressive models and autoregressive moving average models.

The question of determining whether the given model satisfactorily represents the given data still remains open. The validation of the model can be done by hypothesis testing, by constructing confidence bounds, in which case we estimate the length of the observation history needed to get e.g. discrepancy at a given significance level.

To confirm the validity of the model we have to compare the main chatacteristics of the model as eigenvalues, correlograms, extreme values with the corresponding characteristics of the empirical data on the basis of confidence regions.

It is fair to remark that sometimes it is impossible to construct a model with the desired accuracy locally and in generality. E.g., in modeling river flows it is impossible to construct a single model which gives both good one-day forecast and one-year forecast (see Kashyap, A.R. Rao [1]). In such cases separate models may be given for each frequency domain. The situation is even more worse in some economic time series, where the observation history is usually very short. It seems to us that the presence of cyclical behaviour in economic time series (not seasonal or yearly) is fictitious and in most cases for any random sequence, one can detect any frequency depending on his psychologycal makeup. This is one of the reasons why we suggest to use as simple models as possible.

In the last section (Section 1.8.) of Chapter 1 we shall give some illustration of the Kalman filtering in the case when additive noise is present. We give explicite solution of the Riccati equation which means that many of the problems, discussed only in the steady state case (see e.g. Liptser-Shiryaev's book [1] or Basawa, Prakasa Rao [1]) can be handled in finite time interval and even more easily.

This simple and new result enables us to answer directly some statistical problems of Balakrishnan and Liptser-Shiryaev in the presence of noise and calculate the Radon-Nikodym derivatives of stationary processes with rational spectral density, as well.

1.2 The Brownian Motion

In the theory of stochastic processes and its applications a fundamental role is played by the Brownian motion process which provides a model for Brownian motion which was physically observed by a botanist, R. Brown. The same phenomenon can be observed in electric circuits and other physical applications, covering a wide range of fields: kinetic theories in plasma physics, quantum noise, etc. The Brownian motion process was studied by Einstein [1], in a series of papers, who gave an elegant theory, describing the motion of the suspended particles under the influence of a fluctuating force. The Brownian motion from the physical point of view can be regarded as a suspended particle's trajectory in the limit under the effect of finite collisions. Next we give a more detailed description of this phenomenon; for the sake of simplicity we examine only the one-dimensional Brownian motion process.[1)] The stochastic differential equation, which is the counter part of the diffusion equation introduced first by Einstein, leads to the determination of Avogadro's number if we use statistical methods. The diffusion coefficient occuring in our equation is a function of the temperature, friction coefficient of the medium as well as the dimension of the particle.

Let us denote by $\xi(t)$ the trajectory of a particle of, say, radius, $r \simeq 10^{-4}$ cm., with mass m in a fluid of absolute temperature T. Let $v(t) = \xi'(t) = d\xi/dt$ denote the velocity. We first observe that the Brownian particle suffers 10^{21} collisions per second and if the time t is large compared to collision times, the cumulative effect of all the impulses, by the central limit theorem, leads to a normal distribution.

The following equation (Langevin's equation) can be formally derived from Newton's second Law

$$(1.2.1) \qquad dv(t) = -\frac{\lambda}{m}\, v(t)dt + dF(t), \quad (\lambda > 0),$$

where λ is the coefficient of friction (viscosity) figuring in Stoke's law and dF means the forces acting on the particle by the random collisions. From the central limit theorem dF(t) may be regarded as the increment of a Wiener process with unknown local variance σ_F^2 and 0 mean, i.e.,

$$E\, dF(t) = 0, \qquad E\,(dF(t))^2 = \sigma_F^2 dt.$$

This equation (1.2.1) has the form of the well known stochastic differential equation, defining an elementary Gaussian process (see Section 2.2). This is the so called Ornstein - Uhlenbeck process. From the stationarity condition (2.2.3)

$$(1.2.2) \qquad \sigma_F^2 = \frac{2\lambda}{m}\,\sigma_v^2 \;, \quad E\, v^2(t) = \sigma_v^2 \;,$$

---

1) This description was proposed me by A.Krámli.

we get that the solution v(t) of equation (1.2.1) will be stationary if and only if v(0) is normally distributed with parameters $(0, \frac{m\sigma_F^2}{2\lambda})$.

The last condition on initial distribution expresses the heuristic fact that in the case of equilibrium (steady state, or stationarity) the particle loses in average as many energy by the friction as it gets by the random collisions. Using the ergodic theory (see Appendix B2, Theorem 3) the mean kinetic energy equals to

(1.2.3) $$\lim_{t\to\infty} \frac{1}{t} \int_0^t v^2(s)ds = \sigma_v^2 = \frac{m\sigma_F^2}{2\lambda} \cdot$$

The covariance function of the process v(t) has the form

(1.2.4) $$\frac{\sigma_F^2 . m}{2\lambda} \cdot e^{-\frac{\lambda}{m}|t-s|},$$

(see (2.2.2)). By Stoke's law λ is proportional to the radius r, and so $\frac{\lambda}{m}$ has the order r^{-2}, practically a very large number. Using this fact, the covariance function (1.2.4) may be calculated as an approximation of the function

(1.2.5) $$\frac{\sigma_F^2 \cdot m}{2\lambda} \delta(s-t),$$

where $\delta(0) = \infty$, $\delta(x) = 0$, $x \neq 0$, is the Dirac delta function,

$$e^{-x} \sim \frac{1}{x} [1 + \frac{1}{x} + \frac{x}{2} + \frac{x^2}{3!} + . . .]^{-1}, \text{ if } x >> 1.$$

In this case v(t) can be regarded as the approximation of "white" noise in the continuous time case, and $\xi(t) = \int_0^t v(s)ds$ as a Wiener process, with parameters

(1.2.6) $$E\ \xi(t) = 0, \quad E\ (d\xi(t))^2 = \frac{\sigma_F^2 \cdot m^2}{2\lambda^2}\ dt\ .$$

Now on the basis of Lemma 5 in Appendix B1, we have for any h > 0, not too small with respect to the collision time, that

(1.2.7) $$\frac{1}{n} \sum_{i=1}^{n} (\xi(ih) - \xi((i-1)h))^2 \to h \frac{\sigma_F^2 \cdot m^2}{2\lambda^2},$$

with probability 1 and in mean square too. Note, that for too small h, i.e., for $e^{-\frac{\lambda}{m}h} << 1$ it is not true that $\xi(ih) - \xi((i-1)h)$ remains a random variable with variance $\frac{\sigma_F^2 m^2}{2\lambda^2} h$ and independent of the past.

Boltzmann's law on the equipartition of the energy, i.e., the average kinetic and potential energies are both equal to $kT/2$, where k is the Boltzmann's constant ($k = 1.37 \times 10^{-23}$ joule/degree) and T is the absolute temperature, shows that $v(t)$ is normally distributed $N(0,\sigma_v^2)$, where $\sigma_v^2 = \frac{kT}{m}$. From this and (1.2.3), (1.2.4) we get

$$\frac{kT}{m} = \frac{\sigma_F^2 \cdot m}{2\lambda}. \tag{1.2.8}$$

From (1.2.7) and (1.2.8) one can obtain

$$\frac{1}{n}\sum_{i=1}^{n} (\xi(ih) - \xi((i-1)h))^2 \to \frac{\sigma_F^2 m^2}{2\lambda^2} h = \frac{2kT}{\lambda} h = \frac{RT}{\lambda T} h, \tag{1.2.9}$$

where R is the universal gas constant ($R = 8.31 \times 10^7$) and N is the Avogadro number.

(1.2.9) is the famous Einstein - Smoluchowsky formula (see Einstein [1], 1905), which was experimentally used by Perrin (1916) to get the Boltzmann's constant and the Avogadro's number.

In (1.2.9) the right hand side depends on λ(or $\frac{\lambda}{m}$) too, and up to now we assumed that it can be determined exactly. The Langevin's equation (1.2.1) shows that this is not the case, λ, or λ/m, has to be estimated on the basis of realization of $v(t)$, $0 \le t \le T_0$. If $\lambda/m >> 1$ we can use advantage of Theorem 3 in Section 3.3, which states that $T_0 \cdot \lambda/m$ is approximately normally distributed with parameters $(0, T_0 \cdot \lambda/m)$. In case $T_0 \cdot \lambda/m \sim 1$ we have to use Theorem 1 in the same section and to construct confidence limits one can use Table 1 of Chapter 3. If the observation time $T_0 = 10^{-7}$ sec. and r has the order 10^{-4} one can get that $T_0 \cdot \lambda/m \sim 10$. Assuming $T_0 \cdot (\hat{\lambda}/m) = 10$ is an estimate, then the $\beta = 0.98$ confidence bounds (symmetric in probabilities) are the following

$$0.45 < T_0 \left(\frac{\lambda}{m}\right) < 19.5,$$

which means that for k the limits are $0.5 \times 10^{-23} < k < 2 \times 10^{-22}$.

The analogous electrical problem is an L - R circuit described for the current I(t) by the equation

(1.2.10) $$L\,dI(t) + RI(t)dt = de(t),$$

where R is the resistance (in ohms), L is the inductance (in henries) and e(t) is a purely fluctuating electromotive force (a thermal noise source). In case of a capacitance C (in farads) the equation describing the oscillating electrical circuit with thermal noise is (see later Section 1.3),

(1.2.11) $$L\,dI'(t) + [RI'(t) + \frac{1}{C} I(t)]dt = de(t).$$

The fluctuations giving the voltage source e(t) (electromotive force) between the ends of a resistance are due to the electron gas contained in metals, which is subjected to fluctuations about the mean value. This noise caused by the electrons in the resistances is called thermal noise, which increases with temperature. The fluctuation in the oscillating circuit due to thermal noise is very similar to the vibration of a sensitive torsional galvanometer with a damped oscillation, or to the vibration of a pendulum subject to the collisions of air molecules. However, in this case it is more convenient assumption to suppose that the forces are of the nature of impulses, changing the velocity of the galvanometer or $\frac{dI(t)}{dt}$ discontinuously as it is expressed in equation (1.2.11).

To examine the order of magnitude of the fluctuations we apply again the Boltzmann's theorem of equipartition of energy. For an oscillating circuit the average mean kinetic and potential energies are both equal to $\frac{kT}{2}$. We assume that e(t), the fluctuation due to the thermal noise is a Wiener process with parameters $Ee(t) = 0$, $E(de(t))^2 = \sigma_e^2 \cdot dt$.

Naturally we shall have to turn to physics to obtain the values of the constant σ_e^2. In the case when we assume that e(t) is in a series with a resistance R and an inductance L (see Fig. 1) the fluctuating voltage $U_{c,d}(t) = U(t)$, and current I(t)

(1.2.12) $$L \frac{dI_{cd}(t)}{dt} = U_{cd}(t).$$

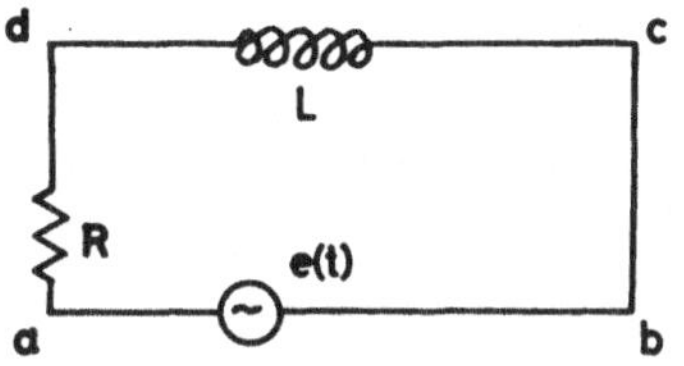

Fig. 1.

In the following we shall examine equation (1.2.10) with $I(t) = I_{ad}(t)$. The solution of (1.2.10) is an elementary Gaussian process (Ornstein - Uhlenbeck) with the covariance function

$$(1.2.13) \qquad B(t) = \frac{\sigma_e^2/L^2}{2R/_L} e^{-\frac{R}{L}|t|} = \frac{\sigma_e^2}{2LR} e^{-\frac{R}{L}|t|},$$

if and only if $I(0)$ is Gaussian $\sim N\ (0, \sigma_I^2)$, where $\sigma_I^2 = \frac{\sigma_e^2}{2LR}$ (see Section 2.2, equation (2.2.2)). This gives for the kinetic energy together with the ergodic theorem (see Appendix B.2, Theorem 3)

$$(1.2.14) \qquad \frac{L}{2} \lim_{T_0 \to \infty} \frac{1}{T_0} \int_0^{T_0} I^2(t)dt = \frac{L}{2} E\, I^2(t) = \frac{\sigma_e^2}{4R}.$$

On the other hand the theorem of equipartition gives

$$(1.2.15) \qquad \frac{1}{2} L\, E\, I^2(t) = \frac{1}{2} kT\ .$$

From (1.2.14) and (1.2.15)

$$(1.2.16) \qquad \sigma_e^2 = 2\ R\ kT$$

which is the well-known Nyquist formula (a factor 2 should be inserted in this formula when for the stationary process $I(t)$ only positive frequencies are considered). The parameter σ_e^2 in (1.2.16) can be calculated from the relation (see (2.2.29) in Section 2.2.1)

$$(1.2.1) \qquad \frac{1}{n} \sum_{i=1}^{n} [I(ih) - I\ ((i-1)h)]^2 \to h\cdot\sigma_e^2\ ,$$

with probability 1, for any $h > 0$. This can be used to estimate k.

Let be given the following values $T = 300^o k$, $L = 10^{-3}$ henry and on the basis of a one second realization the maximum likelihood estimate of reciprocity of the time constant R/L (see Section 3.3)

$$\left(\widehat{\frac{R}{L}}\right) = 2 \text{ ohm}/_{\text{henry}} , \quad \hat{R} = 2 \times 10^3 \text{ ohm},$$

then the following confidence bounds can be gotten if one uses Table 1 in Chapter 3:

Upper bounds: $R_{0.9} = 3750$, $R_{0.95} = 4500$, $R_{0.99} = 5900$,

Lower bounds: $R_{0.1} = 1100$, $R_{0.05} = 900$, $R_{0.01} = 200$.

For the current we get

$$(E\, I^2(t))^{1/2} = \sqrt{\frac{\sigma_e^2}{2LR}} = \sqrt{\frac{2RkT}{2LR}} \sim 2 \times 10^{-9} \text{ amper.}$$

In the case when R is in the vicinity of 2×10^6 ohm the correlation function (1.2.13) is nearly a Dirac delta, i.e., I(t) is approximately a "white noise" process, we should use (1.2.12) and the examination used for the Brownian particle in the first part of this Section.

Let us see a simple example.

Example 1. The relation

$$d\eta(t) + \frac{1}{\tau}\,\eta(t)dt = \frac{1}{R}\,d\xi(t)$$

may represent the voltage transformation $\xi(t)$ into voltage $\eta(t)$ through a capacity-resistance network (see Fig. 2), where $\tau = CR$ is the time

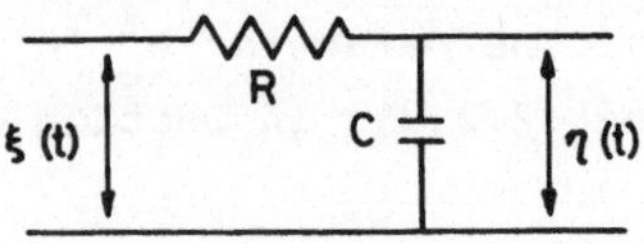

Fig. 2.

constant of the system, as the covariance function of $\eta(t)$(see (2.2.2))

$$B_\eta(t) = \frac{\sigma_\xi^2 \cdot \tau}{2R^2}\, e^{-\frac{1}{\tau}|t|} ,$$

if $\xi(t)$ is Brownian motion process with parameters $E\xi(t) = 0$, $E(d\xi)^2 = \sigma_\xi^2 \cdot dt$. If $\tau = CR \to 0$ the process $\eta(t)$ is approximately a white noise process with covariance function $\frac{1}{2}\sigma_\xi^2 \cdot C^2 \delta(t)$, where $\delta(t)$ is the Dirac delta function.

1.3. The Torsion Pendulum and Electrical Circuits

The mathematical pendulum is represented by a point P of mass m which under the influence of gravity moves on the circumference k of radius ℓ in the vertical plane, where ℓ is said to be the length of the pendulum. The point P is subject to the gravitational force mg, directed vertically downward. The component of this force along the normal to the circumference is balanced by the normal conponent of inertial force and by the reaction of the connection. The component along the tangent to the circumference at the point P in the direction of increase of the angle y is equal to - mg.sin y. Thus, by Newton's second Law, the equation of motion has the form

$$\ell y'' + g \sin y = 0, \tag{1.3.1}$$

which is a nonlinear one. If the coordinate y of P is very close to zero (the equilibrium position) during the motion sin y may be substituted by y and we get the linear equation of the pendulum (equation of the harmonic oscillator)

$$\ell y'' + g y = 0, \tag{1.3.2}$$

with the solution

$$y(t) = r \cos \left(\sqrt{\frac{g}{\ell}} \cdot t + \alpha\right). \tag{1.3.3}$$

The number $\omega = \sqrt{\frac{\ell}{g}}$ is called the pulse of oscillation, and the number of small oscillations of the pendulum per second is given by the frequency ν

$$\nu = \frac{\omega}{2\pi} = \frac{1}{2\pi}\sqrt{\frac{\ell}{g}} \tag{1.3.4}$$

The positive constant r is called the amplitude of the oscillation, and α is the phase. They depend on the initial conditions y(0), y'(0) (location and velocity at t = 0).

Let us consider now a torsion pendulum, suspended in a sealed container, where the only turning forces acting on it are the molecular shocks of the surrounding gas, and the gravitational force. The equation of motion is given in the following form

(1.3.5) $$\ell \cdot m \, d\, y'(t) + [\alpha_1 \, y'(t) + m\, g\, y(t)]dt = dw(t),$$

where the molecular force is resolved into a systematic Stoke's term $\alpha_1 \cdot y'dt$, and a remainder dw(t). The remainder term w(t) is a Brownian motion. As w'(t) does not exist the angular acceleration does not exist too.

The theory of ordinary differential equations with constant coefficients is applied with great success in electrical engineering and in particular in radio engineering. The same is true for the system of stochastic differential equations with constant coefficients, representing the same equations in electrical circuits with the Johson effect due to the thermal motion of the electrons. For elementary two-terminal components see Table 1.

Resistors, inductors (self-induction), and condensers are some of the components from which electrical devices are designed. Each of them is a two-terminal element. The electrical state of a two terminal element is characterized by two values: the current $I_{ab}(t)$, which flows from the pole a to the pole b of the two-terminal element (a,b), and the voltage drop $U_{ab}(t)$ from pole a to pole b, $U_{ab}(t) = V_a(t) - V_b(t)$, where V(t) is the potential.

$I_{ab}(t)$ and $U_{ab}(t)$ are connected by the following laws (see Table 1) governing the performance of the two-terminal element for resistance, inductance and capacitance:

1) Ohm's law with resistance $R_{ab} = R_{ba}$

(1.3.6) $$U_{ab}(t) = R_{ab} I_{ab}(t).$$

2) For a two-terminal element with inductance $L_{ab} = L_{ba}$

$$U_{ab}(t) = L_{ab} \frac{d}{dt} I_{ab}(t) \tag{1.3.7}$$

3) For a condenser with capacitance $C_{ab} = C_{ba}$

$$I_{ab}(t) = C_{ab} \frac{d}{dt} U_{ab}(t). \tag{1.3.8}$$

The function $Q_{ab}(t) = C_{ab}\, U_{ab}(t)$ is called the <u>charge</u> of the condenser.

Table 1

Elementary two-terminal components

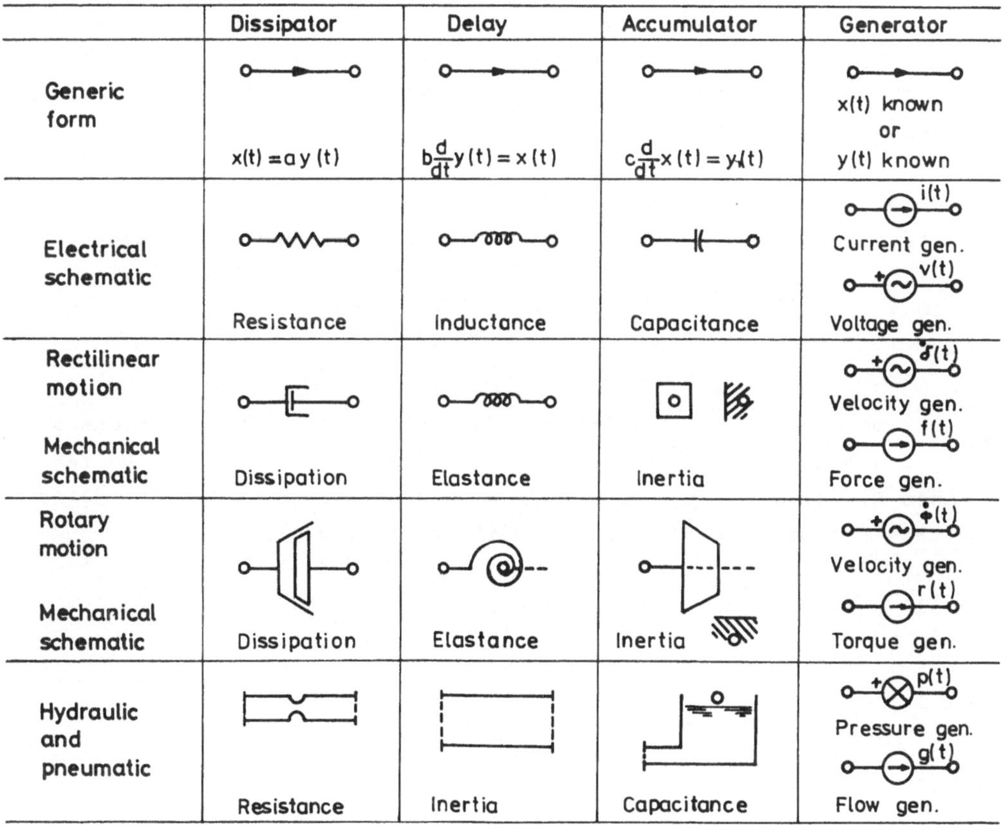

	Dissipator	Delay	Accumulator	Generator
Generic form	$x(t) = a\,y(t)$	$b\frac{d}{dt}y(t) = x(t)$	$c\frac{d}{dt}x(t) = y(t)$	$x(t)$ known or $y(t)$ known
Electrical schematic	Resistance	Inductance	Capacitance	$i(t)$ Current gen. $v(t)$ Voltage gen.
Rectilinear motion Mechanical schematic	Dissipation	Elastance	Inertia	$\dot{\delta}(t)$ Velocity gen. $f(t)$ Force gen.
Rotary motion Mechanical schematic	Dissipation	Elastance	Inertia	$\dot{\phi}(t)$ Velocity gen. $r(t)$ Torque gen.
Hydraulic and pneumatic	Resistance	Inertia	Capacitance	$p(t)$ Pressure gen. $g(t)$ Flow gen.

Two inductances (a_1, b_1) and (a_2, b_2) with values L_1 and L_2 can be in a state of mutual induction which is described by the coefficient of mutual induction M in the following way

$$U_1(t) = U_{a_1 b_1}(t) = L_1 \frac{d}{dt} I_1(t) + M \frac{d}{dt} I_2(t),$$
(1.3.9)
$$U_2(t) = U_{a_2 b_2}(t) = L_2 \frac{d}{dt} I_2(t) + M \frac{d}{dt} I_1(t),$$

where $M^2 \leq L_1 L_2$. All the above elements are called passive.

The active two-terminal elements which serve as the direct cause of electrical currents in a device are voltage sources and current sources.

An electrical network is defined as a finite set of element (in particular, two-terminal elements) with poles that are connected in so-called "junctions" of the network so that at each junction two or more poles of the various elements of the network are joined. The Kirchhoff's laws state the following. Kirchhoff's first law: the sum of all currents entering each junction of a network from all elements connected to this junction is equal to zero. Kirchhoff's second law: The sum of voltage drops around a closed circuit of a network is equal to zero, i.e., if a, b, c, ..., k is a certain sequence of junctions then

(1.3.10) $$U_{ab}(t) + U_{bc}(t) + \ldots + U_{ka}(t) = 0$$
holds.

As an example (see Fig. 1.) let S be a network with four junctions a, b, c, d, where element (a, b) is the inductance L, (b, c) is the resistance R, (c, d) is the condenser with capacitance C, and (a, d) is a voltage source U(t). From Kirchhoff's first law

(1.3.11) $$I_{ab}(t) = I_{bc}(t) = I_{cd}(t) = I_{da}(t) = I(t).$$

As

$$U_{ab}(t) = L \frac{d}{dt} I(t) \ , \ U_{bc}(t) = RI(t),$$
(1.3.12)
$$C \frac{d}{dt} U_{cd}(t) = I(t) \ , \ U_{da}(t) = - U(t),$$

Kirchhoff's second law gives

(1.3.13) $$U_{ab}(t) + U_{bc}(t) + U_{cd}(t) + U_{da}(t) = 0$$

and from (1.3.12), (1.3.13)

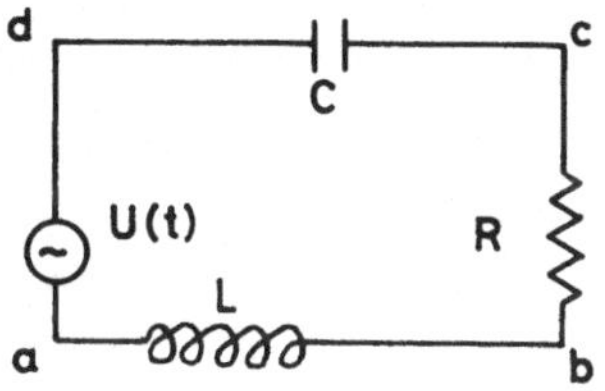

Fig. 1.

(1.3.14) $$[L \frac{d^2}{dt^2} + R \frac{d}{dt} + \frac{1}{C}] I(t) = U'(t)$$

or in charge from

(1.3.14') $$L \frac{d^2}{dt^2} Q_{ab}(t) + R \frac{d}{dt} Q_{bc}(t)+\frac{1}{C} Q_{cd}(t) = U(t),$$

which is a differential equation of the network. If $U(t) = 0$ (or (a, d) is absent) the general solution (assuming that the network has already a current) of equation

(1.3.15) $$L \frac{d^2I(t)}{dt^2} + R \frac{dI(t)}{dt} + \frac{1}{C} I(t) = 0$$

is given by

(1.3.16) $$I(t) = c_1 e^{\lambda_1 t} + c_2 e^{\lambda_2 t}$$

where λ_1, λ_2 are the solution of characteristic equation

(1.3.17) $$L \lambda^2 + R\lambda + \frac{1}{C} = 0, \quad \lambda_{1,2} = - \frac{R}{2L} \pm i \sqrt{\frac{1}{LC} - \frac{R^2}{4L^2}}$$

As L, R, C are positive values the real parts of roots are negative and the electrical process will be damped.

If we take into account the thermal noise the above equation (1.3.15) can be written in the form (see (1.3.14))

$$(1.3.17)\qquad L\,dI'(t) + [RI'(t) + \frac{1}{C} I(t)]dt = dw(t)$$

where dw(t) represents a fictitious voltage source. The known mean particle kinetic energy determines the constant σ_w^2 in $E(dw(t))^2 = \sigma_w^2 dt$.

To calculate the performance of an electrical network consisting of two-terminal elements we must find the current and voltage in each two terminal element in the network. The law governing the performance of each two-terminal element gives one relation and for a network with n two-terminal elements we obtain n relations. The remaining n relations we get from the Kirchhoff's laws. If we use the minimal number of independent currents and after this using Kirchhoff's second law express the voltages also in terms of the currents we get the method of loop currents.

A constant network with n two-terminal elements in terms of loop charges $Q^i(t)$ (as "output") and loop voltage sources $U^i(t)$ (as "input") is described by the loop equations (compare with (1.3.14'))

$$(1.3.18)\qquad \sum_{j=1}^{n} L_{ij}(Q^i(t))'' + \sum_{j=1}^{n} R_{ij}(Q^i(t))' + \sum_{j=1}^{n} \frac{1}{C_{ij}} Q^i(t) = U^i(t),$$

where L_{ij}, R_{ij}, C_{ij}, are the loop inductances, resistances and capacitances, respectively.

To obtain a first order representation let $\underline{X}*(t) = (Q^1(t), \ldots, Q^n(t), (Q^1(t))', \ldots, (Q^n(t))')$ denote the vector of charges and currents, then

$$(1.3.19)\qquad \underline{X}'(t) = A\underline{X}(t) + B_w^{1/2}\,\underline{U}(t)$$

where

$$A = \begin{pmatrix} O & I \\ -L^{-1}G & -L^{-1}R \end{pmatrix}, \quad B_w^{1/2} = \begin{pmatrix} OO \\ OL^{-1} \end{pmatrix}_{2n \times 2n}$$

and $L = (L_{ij})$, $R = (R_{ij})$, $G = (\frac{1}{C_{ij}})$ are symmetric matrices,

L, G are assumed to be nonsingular.

The (1.3.19) description is called state space description (see § 1.5). If $\underline{U}(t)$ is the "white noise" process

(1.3.20) $$E\,\underline{X}(t)\underline{X}^*(t) = B(0)$$

where B(0) is the unique solution of equation (see (2.2.3) and in discrete case (2.1.1))

(1.3.21) $$AB(0) + B(0)\,A^* = -\,B_w$$

and it may be given by

$$B(0) = \frac{1}{2}\begin{pmatrix} G^{-1}\,R^{-1} & 0 \\ 0 & L^{-1}\,R^{-1} \end{pmatrix}$$

or

(1.3.22) $$B(0) = \int_0^\infty e^{-As}\,B_w\,e^{-A^*s}\,ds\ .$$

Particularly interesting case is that of the oscillating circuit with L, R, C described by equation (1.3.17), and assuming that the characteristic equation

$$L\,\lambda^2 + R\lambda + \frac{1}{C} = 0,$$

has complex roots, i.e., $\frac{R^2}{4L^2} - \frac{1}{LC} < 0$, and $\lambda_{1,2} = -\frac{R}{2L} \pm i\sqrt{\frac{1}{LC} - \frac{R^2}{4L^2}}$.

This is the case when LC << 1, or $R \simeq 0$.

Let $I_{ab}(t)$, $U_{cd}(t)$ denote the fluctuating current and voltage (potential difference) between the sides of inductance and condenser, respectively. Let $\xi^1(t) = U_{cd}(t)$, $\xi^2(t) = I_{ab}(t)$, then equation (1.3.17) is equivalent to (1.3.19) and we get

(1.3.23)
$$d\xi^1(t) = \frac{1}{C}\,\xi^2(t)dt,$$
$$d\xi^2(t) = -[\frac{R}{L}\,\xi^2(t) + \frac{1}{L}\,\xi^1(t)]dt + \frac{\sigma_e}{L}\,d\,e(t)$$

or in matrix form

$$(1.3.23')\quad d\underline{\xi}(t) = A\underline{\xi}(t)dt + B_W^{1/2}\, d\underline{e}(t),\ A = \begin{pmatrix} 0 & \frac{1}{C} \\ -\frac{1}{L} & -\frac{R}{L} \end{pmatrix},\ B_W = \begin{pmatrix} 0 & 0 \\ 0 & \frac{\sigma_e^2}{L} \end{pmatrix}.$$

The two dimensional process $\underline{\xi}(t)$ is stationary if and only if (see Theorem 1 in Section 2.2)

$$B(0) = E\underline{\xi}\ (0)\underline{\xi}^*(0)$$

satisfies (1.3.21), which gives

$$(1.3.24)\quad B(0) = \frac{\sigma_e^2}{L^2}\begin{pmatrix} \frac{L^2}{2R} & 0 \\ 0 & \frac{L}{2R} \end{pmatrix} \qquad \lambda = \frac{R}{L},\ \omega = \sqrt{\frac{1}{LC} - \frac{R^2}{4L^2}},$$

and (see Section 4.5, and (4.5.7))

$$(1.3.25)\quad B(t) = E\xi(s+t)\xi^*(s) = e^{A|t|}B(0) =$$

$$= e^{-\lambda|t|}\begin{pmatrix} \cos\omega t + \frac{\lambda}{\omega}\sin\omega|t| & \frac{1}{\omega}\sin\omega|t| \\ -\frac{\lambda^2+\omega^2}{\omega}\sin\omega|t| & \cos\omega t - \frac{\lambda}{\omega}\sin\omega|t| \end{pmatrix}B(0),$$

$$(1.3.26)\quad B^{-1}(0) = \frac{L^2}{\sigma_e^2}\begin{pmatrix} \frac{2R}{L^2} & 0 \\ 0 & \frac{2R}{L} \end{pmatrix}.$$

Note that in the case $R = 0$ we have $\lambda = 0$, $\omega^2 = \frac{1}{LC}$ and

$$A = \begin{pmatrix} 0 & 1 \\ -\omega^2 & 0 \end{pmatrix},$$

$$(1.3.27)\quad e^{A|t|} = \begin{pmatrix} \cos\omega t & \frac{\sin\omega|t|}{\omega} \\ -\omega\sin\omega|t| & \cos\omega t \end{pmatrix},$$

but B(0) is not defined, as in this case stationary solution does not exist, i.e., the linear harmonic oscillator with a white noise on the right hand side has not a stationary solution.

The spectral density functions of the processes $\xi^2(t)$ and $\underline{\xi}(t)$ are (see (2.2.11))

(1.3.28) $$f_{\xi 2}(\lambda) = \frac{\sigma_e^2}{2\pi} \frac{1}{|L(i\lambda)^2 + R(i\lambda) + \frac{1}{C}|^2},$$

and

(1.3.29) $$f_{\xi}(\lambda) = \begin{pmatrix} i\lambda & -1/c \\ 1/L & i\lambda+R/_L \end{pmatrix}^{-1} \begin{pmatrix} 0 & 0 \\ 0 & \sigma_e^2/L^2 \end{pmatrix} \begin{pmatrix} -i\lambda & 1/L \\ -1/C & -i\lambda+R/_L \end{pmatrix}^{-1}.$$

On the basis of the ergodic theorem (Appendix B.2, Theorem 3)

(1.3.30) $$\lim_{T_0\to\infty} \frac{1}{T_0} \int_0^{T_0} (\xi^1(t))^2\, dt = \frac{\sigma_e^2}{L^2} \cdot \frac{L^2}{2R}$$

(1.3.31) $$\lim_{T_0\to\infty} \frac{1}{T_0} \int_0^{T_0} (\xi^2(t))^2\, dt = \frac{\sigma_e^2}{2RL},$$

(1.3.32) $$\lim_{T_0\to\infty} \frac{1}{T_0} \int_0^{T_0} (\xi^1(t)\xi^2(t))dt = 0.$$

We know that $\sigma_e^2 = 2\ R\ kT$ (see 1.2.16)) and from (1.3.30) and (1.3.31)) we get

1.3.33
$$L\cdot 1/2\ E(I_{ab}(t))^2 = L\ \frac{1}{2}\ E\ (\xi^2(t))^2 = \frac{kT}{2},$$
$$C\cdot 1/2\ E\ (U_{cd}(t))^2 = C\ \frac{1}{2}\ E\ (\xi^1(t))^2 = \frac{kT}{2},$$

in agreement with the equipartition principle.

The spectral density function (1.3.28) can be rewritten in the following form, using the notation $\omega_0^2 = \frac{1}{LC}$, which is called the resonance frequency,

(1.3.34) $$f_I(\lambda) = \frac{2RkT}{2\pi L^2} \frac{C^2}{|-\omega_0^2\lambda^2 + RCi\lambda + 1|^2} = \frac{C^2RkT}{\pi L^2} \frac{1}{(1-\omega_0^2\lambda^2)^2 + (RC)^2\lambda^2} =$$
$$= \frac{C^2RkT}{\pi L^2} \cdot \frac{1}{(1-\omega_0^2\lambda^2)^2 + \frac{\lambda^2}{\omega_0^2} \cdot \frac{1}{Q^2}},$$

where $Q = \omega_o \frac{L}{R} = \frac{\omega_o}{2\lambda} = \sqrt{\frac{L}{CR^2}}$ is the overloading (or quality) factor.

Let the elements in the LRC electric circuit be given by R = 0.1 ohm, L = 1 μ henry and C = 25 μμ farad. Then the resonance frequency

$$\omega_o = \frac{1}{\sqrt{LC}} = 2 \times 10^8 \quad , \quad \frac{\omega_o}{2\pi} \approx 31.8 \times 10^6 \text{ cycles/sec,}$$

the quality factor

$$Q = \frac{L\omega_o}{R} = 2000$$

and the "bandwith"

$$\tilde{\omega} = \frac{\omega_o}{Q} = 10^5 \frac{\text{radian}}{\text{sec}} \quad , \quad \frac{\tilde{\omega}}{2\pi} \approx 16 \text{ kilocycles/sec.}$$

$$\omega = \sqrt{\frac{1}{LC} - \frac{R^2}{4L^2}} = 2 \times 10^8 \sqrt{1 - \frac{10^{-7}}{1.6}} \approx \omega_o \ .$$

In order to estimate on the basis of a realization $\xi^1(t)$, $\xi^2(t)$, $0 \le t \le T$, the unknown parameters $\lambda = \frac{R}{2L}$, $\omega_o^2 = \frac{1}{LC}$ in (1.3.23) (or in (1.3.17)) we use advantage of Section 4.5. Both parameters have their physical meaning, decay parameter and resonance frequency, respectively, and further

$$\omega^2 = \omega_o^2 - \lambda^2$$

is the frequency of the covariance function (see (1.3.25)).

For λ and ω_o^2 sufficient statistics exist and the maximum likelihood equations are the following (see (2.3.38)), with respect to λ and $\frac{1}{L}$,

$$\frac{dP_A}{dP}(\xi(t)) = \frac{1}{2\pi} |B(0)|^{-1/2} \exp\{-\tfrac{1}{2}(\underline{\xi}(0), B^{-1}(0)\underline{\xi}(0)) - \frac{1}{2}\frac{L^2}{\sigma_e^2} \int_0^T (\tfrac{R}{L}\xi^2(t) +$$

$$+ \tfrac{1}{L}\xi^1(t))^2 dt - \frac{1}{2}\frac{L^2}{\sigma_e^2}[(\tfrac{R}{L}\xi^2(T) + \tfrac{1}{L}\xi^1(T))^2 -$$

$$-(\tfrac{R}{L}\xi^2(0) + \tfrac{1}{L}\xi^1(0))^2 + \tfrac{1}{2} T \tfrac{R}{L}\} =$$

$$= \frac{1}{\pi\sigma_e}\sqrt{\frac{R}{2L}}\exp\Big\{-\frac{1}{2}\frac{L^2}{\sigma_e^2}\int_0^T[\frac{R}{L}\xi^2(t) + \frac{1}{L}\xi^1(t)]^2\,dt -$$

$$-\frac{1}{2}\frac{L^2}{\sigma_e^2}\quad[\frac{R}{L}\xi^2(T) + \frac{1}{L}\xi^1(T)]^2 + \frac{TR}{2L} - \frac{1}{2}\frac{L^2}{\sigma_e^2}[(\xi^2(0))^2\,(\frac{2R}{2} - \frac{R^2}{L^2}) +$$

$$+ (\xi^1(0))^2\,(\frac{R}{L^2} - \frac{1}{L^2}) - \frac{2R}{L^2}\xi^2(0)\xi^1(0)]\Big\}$$

$$\frac{\partial}{\partial\lambda}\left(\log\frac{dP_A}{dP}\right) = \frac{1}{2}\,\frac{1}{\lambda}\cdot\frac{\sigma_e^2}{L^2} - 4\lambda\int_0^T(\xi^2(t))^2\,dt - \frac{2}{L}\int_0^T\xi^2(t)\xi^1(t)\,dt =$$

$$-4\lambda\,(\xi^2(T))^2 - \frac{2}{L}\,\xi^2(T)\xi^1(T) + T\,\frac{\sigma_e^2}{L^2} - (\xi^2(0))^2\,(2 - 2\lambda) +$$

$$-(\xi^1(0))^2\,\frac{1}{L} + \frac{2}{L}\,\xi^2(0)\xi^1(0) = 0,$$

$$\frac{\partial}{\partial(\frac{1}{L})}\left(\log\frac{dP_A}{dP}\right) = L\,\frac{\sigma_e^2}{L^2} - 2\lambda\int_0^T\xi^2(t)\xi^1(t)\,dt - \frac{1}{L}\int_0^T(\xi^1(t))^2\,dt -$$

$$-2\lambda\,\xi^2(T)\xi^1(T) - \frac{1}{L}\,(\xi^1(T))^2 - (\xi^1(0))^2(\lambda - \frac{1}{L}) +$$

$$+ 2\lambda\xi^2(0)\xi^1(0) = 0.$$

Neglecting the terms of $O(\frac{1}{T})$ we get as a first approximation

$$\frac{4\lambda}{T}\int_0^T(\xi^2(t))^2\,dt + \frac{2}{LT}\int_0^T\xi^2(t)\xi^1(t)\,dt - \frac{\sigma_e^2}{L^2} = 0,$$

$$\frac{2\lambda}{T}\int_0^T\xi^2(t)\xi^1(t)\,dt + \frac{1}{LT}\int_0^T(\xi^1(t))^2\,dt = 0$$

with the solution

$$\hat{\lambda} = \frac{\sigma_e^2}{4L^2}\;\frac{\frac{1}{T}\int_0^T(\xi^1(t))^2\,dt}{\frac{1}{T}\int_0^T(\xi^2(t))^2\,dt\,\frac{1}{T}\int_0^T(\xi^1(t))^2\,dt - \frac{1}{T}\int_0^T\xi^2(t)\xi^1(t)\,dt},$$

$$\left(\hat{\frac{1}{L}}\right) = -2\,\frac{\frac{1}{T}\int_0^T\xi^2(t)\xi^1(t)\,dt}{\frac{1}{T}\int_0^T(\xi^2(t))^2\,dt\,\frac{1}{T}\int_0^T(\xi^1(t))^2\,dt - \frac{1}{T}\int_0^T\xi^2(t)\xi^1(t)\,dt},$$

in agreement with (1.3.30) - (1.3.32) for large T.

In the case when we are measuring only $\xi(t)$, then from (1.3.17) and (4.5.6) we get the following estimations for $\lambda = \frac{R}{2L}$ and $\omega_o^2 = \frac{1}{LC}$.

The likelihood function is the following (see (4.5.6))

$$\frac{dP_{\lambda\omega_o}}{dP}(\xi(t)) = \frac{L^2}{\sigma_e^2}\frac{2\lambda\omega_o}{\pi}\exp\Big\{-\frac{\omega_o^4}{2\sigma^2}\int_o^T(\xi(t))^2\,dt -$$

$$-\frac{(2\lambda)^2-2\omega_o^2}{2\sigma^2}\int_o^T(\xi'(t))^2\,dt + \lambda T - \frac{\omega_o^2\lambda}{\sigma^2}[\xi^2(T)+\xi^2(0)] -$$

$$-\frac{\lambda}{\sigma^2}[(\xi'(T))^2+(\xi'(0))^2] - \frac{\omega_o}{\sigma^2}[\xi(T)\xi'(T)-\xi(0)\xi'(0)]\Big\}$$

and from here, $(\sigma^2 = \frac{\sigma_e^2}{L^2})$,

$$\frac{\partial}{\partial(\omega_o^2)}\left(\log\frac{dP_{\lambda\omega_o}}{dP}\right) = \frac{1}{2\omega_o^2} - \frac{\omega_o^2}{\sigma^2}\int_o^T(\xi(t))^2\,dt + \frac{1}{\sigma^2}\int_o^T(\xi'(t))^2\,dt -$$

$$-\frac{\lambda}{\sigma^2}[(\xi(T))^2+(\xi(0))^2] - \frac{1}{\sigma^2}[\xi(T)\xi'(T)-\xi(0)\xi'(0)] = 0$$

$$\frac{\partial}{\partial\lambda}\left(\log\frac{dP_{\lambda\omega_o}}{dP}\right) = \frac{1}{\lambda} - \frac{4\lambda}{\sigma^2}\int_o^T(\xi'(t))^2\,dt + T - \frac{\omega_o^2}{\sigma^2}[(\xi(T))^2+(\xi(0))^2] -$$

$$-\frac{1}{\sigma^2}[(\xi'(T))^2+(\xi'(0))^2] = 0$$

with the approximate solutions

$$\tilde{\omega}_o^2 = \frac{\int_o^T(\xi'(t))^2\,dt}{\int_o^T(\xi(t))^2\,dt}, \qquad \tilde{\lambda} = \frac{T\sigma^2}{4\int_o^T(\xi'(t))^2\,dt}.$$

The discrete time process $\xi_\Delta(n) = \xi(n\Delta)$ fulfils the equation

(1.3.35) $$\underline{\xi}_\Delta(n+1) = Q\xi_\Delta(n) + B_\varepsilon\,\underline{\varepsilon}(n+1)$$

where (see Section 4.5)

$$(1.3.36) \qquad Q = e^{A\Delta} = e^{-\lambda\Delta}\begin{pmatrix} \cos\omega\Delta + \frac{\lambda}{\omega}\sin\omega\Delta & \frac{1}{\omega}\sin\omega\Delta \\ -\frac{\lambda^2+\omega^2}{\omega}\sin\omega\Delta & \cos\omega\Delta - \frac{\lambda}{\omega}\sin\omega\Delta \end{pmatrix}$$

and

$$(1.3.37) \qquad B_\varepsilon = B(0) - QB(0)Q^* =$$

$$= \begin{pmatrix} \sigma_1[1-e^{-\lambda\Delta}(\cos\omega\Delta - \frac{\lambda}{\omega}\sin\omega\Delta)] & -\frac{\sigma_2}{\omega}\sin\omega\Delta \\ \sigma_1\frac{\lambda^2+\omega^2}{\omega}\sin\omega\Delta & \sigma_2[1-e^{-\lambda\Delta}(\sin\omega\Delta - \cos\omega\Delta)] \end{pmatrix},$$

$$B(0) = \begin{pmatrix} \sigma_1 & 0 \\ 0 & \sigma_2 \end{pmatrix} = \begin{pmatrix} \frac{\sigma_e^2}{4\lambda(\lambda^2+\omega^2)} & 0 \\ 0 & \frac{\sigma_e^2}{4\lambda} \end{pmatrix},$$

$$\omega = \sqrt{\frac{1}{LC} - \frac{R^2}{4L^2}}, \qquad \lambda = \frac{R}{2L}.$$

The eigenvalues of Q are $e^{-\frac{R}{2L} \pm i\omega}$. The covariance function of $\underline{\xi}_\Delta(n)$ has the form

$$(1.3.38) \qquad B(n) = Q^n \cdot B(0)$$

and $\xi^2_\Delta(n)$ has the one-dimensional covariance function

$$(1.3.38') \qquad B_{\xi 2}(n) = e^{-\lambda\Delta\cdot n}\frac{\sigma_e^2}{4\lambda}[\cos\omega\Delta n - \frac{\lambda}{\omega}\sin\omega\Delta n] =$$

$$= \frac{\sigma_e^2}{4\lambda}e^{-\lambda\Delta n}\frac{\cos(\omega\Delta n + \Psi)}{\cos\Psi}, \quad \operatorname{tg}\Psi = \frac{\lambda}{\omega}.$$

Let us take the following results: The time interval T = 1 sec.,

$$\sigma^2 = \frac{\sigma_e^2}{L^2} = 4 \times 10^{-16},$$

and further

$$\int_0^1 (\xi'(t))^2\,dt = \frac{4}{9}10^{-15}, \quad \int_0^1 (\xi(t))^2\,dt = 0.25 \times 10^{-17},$$

then

$$\tilde{\omega}_o^2 = 9 \times 10^2 \quad , \quad \tilde{\lambda} = 0.9$$

$$\tilde{\omega}^2 = 9 \times 10^2 - 0.81 \quad , \quad \frac{\tilde{\omega}}{2\pi} \sim 4.77 \text{ cycles/sec.}$$

On the basis of Remark 1 in Section 4.5 the confidence bounds for λ are the following (see Table 1 and Theorem 2 in Section 3.3)

$$\lambda_{0.9} = 1.85, \qquad \lambda_{0.95} = 2.36, \qquad \lambda_{0.99} = 3.25,$$

$$\lambda_{0.1} = 0.04, \qquad \lambda_{0.05} = 0.005, \qquad \lambda_{0.01} = 0.001,$$

while using the normal approximation $D^2\ (\tilde{\lambda}) \simeq \lambda$ we get

$$\lambda_{0.9} = 1.94, \qquad \lambda_{0.95} = 2.23, \qquad \lambda_{0.99} = 2.79,$$

$$\lambda_{0.1} = -0.14, \qquad \lambda_{0.05} = -0.43, \qquad \lambda_{0.01} = -0.99.$$

The confidence bounds at the same levels for ω_o are the following ($D^2\ (\omega_o^2) = 2\ \frac{\lambda}{2}\ \omega_o^2 \simeq 810$, $D^2\ \tilde{\omega}_o \simeq 13.3$)

$$\omega_{0.9} = 34.67 \qquad \omega_{0.95} = 36.00 \qquad \omega_{0.99} = 38.50$$

$$\omega_{0.1} = 25.33 \qquad \omega_{0.05} = 24.00 \qquad \omega_{0.01} = 21.50$$

Denote that the normal approximation cannot be used for λ as the lower bounds are all negative, which is meaningless.

Remark 1. Note that in the case $\frac{1}{LC} \sim 0$ we have a first order equation with one unknown λ the maximum likelihood estimator of which is known, and its distribution is also given (see Table 1 in Chapter 3). If $\frac{R}{2L}$ is small it can be poorly estimated. If we assume that ω_o^2 is known and given, λ has the same estimate as in case $\omega_o^2 = 0$ and the distribution is also known.

Remark 2. If $\frac{R}{L} \sim 0$ and $\frac{1}{LC} = \omega_o^2$ is fixed we get the harmonic oscillator. If we assume $\frac{\sigma_e^2}{L^2}$ is fixed then in (1.3.34) we get the "spectrum" of harmonic oscillator, where $Q \to \infty$. The estimator of $\omega_o^2 = \frac{1}{LC}$ is the following

$$\hat{\omega}_o^2 = \frac{\int_o^T (I'(s))^2 \, ds}{\int_o^T (I(s))^2 \, ds} .$$

On the other side $\lambda = \frac{R}{2L}$ can be estimated very poorly (see Section 3.3, Theorem 2).

Remark 3. If $EI'(t) = m$ and λ are unknown, ω_o^2 is known we have the situation described in Section 3.4. In this case for m there exist only infinite confidence limits and λ cannot be distinguished from 0 (see Theorems 2 and 3 in Section 3.4).

1.4. The Chandler Wobble

1.4.1. The Rotation of the Earth

The earth's rotation has occupied the interest of scientists for at least the last 300 years. The earth's rotation is conveniently separated into three parts: precession and nutation, polar motion and changes in length-of-day. Precession and nutation describes the rotational motion of the earth in space and is a consequence of lunar and solar gravitational attraction on the earth's equatorial bulge. Polar motion, or wobble, is the motion of the rotation axis with respect to the earth's crust. Changes in the length-of-day are a measure of a variable speed of rotation about the instantaneous pole.

The instantaneous axis of the earth's rotation constantly changes its position with respect to the minor axis of the ellipsoidal earth, this is called as free nutation. This motion has the following interesting property: there is a one-year period; if it is removed, there remains the so called Chandler wobble, which has a period of about 435 days (14 months). This latter oscillation is not an exact period motion; moreover, its amplitude varies over a ten-to-twenty-year period. The deviation of the North pole from its mean position is about within a region approximately the size of a tennis court, 60 x 60 feet, (see Fig. 1.).

The wandering of the pole was predicted by Euler in 1765 (see Fig. 2 and 3.) on the basis of the free nutation of a rigid body with period of 305 days. The Chandler wobble, discovered in 1891 after a long and fruitless search for a 10-months period in astronomical latitude observations, is still associated with almost as much controversy today as it was then, and many of the questions that were raised by Chandler, Newcomb, Kelvin and others are still open. The questions relate to the three essential problems associated with the Chandler wobble. 1. Can the lengthening of the period, from 305 days predicted by Euler for a rigid earth to the observed 435 days, be explained quantitatively? 2. Being a free motion, the Chandler wobble will ultimately be damped out but the astronomical record of near 150 years does not show any indication of a gradually diminishing amplitude. What maintains the motion against damping? 3. If damping occurs, where is the rotational energy dissipated?

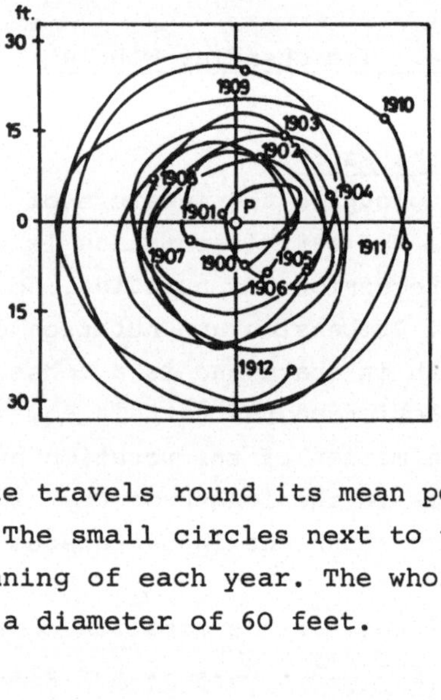

Fig. 1. The North Pole travels round its mean position (data for 1900-1912, after Wanach). The small circles next to the dates show the pole's position at the beginning of each year. The whole process takes place within a circle with a diameter of 60 feet.

It seems to us that the description by stochastic differential equations of the Chandler wobble admits statistical examination, gives a new aspect of the whole phenomena.

As noted by Newcomb in 1892 (soon after Chandler's discovery) the discrepancy between the observed 435-day period and the 305-day period can be attributed to the elastic yielding of the earth subject to a variable centrifugal force. However, in view of the quite different roles played by the mantle, core and oceans, the resulting Chandler's wobble is not readily interpreted and cannot be directly compared with Love number estimates based on other geophysical observations. Jeffreys,

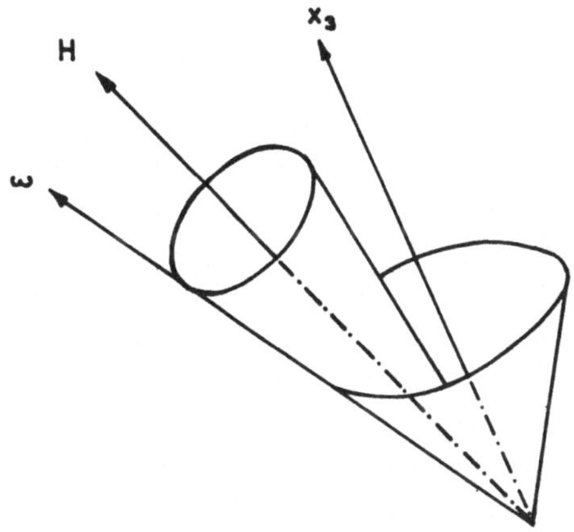

Figure 2. The Poinsot representation of rigid body rotation. H is the angular momentum axis, ω the instantaneous rotation axis, and x_3 the principal axis. In the absence of external torques, H is fixed in space; ω describes a periodic nearly diurnal motion in space about H and a much larger motion, the free Eulerian nutation, about x_3.

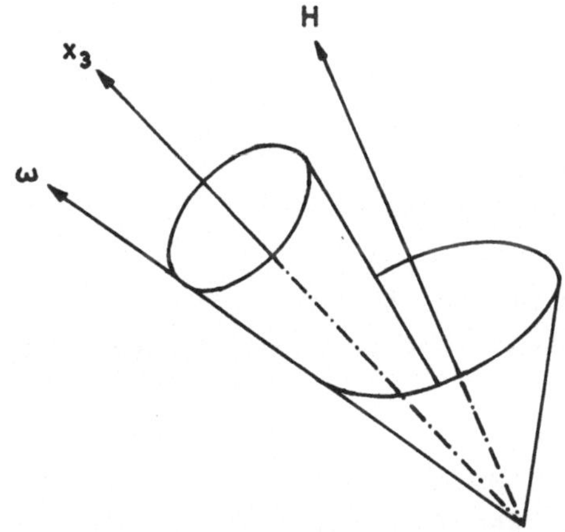

Figure 3. The Poinsot representation of the nearly diurnal free wobble for a body containing a spheroidal, liquid-filled cavity. The motion of ω about x_3 describes a nearly diurnal free wobble, and the much larger motion of ω about H describes the associated nutation in space.

in 1949, recognized that the actual core effect on the Chandler period would be modified if the shell is not rigid, and pointed out the need for a unified approach, allowing for both the elastic deformations in the mantle and the fluid motions in the core. He used first the stochastic difference equation method to estimate the period and damping parameters of the motion.

Chandler suggested (in 1891) that the polar motion was made up to two principal components with periods one year and 14 months (428 days) respectively. Today it is accepted that the period of about 435 sideral days is the best estimate. The exactly normal distribution of this estimator (proposed by Kolmogorov in 1961) is very suggestive and, if a linearly damped oscillation of the covariance function is assumed, the relaxation time is of the order of 15-30 years; the amplitude of covariance function would decay to e^{-1} of its original value is something like 15-30 years.

Munk and MacDonald, in reviewing the Chandler wobble problem in 1960 [1], concluded that, of the three problem areas, mentioned above, only the lengthening of the period could be explained satisfactorily. They examined several excitation mechanisms only to discard them. They concluded also that there is no shortage of energy sinks, the core, mantle and oceans all being possible contenders. Excitation of the Chandler wobble now appears to be a consequence of changes in the earth's inertia tensor associated with both earthquakes and the atmospheric mass redistribution.

Theoretical estimates of the parameters are based nowadays on Kolmogorov's ideas, proposed in 1960, they lead to refine and introduce more realistic core models and to the re-evaluation of the ocean contribution, with the result that the observed and computed periods agree to within a few days.

The 14-months motion of the rotation axis is referred to synonymously as the Chandler wobble, the free nutation or Eulerian precession of the earth. Our discussion will show that both parameters, the period $T = \frac{2\pi}{\omega}$, and decay λ respectively can be exactly handled. The distribution of their estimates are also given.

The x and y coordinates of the deviation of the North Pole are measured in units of 0".001 = .101 ft. The values of x(t) and y(t) fall in the intervals - 0".40, 0".50 and - 0".30, 0".50 respectively (see Fig. 4., Yumi and Vicente [1], Orlov [1], Munk and MacDonald [1]).

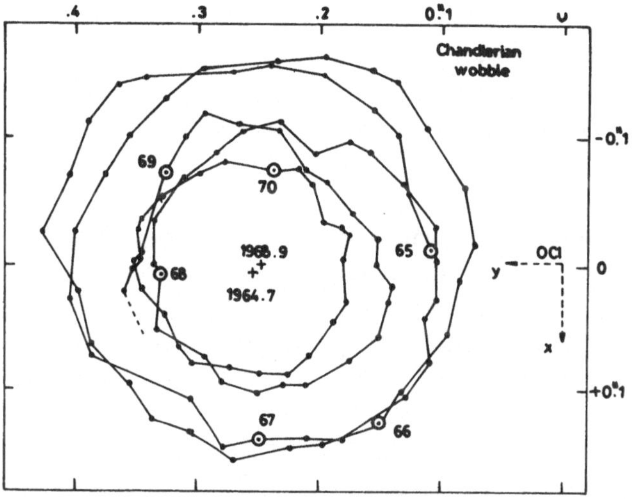

Fig. 4. Chandlerian wobble obtained by removal of the mean annual pole path.

Since 1899 the International Latitude Service (ILS) has measured the variation of latitude at five stations spread along 39° 08′ north latitude. A conventional pole of rotation (the CIO) has been adopted. Observations, using visual zenith telescopes, each night consist of six to eight pairs of stars, and throughout the year 12 (or 10) such groups are observed. To minimize the errors associated with the micrometer readings, these groups are selected so that the sum of the micrometer measures for a group is very nearly zero. But as precession slowly moves the stars out of the zeniths of the latitude stations, this condition is no longer satisfied after some years, and a new group of stars must be selected. Such changes in the star catalogues were made in 1906.0, 1912.7, 1935.0, 1955.0 and 1967.0. Uncertainties in the star positions may then introduce discontinuities in the pole path. To allow for catalogue errors, for errors in the fundamental constants defining nutation and aberration and for the neglect of certain parallax terms, the equation expressing the relation between a change in latitude $\Delta\phi(t)$ and the pole coordinates is modified to

$$\Delta\phi_i(t) = m_1 \cos \lambda_i + m_2 \sin \lambda_i + z(t),$$

where the correction term $z(t)$ absorbs all errors that are common to the stations. Despite these efforts, the data are still in an unsatisfactory state. The uniform revision of all ILS results by Melchior, announced by Munk and MacDonald in 1960, is still awaited, although some important preliminary steps have been taken (Melchior (1972)). S. Yumi and R. Vicente anticipated that the final revisions will be completed by early 1980.

Since 1955, the Bureau International de l'Heure (BIH) has routinely computed the position of the instantaneous rotation axis in an adjustment that also determines the rate of rotation. In 1975 a total of 38 stations contributed to the pole position solutions with weights ranging from 1 to 100. A comparison of the BIH and ILS results reveals significant differences. Typical results are illustrated in Figure 5. Part of the discrepancy is a consequence of the uncertainty in the ILS seasonal terms. Non-seasonal differences of 0".1 or more also occur and may persist for several months. Furthermore, there appear to be systematic differences: annual mean differences between the ILS and

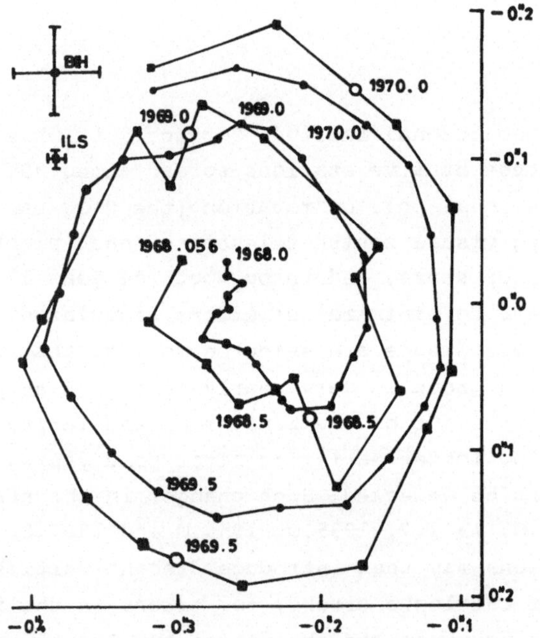

Figure 5. Pole paths as determined by the ILS and by the BIH from 1968.0 to 1970.0. The ILS data are the unsmoothed values given at successive intervals of 0.0833 yr. For convenience in comparing with the BIH results, interpolated values at intervals of 0.5 yr are also indicated. BIH values are the unsmoothed values at intervals of 0.05 yr. Error estimates for the two data sets are indicated at top left-hand side.

BIH pole positions from 1962 to 1975 show fluctuations of the order 0".02 in both m_1 and m_2.

1.4.2. The Mathematical Description and Statistical Investigation.

The position of the pole of rotation at time t is conveniently described by the complex stochastic process

$$z(t) = x(t) + iy(t),$$

where x(t), y(t) are the displacements from the CIO towards Greenwhich and towards 90° West of Greenwhich respectively. Further, we assume that

$$z(t) = me^{i2\pi t} + \zeta(t),$$

where the first term is a periodical component. From the Eulerian equations of motion with respect to the rotating reference-axis it may be deduced that $\zeta(t)$ satisfies a linear stochastic differential equation

(1.4.1) $$d\,\zeta(t) = -\,\gamma\zeta(t) + d\chi(t),$$

with $\gamma = \lambda - i\omega$, $\lambda > 0$, $\zeta(t) = \xi^1(t) + i\xi^2(t)$, $\chi(t) = \phi^1(t) + i\phi^2(t)$

is a complex Wiener process with $Ed\phi^j = 0$, $E(\phi^j(t))^2 = a \cdot t$; $j = 1,2$.

(1.4.1) is equivalent to

$$d\xi^1(t) = -\,\lambda\xi^1(t)dt - \omega\xi^2(t)dt + d\phi^1(t),$$

(1.4.1′)

$$d\xi^2(t) = +\,\omega\xi^1(t)dt - \lambda\xi^2(t)dt + d\phi^2(t).$$

The process $\phi^j(t)$ is called excitation process, $d\phi^j(t)$ describes the change in the earth's inertia tensor in the time interval (t, t + dt). If $\phi^j(t)$, j = 1,2, would be equal to zero the solution of (1.4.1) is provided by

$$\bar{e}^{\gamma|t|} = \bar{e}^{\lambda|t|} (\cos \omega t + i \sin \omega t),$$

i.e. a motion with damped oscillation of frequency ω. In the stochastic case $\bar{e}^{\gamma|t|}$ is the correlation function of the process $\zeta(t)$ (see (4.1.2) and (4.1.11)).

This model with continuous time representation was proposed by A. Kolmogorov in 1960 at the Moscow State University (see Arató, Kolmogorov, Sinay (1962)). The empirical correlation function of the Chandler's component ζ was calculated by us and it is given in Fig. 8. It was obtained as a result of the data processing of Table 6 from Orlov's paper [1].

Table 1. Seasonal components of the rotation pole, in units of 0".01.

Source	Interval	m_1		m_2	
Jeffreys, 1952	1892-1938	-3.6 cos ·	-8.5 sin ·	7.0 cos ·	-2.9 sin ·
Pollak, 1927	1890-1924	-3.7	-8.9	7.0	-3.9
Rudnick, 1956	1891-1945	-3.2	-8.2	6.7	-2.8
Walker and Young, 1957	1899-1954	-6.4	-7.1	7.0	-4.6
	1900-1934	-5.5	-7.0	7.5	-4.6
	1900-1920	-4.8	-6.0	6.6	-3.7
Jeffreys, 1940 Markowitz, 1942	1912-1935 1916-1940	-3.2	-7.8	5.6	-1.6
Arató, Kolmogorov, Sinay	1891-1951	-3.5	-8.5	7.0	-2.8
Walker and Young, 1957	1899-1954†	-0.1 cos 2	+0.6 sin 2	-0.5 cos 2	+0.0 sin 2
	1900-1934	-0.2	0.7	-0.6	-0.3
	1900-1920	-0.3	0.8	-0.8	-0.6

The annual variation $m\ e^{i2\pi t}$ in rotation is due to meteorologic events. These are treated by considering the balance of momentum and mass for the planet earth. The semi-annual changes were not significant and they were omitted. Table 1. summarizes the various results with respect to the annual and semiannual terms in the latitude observations.

Table 2.

T	Period of Analysis	m_1 a_1	b_1	m_2 c_1	d_1
ILS	1956-1970	-5.4	-9.5	-7.0	-3.8
BIH	1956-1970	-3.2	-8.7	-7.4	-2.7
Gaposchkin (1972) ILS	1891-1970	-4.3	-8.0	5.8	-3.4
Jeffreys (1968) ILS	1899-1961	-6.6	-6.2	6.0	-4.5
	1899-1905	-2.6	-5.0	5.9	-1.0
	1906-1912	-5.0	-6.1	6.0	-3.3
	1913-1919	-6.1	-6.2	7.7	-6.6
	1920-1926	-5.4	-8.9	8.6	-4.9
	1927-1933	-7.8	-8.00	94	-7.0
	1934-1940	-7.4	-5.6	6.0	-5.0
	1941-1947	-4.0	-5.8	5.2	-1.8
	1948-1954	-10.7	-6.7	5.7	-6.2
	1954-1961	-10.2	-3.9	-7	-4.5

To get reliable estimations for m (and for the CID) we have to mention that the maximum likelihood estimates of m and the CID assumes the knowledge of γ, i.e., a preliminary estimation of λ and ω (see Section 4.4). In Table 2. the dependence of the annual components on the observation interval is pointed out.

The latitude component z(t) before and after removal of the seasonal terms is shown in Fig. 6. The non-seasonal residue reveals 14 months oscillations in wave "packets." The data are from the book of Munk & MacDonald. This 14-months oscillation is brought out more clearly from the plot of covariance function (see Fig. 8).

The spectral density of the complex process $\zeta(t)$ has the form

$$f_\zeta(s) = \frac{a}{\pi}\frac{1}{|is+\gamma|^2} = \frac{a}{\pi}\frac{1}{[\lambda^2+(\omega-s)^2]^2}, \qquad \sigma^2 = \frac{a}{\lambda}.$$

Spectral estimates are given in Fig. 7. Peaks are present at 0.85 and 1 cycles/per year corresponding to rotations in a negative direction with periods of 14.3 months and 12 months respectively.

A specific dissipation function Q^{-1} is used as the measure of the rate at which energy is dissipated in a vibrating system. It corresponds to the usual definition of overloading (quantity) factor in electrical circuits (see (1.3.34)). It is defined as: $Q = \frac{\omega}{2\lambda}$.

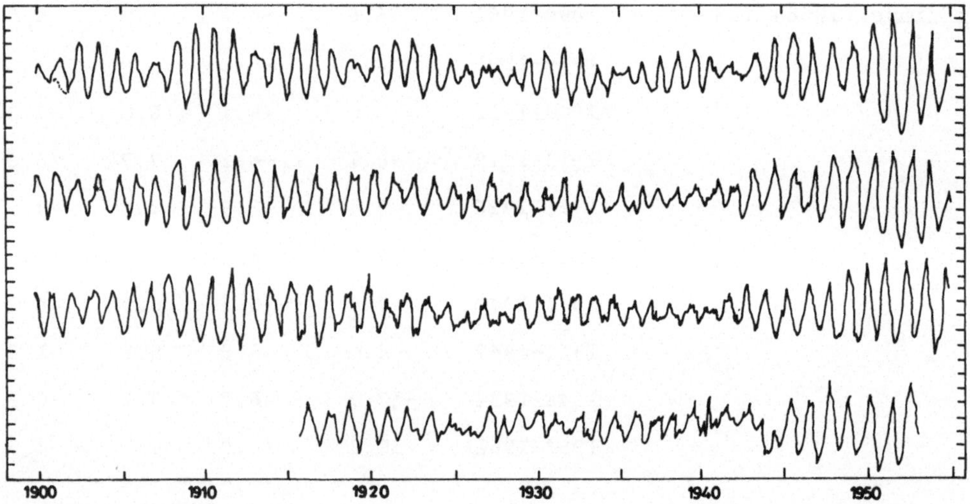

Fig. 6. The component of the unsmoothed ILS observations, before (top) and after (second curve) removal of the seasonal variation; the component of the unsmoothed ILS observation after removal of the seasonal variation (third curve) and the corresponding non-seasonal variation in the latitude of Washington, (bottom).

We shall consider a two-dimensional stationary stochastic process whose components $\xi^1(t)$ and $\xi^2(t)$ satisfy the following stochastic differential equations (1.4.1) where $\phi^1(t)$ and $\phi^2(t)$ are two independent Wiener processes with

$$E\,d\phi^1 = E\,d\phi^2 = 0\ , \quad E(d\phi^1)^2 = E(d\phi^2)^2 = a.dt.$$

Putting

$$\zeta(t) = \xi^1(t) + i\xi^2(t),\ \chi = \phi^1 + i\phi^2\ , \quad \gamma = \lambda - i\omega,$$

we can write the system (1.4.1) in a form of one equation.

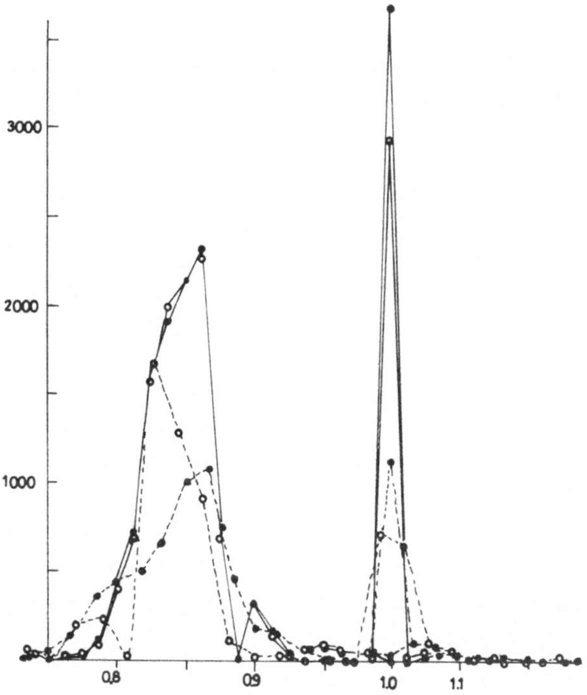

Fig. 7. Power spectra of latitude. Solid lines: $\overline{S_{11}}$ and $\overline{S_{23}}$ 1900 to 1954, from ILS. Dashed lines, open circles: 1/2 S^{+}_{11}, 1891 to 1945, from Kulikov's complication, according to Rudnick. Dashed lines, solid circles: Washington latitude for 1916-1952. The ordinate gives power density.

The complex covariance function of our process is of the form

$$C(\tau) = A(\tau) + iB(\tau) = E(\zeta(t)\overline{\zeta(t + \tau)}) = \sigma^2 e^{-\lambda|\tau|-i\omega\tau}, \qquad (1.4.2)$$

where $\sigma^2 = a/\lambda$.

If the process is observed on the interval [0, T], then it is possible to determine its empirical covariance function

$$c(\tau) = a(\tau) + ib(\tau) = \frac{1}{T-\tau}\int_0^{T-\tau} \zeta(t)\overline{\zeta(t + \tau)}\, dt. \qquad (1.4.3)$$

With probability one the empirical covariance function has its right derivative at zero (see Lemma 1 in Section 4.1)

$$c'(0) = -a - \frac{1}{T} s_1^2 + \frac{1}{T} s_2^2 - ir,$$

where a is a parameter, introduced above, which characterizes the intensity of the "white noises" $\phi^1(t)$ and $\phi^2(t)$, and

$$s_1^2 = \frac{1}{2}[|\zeta(0)|^2 + |\zeta(T)|^2], \quad s_2^2 = \frac{1}{T}\int_0^T |\zeta(t)|^2 dt, \quad r = \frac{1}{T}\int_0^T |\zeta(t)|^2 d\theta.$$

The integration in the expression for r is performed with respect to the angular variable θ, which is determined from the equation:

$$\zeta(t) = |\zeta(t)| e^{i\theta(t)}.$$

As we mentioned in Figure 8 we give the empirical covariance function $c(\tau)$ for Chandler's variations of the coordinates of the earth's pole. It can be immediately seen from the figure that the period $2\pi/\omega$ is approximately equal to 14 months. A regular character of the obtained spiral may suggest a supposition that the parameter λ also can be estimated in a very precise way. This, however, is not the case, as will be explained. These oscillations, however, are not strictly periodic but have large, primarily smooth changes in the amplitude (the waves being of the order of 10-20 years, see Fig. 9.) Figure 8 shows that this Chandler's component of polar displacement satisfies quite well the hypothesis with respect to equation (1.4.1).

The parameter a is precisely determined (see Section 4.1) from the complete realization of the process ($a \sim 0''.035$). It remains to investigate the problem of estimation of the parameters λ and ω. Let us use advantage of Chapter 4 and denote by P the probability measure in the space of the sample functions of our process for the interval $[0,T]$. In the space we also introduce the standard measure

$$V = L \times W$$

where L is the usual Lebesgue measure on the plane $\zeta(0)$, and W is the two-dimensional Wiener measure on the space of increments $\zeta(t) - \zeta(0)$ with the same characteristics which are assumed for the stochastic process $\chi(t)$.

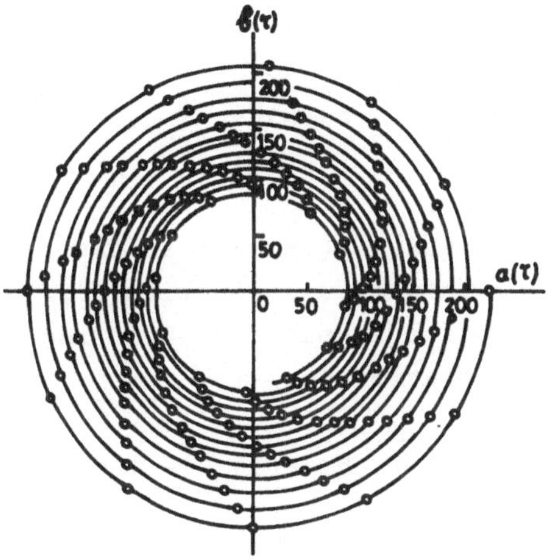

Figure 8 $\tau = 0, 1 \cdot n$; $n = 0, 1 \ldots, 156$

We have (see (4.1.7))

(1.4.4) $$\frac{dP}{dV} = C \exp \left[- \frac{\lambda^2 + \omega^2}{2a} T s_2^2 - \frac{\lambda}{a} s_1^2 + \lambda T + \frac{\omega}{a} Tr\right],$$

where $C = \frac{\lambda}{\pi a^2}$. The formula (1.4.4) shows that the system of three statistics is a sufficient system of statistics for this problem.

Differentiating the expression

$$L = \log \frac{dP}{dV} = - \log \pi a^2 + \log\lambda - \frac{\lambda^2 + \omega^2}{2a} T s_2^2 - \frac{\lambda}{a} s_1^2 + \lambda T + \frac{\omega}{a} T_r$$

with respect to ω and λ, we get the equations

(1.4.5) $$\frac{\partial L}{\partial \omega} = - \frac{\omega}{a} T s_2^2 + \frac{T}{a} r = 0,$$

(1.4.6) $$\frac{\partial L}{\partial \lambda} = \frac{1}{\lambda} - \frac{\lambda}{a} T s_2^2 - \frac{s_1^2}{a} + T = 0.$$

These equations serve for the determination of the maximum likelihood estimates $\hat{\omega}$ and $\hat{\lambda}$. From equation (1.4.5) we obtain

$$\hat{\omega} = \frac{r}{s_2^2} \cdot$$

It is shown that

$$\frac{\hat{\omega} - \omega}{\sigma(\hat{\omega})}, \quad \sigma^2(\hat{\omega}) = \frac{a}{Ts_2^2},$$

is normally (0,1) distributed (this is an exact result and not an asymptotic one!), see Theorem 3 in Section 4.3. The equation (1.4.6) always has a unique positive solution. Denoting $\lambda T = \kappa$, $\hat{\lambda}T = \hat{\kappa}$, we obtain the following equation for $\hat{\kappa}$:

$$h_2\hat{\kappa}^2 + (h_1 - 1)\hat{\kappa} - 1 = 0,$$

where $h_1 = s_1^2/aT$, $h_2 = s_2^2/aT$.

The distribution of statistics h_1 and h_2 and, hence, also the distribution of $\hat{\kappa}$ depends only on the parameter κ. Since $\hat{\kappa}$ has a continuous distribution, it is possible to find α, k such that for any α, $0 < \alpha < 1$ and any κ, $0 < \kappa < \infty$,

(1.4.7) $$P_\kappa(\hat{\kappa} > k) = \alpha.$$

Inverting the function

$$k = k_\alpha(\kappa),$$

we get the function

$$\kappa = \kappa_\alpha(k).$$

(It has been established that the function $k_\alpha(\kappa)$ varies monotonically from 0 to ∞ as κ varies from 0 to ∞; thus, the inversion is possible and is unique.)

Clearly

(1.4.8) $$P_\kappa(\kappa < \kappa_\alpha(\hat{\kappa})) \equiv \alpha.$$

We have organized (see Section 4.2) the calculations of the function $\kappa_\alpha(\hat{\kappa})$ for α = 0.1; 0.05; 0.025; 0.01; 0.001; 0.9; 0.95; 0.975; 0.99; 0.999.

For small values of $\hat{\kappa}$ the equation (1.4.7) is equivalent to (see Theorem 2 in Section 4.2)

(1.4.9) $$P_\kappa(\hat{\kappa} < x\,\kappa) = \exp(-\frac{1}{x}) \; ,$$

i.e., the ratio $\hat{\kappa}/\kappa$ has χ^2 distribution with two degrees of freedom.

For large $\hat{\kappa}$ the equation (1.4.8) is equivalent to (see Theorem 3 in Section 4.2)

(1.4.10) $$P\;(\kappa < \hat{\kappa} + x\sqrt{\hat{\kappa}}) \approx \frac{1}{\sqrt{2\pi}} \int_{-\infty}^{x} e^{-t^2/2}\,dt,$$

i.e., the estimate of $\hat{\kappa}$ is asymptotically normal with variance

(1.4.11) $$\sigma^2(\hat{\kappa}) \,\iota\, \hat{\kappa}.$$

The introduction of the Wiener functions ϕ^1 and ϕ^2 in the equation (1.4.1), i.e., of the perturbations of "white noise" type is, of course, a crude idealization for the case of the earth's pole displacement. It would have been more correct to write:

$$d\xi^1 = (-\lambda\xi^1 - \omega\xi^2)dt + f, \quad d\xi^2 = (-\lambda\xi^2 + \omega\xi^1)dt + g.$$

However, the data of Orlov shows that the values of functions f(t) and g(t) at the periods of time t, which are several years apart, are practically independent. Thus, the substitution of functions f and g by the "equivalent white noise" is legitimate. The error introduced in the determination of the intensity a of this equivalent white noise is, apparently, sufficiently small that it does not significantly influence the results of the estimation of parameter λ. The value $\hat{\omega}$ is calculated from a discrete analogy which was obtained using the maximum likelihood method for the "system with discrete time," (see (4.1.5')).

Concerning the estimation of parameters λ and ω for the case of the displacement of the earth's pole, Munk and MacDonald obtained something close to our results: $\lambda = 1/15$, $2\pi/\omega = 1.193$. Similar values also have been given by Jeffreys(1942). However, in Walker, Young and Pancenko's papers, sharply differing values are indicated: $\lambda = 0.3$ and $\lambda = 0.01$, see Table 3.

We get: $\hat{\omega} = 5.274$, $\hat{\kappa} = 3.6$, $2\pi{:}\ \hat{\omega} = 1.191$, $\sigma(2\pi{:}\ \hat{\omega}) = 0.006$, which gives a 435.0 ± 2.19 days period. The asymptotical formula (1.4.11) gives

$$\sigma^2(\hat{\kappa}) = 3.6$$

Table 3. Estimates of the Chandler wobble parameters: period, relaxation time and Q

Author	Data	Period (sidereal day) $(\frac{2\pi}{\omega}\times 365.25)$	Frequency (cycle yr^{-1}) $(\frac{\omega}{2\pi})$	Relaxation time (yr) $(\frac{1}{\lambda})$	Q
Mandelbrot & McCamy (1970)	ILS 1900-1954			11-13	30-5
Arató, Kolmogorov & Sinay (1962)	(Orlov, 1891-1951)	435.0±2.15 & distribution	0.840±0.996	16,7 with exact distribution!	
Jeffreys (1968)	ILS 1899-1961	434.3±2.2	0.843±0.004	23 (14-73)	60 (40-190)
Brillinger (1973)	ILS 1900-1970	433.4±1.6	0.845±0.025	17 (11-30)	45 (30-80)
Currie (1974)	ILS 1900-1973	434.1±1.0	0.844±0.002	14 (10-18)	36 (24-46)
Wilson & Haubrich (1976)	ILS 1901-1970	435.2±2.6	0.841±0.005	38 (20-150)	100 (50-400)
Guinot (1972)	ILS+8others 1900-1970	436.9±0.7	0.838±0.001	16 (15-17)	40
Ooe (1978)	ILS 1900-1975	436.2±2.2	0.8400±0.004	38	100 (50-300)
Kulikov	ILS (smoothed) (1.195)	435.7±4	0.838	11	30
Munk, MacDonald	ILS (unsmoothed)	436.5±5	0.836	(11-22)	(30-60)
Jeffreys	ILS (1.189)	434.3±2.2	0.841	(15.2±1.6)	
Walker, Young	ILS (1.193)	435.7±2	0.838	(2-3)	11
Pancenko	ILS (1.193)	435.7	0.838	100	

Since κ is a fortiori positive, while the formula (1.4.10) gives for level < 0.03 a negative estimate k_α, it is evident that the asymptotics of the formula (1.4.10) is not yet suitable.

To explain the reasons for discrepancies in estimates of the decay parameter λ obtained in various papers (see Table 3) we use advantage of Table 1 in Chapter 4, and give the confidence limits of $\hat{\lambda}$ and $\hat{\kappa} = \hat{\lambda}T$ (on the basis of observations over T = 60 years). The following lower and upper confidence limits are obtained from our table according to the observed value $\hat{\kappa} = 3.6$ ($\hat{\lambda}$=0.06):

$\kappa_{0.999}=8.88$; $\kappa_{0.99}=7.46$; $\kappa_{0.975}=6.75$; $\kappa_{0.95}=6.18$; $\kappa_{0.90}=5.50$;

$\kappa_{0.10}=1.275$; $\kappa_{0.05}=0.818$; $\kappa_{0.025}=0.496$; $\kappa_{0.01}=0.232$; $\kappa_{0.001}=0.041$;

This corresponds to the following estimates for λ:

$\lambda_{0.999}=0.148$; $\lambda_{0.99}=0.125$; $\lambda_{0.975}=0.112$; $\lambda_{0.95}=0.103$; $\lambda_{0.90}=0.091$;

$\lambda_{0.10}=0.021$; $\lambda_{0.05}=0.014$; $\lambda_{0.025}=0.008$; $\lambda_{0.01}=0.0039$; $\lambda_{0.001}=0.0007$.

Hence it is seen that the discrepancy between the value $\hat{\lambda} = 0.06$ obtained by us and the value $\hat{\lambda} = 0.3$ obtained by Walker, Young, is significant even at the level p = 0.999, and this means that the method of estimation used is not consistent. At the same time, the value $\hat{\lambda} = 0.01$ does not differ significantly from $\hat{\lambda} = 0.06$ already at the level p = 0.05.

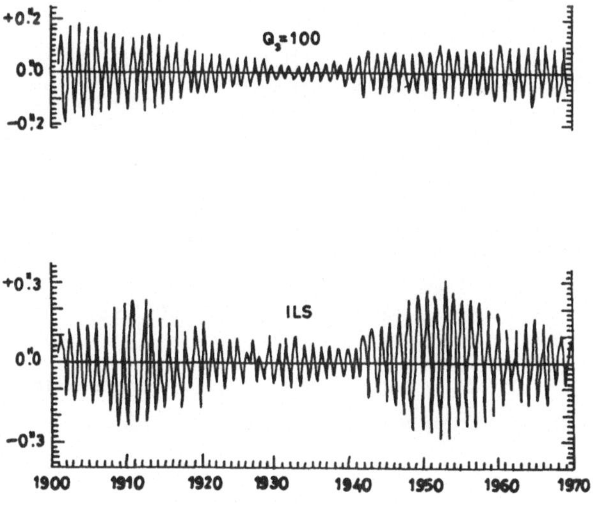

Figure 9. Changes of amplitude of the Chandler wobble from 1900 to 1970 due to the seismic excitation function of O'Connel & Dziewonski for $Q_w = 100$, compared with the astronomical results (lower curve).

1.5. System Description (State Space Theory of Linear Systems)

The terms input, output, systems, control, feedback, for engineers have an operational meaning even without precise definitions. They are using block diagrams in which e.g. a computer, aircraft, electrical network, is a "black box" with inputs and outputs

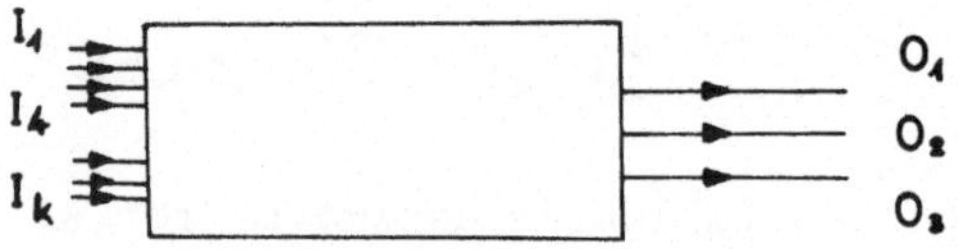

Figure 1.

labelled with apropriate arrows. $I_1, I_2, \ldots, I_k$ are the "inputs" and O_1, O_2, O_3, the outputs. Assuming linearity and time-invariant case, we have dynamic equations relating the outputs e.g. in the following way

$$\frac{d^p}{dt^p} \sigma_1(t) + \ldots + c_p \frac{d}{dt} \sigma_1(t) = \sum_{j=1}^{3} a_{1j}\sigma_j(t) + \sum_{j=1}^{k} b_{1j} I_j(t),$$

$$(1.5.1) \quad \frac{d}{dt} \sigma_2(t) = \sum_{j=1}^{3} a_{2j}\sigma_j(t) + \sum_{j=1}^{k} b_{2j} I_j(t),$$

$$\frac{d}{dt} \sigma_3(t) = \sum_{j=1}^{3} a_{3j}\sigma_j(t) + \sum_{j=1}^{k} b_{3j} I_j(t).$$

We say that (1.5.1) is the "input-output" system description. If $\underline{x}(t) = (\sigma_1(t), \sigma_2(t), \sigma_3(t), \sigma'_1(t), \ldots, \sigma_1^{(p-1)}(t))$ then (1.5.1) can be rewritten in the form ($\underline{i}^* = (I_1, I_2, \ldots, I_k)$)

$$(1.5.1') \qquad d\underline{x}(t) = A\,\underline{x}(t)dt + B\,\underline{i}(t)dt.$$

Let $\underline{y}(t) = C\underline{x}(t) + D\underline{i}(t)$ where $\underline{x}(t)$ is an ℓ dimensional vector $C = C_{\ell x p+2}$, $D = D_{\ell x k}$, then we can write

$$(1.5.2) \qquad \frac{d\underline{x}(t)}{dt} = A\,\underline{x}(t) + B\underline{i}(t),$$

$$\underline{y}(t) = C\,\underline{x}(t) + D\underline{i}(t).$$

We have said (1.5.2) to be the State Space description of the system. We call $\underline{x}(t)$ the "state" of the system (1.5.2), which is an important concept and equation (1.5.2) as a first order system of linear differential equations in most cases is not simply the rewritten form of (1.5.1). The first equation in (1.5.2) is called the "input-state" equation with the solution (see Appendix (A2.10))

$$(1.5.3) \qquad \underline{x}(t) = e^{A(t-t_o)}\underline{x}(t_o) + \int_{t_o}^{t} e^{A(t-s)}B\underline{i}(s)ds, \; t \geq t_o.$$

The second equation in (1.5.2) is called the "state-output" equation. In the black-box diagram (Fig. 1) the state of the system is only implicitly given. We call a system linear and time invariant if it can be represented in the form (1.5.2) with given A, B, C, D. Note that in older engineering literature on linear systems it was understood that

$$(1.5.4) \qquad \underline{y}(t) = \int_{-\infty}^{t} w(t-u)\underline{i}(u)du$$

where $\underline{y}(t)$ is the output, $\underline{i}(t)$ is the input, and $w(t)$ is the impulse response or weighting matrix function. (1.5.4) can be expressed as

$$\underline{y}(t) = \int_{-\infty}^{t_o} w(t-u)\underline{i}(u)du + \int_{t_o}^{t} w(t-u)\underline{i}(u)du.$$

In (1.5.2) we do not assume that $\underline{y}(t)$ may be expressed in the form (1.5.4). The state space representation is more general as it allows to consider a larger class of linear systems.

Representations (1.5.2) and (1.5.4) are equivalent if the state space $\underline{x}(t)$ is controllable. Controllability of $\underline{x}(t)$ means, that for any two states $\underline{x}_1$ and $\underline{x}_2$ one can find an input $\underline{I}(t)$ to transfer the state from $\underline{x}_1$ to $\underline{x}_2$ for some time period [O,t], i.e.,

$$\underline{x}_2 = e^{At}\,\underline{x}_1 + \int_0^t e^{A(t-u)}BI(u)du.$$

A condition for controllability is that the matrix

$$B(0) = \int_0^T e^{At}\,BB^*e^{A^*t}dt$$

be non-singular for any $T > 0$ (see § 2.2. Theorem 1, (2.2.3) and (2.2.4)) or in algebraic form that the matrix $[B \ AB \ \dots \ A^{n-1} B]$ has rank n where $A = A_{nxn}$, $B = B_{nxn}$ (see § 2.1 Lemma 1).

We say that system (1.5.2) is <u>stable</u> if

(1.5.5) $$\lim_{t\to\infty} \| e^{At} \underline{x} \| = 0,$$

for every initial state $\underline{x}$. A necessary and sufficient condition for stability is that all the eigenvalues of A have negative real parts (see Lemma 2 in Section 2.1 in discrete time case and Theorem 1 in Section 2.2 in continuous case).

We call the system $\underline{x}(t)$ described by (1.5.2) observable if

(1.5.6) $$C\, e^{At}\, \underline{x} = 0, \quad t > 0,$$

implies that $\underline{x} = 0$. In this case observing $\underline{y}(t)$, $t \geq 0$, and knowing $\underline{I}(t)$, $t \geq 0$, we can determine $\underline{x}(0)$. Indeed from (1.5.2)

$$\underline{y}(t) = C\, e^{At}\, \underline{x}(0) + \int_0^t e^{A(t-u)} B \underline{I}(u) du + D \underline{I}(t),$$

where the second and third term are known. Condition (1.5.6) is equivalent to the condition that the matrix

$$\begin{pmatrix} C \\ C\,A \\ \cdot \\ \cdot \\ \cdot \\ C\,A^{n-1} \end{pmatrix}$$

has rank n, as $e^{At} = \sum_{k=0}^{n} C_k(t)\, A^k$, (see (A1.9)).

If $\underline{x}(n) = \underline{x}(n\,\Delta)$ and $\underline{y}(n) = \underline{y}(n\,\Delta)$ are the discrete versions of processes $\underline{x}(t)$ and $\underline{y}(t)$ (they are sampled at time points $n\,\Delta$, $n = 0, \pm 1, \pm 2, \dots$), then we have

$$\underline{x}(n+1) = \underline{x}((n+1)\Delta) = e^{A\Delta}\,\underline{x}(n) + \int_{n\Delta}^{(n+1)\Delta} e^{A[(n+1)\Delta-s]}\,B\underline{I}(s)ds =$$

$$= Q\underline{x}(n) + \int_{0}^{\Delta} e^{A(\Delta-s)}\,B\underline{I}(n\Delta+s)ds,$$

where the second term may be approximated as $B_\Delta \cdot I((n+1)\Delta)$ and we obtain (see Theorem 2 in Section 2.1)

$$\underline{x}(n+1) = Q\underline{x}(n) + B_\Delta \underline{I}\,(n+1), \qquad B_\Delta = B(0) - Q\,B(0)Q^*,$$

(1.5.7)

$$\underline{y}(n) = C\underline{x}(n) + D\underline{I}(n).$$

Equations (5.1.7) have the same form as (5.1.2) and they can be handled in the same manner.

One of the main field of applications of state space description is the signal theory. A signal $\underline{\zeta}(t)$ is a p-dimensional vector function (in most cases stochastic) $-\infty < t < \infty$. We assume that $\underline{\zeta}(t)$ has a spectral representation (in deterministic case the Fourier transform).

(1.5.8) $$\underline{\zeta}(t) = \int_{-\infty}^{\infty} e^{i\lambda t}\,\underline{\Phi}(d\lambda)\ ,$$

where $\underline{\Phi}$ is a random spectral measure with $E\,\underline{\Phi}(d\lambda) = 0$, $E\,\underline{\Phi}(d\lambda)\,\underline{\Phi}^*(d\lambda) = F_\zeta(d\lambda) = f_\zeta(\lambda)d\lambda$.

A signal is said to have finite energy if with probability 1

(1.5.9) $$\int_{-\infty}^{\infty} \|\underline{\zeta}(t)\|^2\,dt < \infty,$$

(1.5.10) $$\int_{-\infty}^{\infty} \|\underline{\zeta}(t)\|^2\,dt < \infty \qquad \text{(in deterministic case).}$$

The signal $\underline{\zeta}(t)$ is said to have finite power if

(1.5.11) $$\lim_{T\to\infty} \frac{1}{2T} \int_{-T}^{T} \|\zeta(t)\|^2 < \infty.$$

Final power means that the ergodic property holds (see Theorem 3 in Appendix 2). In most cases the signals are considered as stationary Gaussian processes. If we assume that the real signal $\zeta(t)$ has a rational spectral density, then it is always a component of a multivariate elementary Gaussian process (see § 2.2 Theorem 4), i.e

(1.5.12) $$\zeta(t) = \int_{-\infty}^{\infty} e^{i\lambda t} \frac{Q(i\lambda)}{P(i\lambda)} \Phi(d\lambda)$$

where $E\Phi(d\lambda) = 0$, $E|\Phi(d\lambda)|^2 = \frac{d\lambda}{2\pi}$ and $\zeta(t)$ is the first component of

(1.5.13) $$\underline{\zeta}(t) = \int_{-\infty}^{t} e^{A(t-u)} d\underline{w}(u)$$

where

(1.5.14) $$\underline{w}(t) = \int_{-\infty}^{\infty} \frac{e^{i\lambda t}-1}{i\lambda} \underline{\Phi}(d\lambda)$$

is a Wiener process (see § 2.1, Lemma 5), and A is given.

If we consider signals for which the spectral density is rational we need the model

(1.5.15) $$\begin{aligned} \underline{\zeta}(t) &= C\,\underline{x}(t) \\ d\,\underline{x}(t) &= A\,\underline{x}(t)\,dt + B_W^{1/2}\,d\underline{w}(t) \end{aligned}$$

where $\underline{x}(t)$ is the "state", $\underline{w}(t)$ is the standard Wiener process and A_{nxn}, B_{nxn}, C_{kxn} are given. The solution of (1.5.15) we have in the form (see Theorem 1 in Section 2.2)

(1.5.16) $$\underline{\zeta}(t) = C\,e^{At}\,\underline{x}(0) + \int_0^t C\,e^{A(t-u)}\,B_W^{1/2}\,d\underline{w}(u).$$

If $\underline{x}(0)$ is Gaussian and independent of $\underline{w}(t)$, $t \geq 0$, then $\underline{\zeta}(t)$ is Gaussian. The $\underline{x}(t)$ process is Markov. It is stationary if and only if $\underline{x}(0)$ has covariance matrix $B(0)$ which is the solution of

(1.5.17) $$A\,B(0) + B(0)\,A^* = -B_W.$$

The above statements are expressed in Theorem 1 of § 2.2. Equation (1.5.17) has at most one solution if A has eigenvalues with negative real parts. In this case the system is stable and

(1.5.17') $$B(0) = \int_0^{\infty} e^{Au}\,B_W\,e^{A^*u}\,du.$$

Any digital computer processing of signals would involve a discretization operation (so called "sampling" operation in the analog-digital, called A-D converter). The discrete sequence of signals $\underline{\zeta}(n)$, $n = 0, \pm 1, \pm 2, \ldots$, is an elementary Gaussian process if we assume that the components have rational spectral density in $e^{i\lambda}$ (see Theorem 6 in § 2.1) and $\underline{\zeta}(n)$ is the solution of

$$\underline{\zeta}(n) = Q\underline{\xi}(n-1) + \underline{\varepsilon}(n)$$

where $\underline{\varepsilon}(n)$ is a discrete Gaussian white noise process (independent, identically distributed sequence of Gaussian random vectors with $E\,\underline{\varepsilon}(n) = \underline{0}$, $E\,\underline{\varepsilon}(n)\,\underline{\varepsilon}^*(n) = B_\varepsilon$). If $\underline{\zeta}(0)$ has Gaussian distribution with $E\,\underline{\zeta}(0) = \underline{0}$, cov $(\underline{\zeta}(0), \underline{\zeta}(0)) = B(0)$, where

$$B(0) = Q\,B(0)\,Q^* + B_\varepsilon$$

and $\underline{\zeta}(0)$ is independent of $\underline{\varepsilon}(n)$, $n \geq 1$, then $\underline{\zeta}(n)$ is stationary with spectral density (see (2.1.24))

$$f_\zeta(\lambda) = \frac{1}{2\pi}(I - e^{-i\lambda}Q)^{-1}\,B_\varepsilon\,(I - e^{i\lambda}\,Q)^{*-1}$$

where $(I - e^{-i\lambda}\,Q)^{-1}\,B^{1/2}_\varepsilon$ is called the transfer function.

Other typical example for discretization one can take from the analysis of hydrologic time series and construction of models for them. The application of stochastic models of run off series to water resources system design has been treated in detail in many books. The following consideration was many times used in the lectures of A. Kolmogorov at the Moscow State University. Let us denote the level of a water reservoir at the nth year by Z_n, then we have the following balance equation

$$Z_{n+1} = Z_n - K \cdot S(Z_n) + \Sigma_{n+1}\,,$$

where Σ_{n+1} is the input in the $(n + 1)$th year, $S(Z)$ means the surface of the reservoir at level Z, K is the coefficient of evaporation. Let $E\,Z_n = m$ and assuming that $S(Z) = S(m) + c(Z - m)$ by the notation

$$Z_n - m = \xi_n$$

we get the relation

$$\xi_{n+1} = (1 - c\,k)\xi_n + \Sigma_{n+1} - KS(m) =$$

$$= \alpha\xi_n + \varepsilon_{n+1}\,, \qquad 0 < \alpha < 1.$$

In many practical cases it is natural to assume that ε_n is a Gaussian white noise process. Depending on the yearly flows of rivers ε_n may be taken as an A R process.

The models for monthly and yearly flows of rivers are relatively simple (see e.g. Kashyap, A. Rao [1]). In daily flow models the mean and variance seem time varying in many years history. Using the following table (see Kashyap, A. Rao [1]) of empirical correlation coefficients of observed flows of the rivers Krishna, Godavari and Mississippi, we construct by the help of Table 1 in Section 3.3 confidence limits, which may strengthen the validation of the A R models. The number of observation N = 59, and $\sigma_\xi^2 = 1$ in all cases.

Table 1. Empirical correlation coefficients of yearly flows.

	0	1	2	3	4	5	6	7	8
Krishna	1	.2696	.0834	.1063	-.168	-.0847	.0169	.0786	-.0769
Godavari	1	.0019	.2394	.0261	.0383	-.1795	-.1332	.0435	-.1716
Mississippi	1	.2967	.0185	.0426	-.033	.0088	-.0093	-.0104	-.0766

For the Mississippi river we get

$$\xi(n) = \varrho\xi(n-1) + \varepsilon(n),$$

with the maximum likelihood estimator

$$\hat{\varrho} = 0.2967$$

and confidence limits (in parantheses the normal approximation are given):

$$\varrho_{0.9} = 0.39\ (0.45), \quad \varrho_{0.95} = 0.41\ (0.49), \quad \varrho_{0.99} = 0.44\ (0.56),$$

$$\varrho_{0.1} = 0.23\ (0.13), \quad \varrho_{0.05} = 0.21\ (0.09), \quad \varrho_{0.01} = 0.19\ (0.0).$$

For the Godavari river we get

$$\xi(n) = a_1\xi(n-1) + a_2\xi(n-2) + \varepsilon(n),$$

where (see Section 4.6)

$$\hat{a}_1 = 2\, e^{-\lambda} \cos\omega = \frac{\hat{\rho}(1)(1-\hat{\rho}(2))}{1-\hat{\rho}(1)^2} = 0.0001$$

$$\hat{a}_2 = e^{-2\lambda} = \frac{\hat{\rho}(2)-\hat{\rho}(1)}{1-\hat{\rho}(1)^2} = 0.2375 ,$$

and the confidence limit for a_2

$$(a_2)_{0.9} = 0.37\ (0.43),\ (a_2)_{0.1} = 0.16\ (0.03).$$

For the Krishna river the results are the following:

$$\xi(n) = a_1\ \xi(n-1) + a_2\ \xi(n-2) + \varepsilon(n)$$

where

$$\hat{a}_1 = 0.2667 ,\quad \hat{a}_2 = -\ 0.2009,$$

with confidence limits

$$(a_2)_{0.9} = -\ 0.30,\ (-0.36) ,\quad (a_2)_{0.1} = -\ 0.13,\ (-0.04).$$

Typical economical applications are discussed in the classical book of Kendall-Stuart ([1], Vol 3). The trend-free wheat-price index (European prices) compiled by Beveridge is a second order autoregressive process with small ρ, where for ρ we can construct confidence regions. This series extends over 370 years, a phenomenal length of time for economic series.

The annual yields per acre of barley in England and Wales (from 1884 to 1939) is a first order A R process. The sheep population (from 1867 to 1939) in England and Wales, after eliminating the trend is again a first order A R process (which can be approximated by a high order moving average process).

In the above examples the connection between time series analysis and system description without error was discussed. The statistical problems of systems description with error the reader will find in § 1.8.

1.6. Measurement Analysis in Computer Systems

1.6.1. Measurement of Performance.

This part discusses some statistical problems which arise in analyzing the results of experiments involving the measurement evaluation and comparison of the performance of computing systems. The sequences are generally correlated and in most cases contain a portion which is non-stationary. The computer system is operating under a stochastic load and generates stochastic response sequences which are assumed stationary. Such sequences include system response times, utilizations, troughputs (e.g. transactions/sec.), device waiting times, etc. The properties of these output sequences are unknown and the system is being measured in order to estimate characteristics of the specific sequences. As an example the experimenter might be interested in the mean, covariance function of the response times (or response time distribution) and in the utilizations of the major system components (CPU, memory, disks, etc.). Furthermore, the experimenter is often interested in estimating the above quantities as a function of some input parameter such as the number of terminals or transaction rate and in comparing these estimated functions for alternative system configurations. The output sequences are correlated (often strongly) and hence the usual statistical procedures which assume independent observations do not apply.

Let us consider a database system (Heidelberger P. (1981)), where transaction response time and transaction rate are particularly important. These have been chosen as the major criteria for evaluating an alternative system. There were made modifications to the operating system so that certain supervisory functions which account for a substantial amount of processor utilisation are executed on a separate processor.

A typical sample correlation function is given in Figure 1.

The transaction response time is defined as the time between receipt of the input message to the time when all output messages have been sent. This includes the queuing delay before the transaction is scheduled to run in an address space, the cpu execution within the program and the database call handling component, the time for all I/O operations, the time in cpu dispatch queue and in the time in output

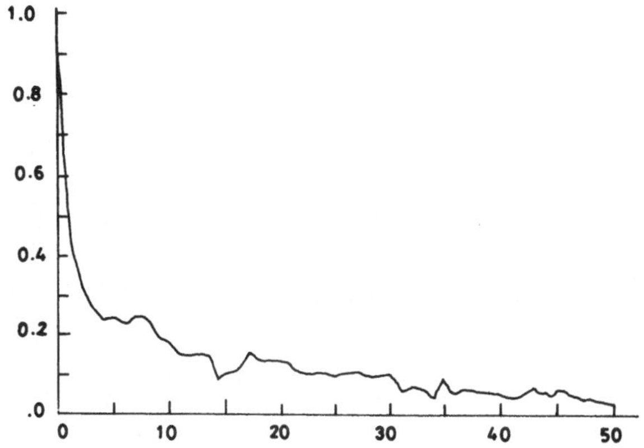

Fig. 1. Sample covariance function of transaction response time.

message queue. Even though in simulation experiments there are many sources of random fluctuations. The database system was running under the operating system MVS on an IBM computer.

On the basis of the sample covariance function we assume that the observations come as a realisation of a one dimensional elementary Gaussian process $\xi(n)$ with unknown parameters $m = E\ \xi(n)$, $\sigma_\xi^2 = D^2\ \xi(n)$ and $\text{corr}\ (\xi(n),\ \xi(n-1)) = \varrho$, i.e.,

(1.6.1) $$(\xi(n) - m) = \varrho\ (\xi(n-1) - m) + \varepsilon(n),$$

where $\varepsilon(n)$ is a Gaussian white noise with $E\ \varepsilon(n) = 0$, $\sigma_\varepsilon^2 = (1 - \varrho^2)\sigma_\xi^2$.

First we are interested in the construction of confidence limits for the parameter m. If we denote the process of the base system by $\xi_1(n)$ and by $\xi_2(n)$ the alternative system with certain functional redistribution then the main question is that the difference of sample means

$$\overline{X}_{N,1} - \overline{X}_{N,2}$$

differs significantly from 0 or not, where N is the sample size and

(1.6.2) $$\overline{X}_{N,i} = \frac{1}{N} \sum_{n=1}^{N} X_{n,i}\ , \quad i = 1,2.$$

The question we shall investigate is the following: how many observations are needed to constract a positive lower confidence bound for $\overline{X}_{N,1} - \overline{X}_{N,2}$ on significance level 90% and 95% ?

For N = 2000 and for two independent realisations of the base system and modified system respectively we get the following estimates

$$(1.6.3) \qquad \hat{\sigma}^2_{\xi_1} = 5.33 \ , \quad \hat{\varrho}_1 = 0.875,$$

$$\hat{\sigma}^2_{\xi 2} = 6.25 \ , \quad \hat{\varrho}_2 = 0.912.$$

Using the sample meals (1.6.2), they are normally distributed with the approximate variance

$$(1.6.4) \qquad D^2(\overline{X}_{N,1}) = \hat{\sigma}^2_{\xi_1} \frac{1 + \hat{\varrho}_1}{1 - \hat{\varrho}_1} \cdot \frac{1}{N} \approx \frac{79.95}{N} ,$$

$$D^2(\overline{X}_{N,2}) = \hat{\sigma}^2_{\xi_2} \frac{1 + \hat{\varrho}_2}{1 - \hat{\varrho}_2} \frac{1}{N} \approx \frac{135.8}{N} ,$$

(see Section 3.5, Equations (3.5.53)). Note that $\frac{1 + \hat{\varrho}}{1 - \hat{\varrho}}$ is the amount of additional sampling required for the mean of a correlated sequence to have the same variance as the mean of an uncorrelated sequence, assuming σ^2_ξ is the same.

Now fixing $(\hat{\sigma}^2_{\xi_1}, \hat{\varrho}_1)$, $(\hat{\sigma}^2_{\xi_2}, \hat{\varrho}_2)$ for different N we get the following upper and lower bounds of ϱ_i on the basis of Table 1 in Section 3.5 (see Example 2 in the same Section), $\tilde{\lambda} = - N \log \hat{\varrho}$

		p	N = 100	200	500	1000	10,000
$\hat{\varrho}_1$	lower	0.025	0.768	0.810	0.836	0.860	0.870
		0.05	0.810	0.830	0.865	0.863	0.871
$\hat{\varrho}_1$	upper	0.975	0.970	0.944	0.917	0.890	0.880
		0.950	0.955	0.935	0.905	0.887	0.879
$\hat{\varrho}_2$	lower	0.025	0.853	0.865	0.884	0.902	0.909
		0.050	0.861	0.869	0.887	0.903	0.909
$\hat{\varrho}_2$	upper	0.975	0.990	0.972	0.951	0.922	0.915
		0.950	0.980	0.962	0.945	0.921	0.915

Here we need not use Tables 2 - 4 in Section 3.5 as N is sufficiently large. But using the upper confidence bounds $\hat{\varrho}_{i,0.95}$ in (1.6.4) we shall get fairly wide confidence intervals for m_1 and m_2. Shorter confidence intervals can be gotten if we use the lower limit for ϱ. Notice that in this case we cannot use the confidence intervals for independent observations, i.e., if ϱ were to be zero.

The confidence intervals for the means are overlapping if $N \leq 1000$. Thus it might be desirable to have more than some thousand observations for getting positive confidence intervals for the difference of the two means.

In the case when ϱ, σ, m are all unknown, it does not exist such a statistic with known distribution as Student's t in the independent observation case. For the difference of the two mean values we can use inequalities and approximations.

Measuring the base system and the modified one at 4.86, 5.26, 5.58 and 5.87 transactions/second after 10,000 observations, for each measurement, there may be found a significant difference between the two systems. In this case the response time may be regarded as a linear function of the troughput.

The method for placing confidence bounds on the mean of an output sequence in the stationary case is generally used in the literature in discrete event simulation and the stopping times in a run length control procedure can be handled in the case when it uses the relative width of the confidence intervals. The joint estimate of the variance and correlation is used in the Gaussian case, when sufficient statistics exist and the computational requirements and storage remain low. Our method is based on the direct calculation of covariances and the approximation of the discrete process by a continuous process.

Instead of using the spectral analysis techniques, which assume indirectly the asymptotic normality, we are using the stochastic difference and differential equation method, which enables us to calculate the confidence limits in advance, to get exact results in the Gaussian case and, at the same time, good approximations for non-Gaussian sequences.

The results are in good agreement with those of the simulation, though the calculations can be carried out on a small calculator, using the tables of the known exact distribution of the maximum likelihood estimator of the damping parameter of an autoregressive (AR) process.

The main novelty in our method is not only it's simplicity, but in the direct estimation of the correlation and giving sufficient statistics. Indeed, instead of the tedious calculations of spectral densities we are using only the first covariances and the end variables which keeps the storage requirements of the method extremely low.

The spectral density function, $f_\xi(\lambda)$, of the process $\xi(n)$ has the form

$$f_\xi(\lambda) = \frac{1}{2\pi}\frac{\sigma_\varepsilon^2}{|1-\varrho e^{-i\lambda}|^2} = \frac{(1-\varrho^2)\sigma_\xi^2}{2\pi}\frac{1}{(1-\varrho\cos\lambda)^2+\varrho^2\sin^2\lambda})$$

$$f_\xi(0) = \frac{\sigma_\xi^2}{2}\cdot\frac{1+\varrho}{1-\varrho}, \quad 0 \le \varrho < 1.$$

If ϱ and σ_ξ^2 are known the maximum likelihood estimator of $\mu = E\xi(n)$ is the following (where $\frac{X_1+X_N}{2}$, $\sum_1^N X_i$ are the sufficient statistics),

$$\hat{\mu} = \frac{X_1 + X_N + (1-\varrho)\sum_2^{N-1} X_i}{2+(1-\varrho)(N-2)}$$

which is normally distributed with parameters

$$(\mu, \sigma_\xi^2\frac{1+\varrho}{2+(1-\varrho)(N-2)}) .$$

Assuming that $\xi(n)$ is the discrete variant of the continuous process $\xi(t)$ with the differential

$$d\xi(t) = -\lambda\xi(t)dt + \sigma_w \cdot dw(t), \quad \varrho = e^{-\lambda\Delta t},$$

where $w(t)$ is the standard Wiener process, then it is known that σ_w can be estimated exactly and $2\lambda\sigma_\xi^2 = \sigma_w^2$. The damping parameter λ (and so ϱ, too) can be estimated poorly and this is the reason why μ has fairly wide confidence intervals. The maximum likelihood estimator of λ is approximately normally distributed if $\lambda T \simeq 1000$. In the continuous time case the sufficient statistics of the unknown parameter μ are

$\xi(0) + \xi(T)$, $\int_0^T \xi(t)\,dt$ and the maximum likelihood estimator has the form

$$\hat{\mu} = \frac{\xi(0) + \xi(T) + \lambda \int_0^T \xi(t)dt}{2 + \lambda T},$$

with variance $2\sigma_\xi^2/(2 + \lambda T)$. Note that for $T = 1$, $\sigma_w^2 = 1$

$$D^2 \left(\frac{\xi(0) + \xi(1)}{2}\right) = \frac{1 + e^{-\lambda}}{4\lambda} < D^2 \left(\int_0^1 \xi(t)dt\right) = \frac{\lambda + e^{-\lambda} - 1}{\lambda^3}, \text{ if } \lambda < 2,$$

i.e., depending on λT the mean of two observations can be a better estimate for μ than $\frac{1}{T}\int_0^T \xi(t)dt$, and of course better than

$$\frac{1}{N + 1} \sum_{i=0}^{N} \xi\left(\frac{Ti}{N}\right) .$$

Using advantage of the Table 1 given in Chapter 3. and the approximate variance of $\hat{\mu}$

$$\frac{\sigma_\xi^2}{N} \cdot \frac{1 + \varrho}{1 - \varrho},$$

we get the following approximate confidence intervals at level $p = 0.9$ for $\frac{\mu}{\sigma_\xi}$, using the upper $\hat{\varrho}_{.95}$ and lower $\hat{\varrho}_{.05}$ confidence bounds for ϱ at the levels 0.95 and 0.05

$$-1.645 \sqrt{\frac{1 + \hat{\varrho}}{(1 - \hat{\varrho})N}} < \frac{\mu}{\sigma_\xi} - \frac{\hat{\mu}}{\sigma_\xi} < 1.645 \sqrt{\frac{1 - \hat{\varrho}}{(1 - \hat{\varrho})N}} = \tilde{\sigma}_{\hat{\varrho}}(.9),$$

and we call $\tilde{\sigma}_{\hat{\varrho}}(0.9)$ the half confidence interval width at the level $p = 0.9$.

Table 1. contains the lower and upper estimates of ϱ for different sample size and the half confidence interval width at level $p = 0.9$ and for all the values ϱ, $\hat{\varrho}_{.95}$, $\hat{\varrho}_{.05}$.

From Table 1 one can get estimation for the run length control too, in the sense that required half-width is attained or not. At given ϱ and ε (half-width) with $\varrho_{.05}$ one can get the maximum value $N(\varrho)$ for which

TABLE 1

N	100	500	1000	5000	10000	50000
		$\rho=0.98$				
λ	2.020	10.101	20.203	101.014	202.03	1010.14
$\hat{\rho}_{.95}$	0.9996	0.995	0.991	0.985	0.984	0.981
$\hat{\rho}_{.05}$	0.956	0.969	0.971	0.976	0.977	0.979
$\hat{\sigma}_{.9}(\rho)$	1.637	0.732	0.518	0.231	0.164	0.979
$\hat{\sigma}_{.9}(\hat{\rho}_{.95})$	11.630	1.469	0.774	0.268	0.183	0.075
$\hat{\sigma}_{.9}(\hat{\rho}_{.05})$	1.097	0.586	0.429	0.211	0.153	0.071
		$\rho=0.99$				
λ	1.010	5.025	10.050	50.252	100.50	502.52
$\hat{\rho}_{.95}$	0.9999*	0.9993	0.9976	9.9934	0.9924	0.9911
$\hat{\rho}_{.95}$	0.9750	0.9816	0.9841	0.9869	0.9879	0.9891
$\hat{\sigma}_{.9}(\rho)$	2.321	1.038	0.734	0.328	0.232	0.104
$\hat{\sigma}_{.9}(\hat{\rho}_{.95})$	23.263*	3.032	1.501	0.404	0.266	0.110
$\hat{\sigma}_{.9}(\hat{\rho}_{.05})$	1.462	0.763	0.581	0.287	0.211	0.099
		$\rho=0.995$				
λ	0.5012	2.506	5.013	25.063	50.125	250.63
$\hat{\rho}_{.95}$	0.9999*	0.9999	0.9996	0.9973	0.9967	0.9959
$\hat{\rho}_{.05}$	0.9852	0.9893	0.9908	0.9928	0.9934	0.9942
$\hat{\sigma}_{.9}(\rho)$	3.286	1.469	1.039	0.465	0.329	0.147
$\hat{\sigma}_{.9}(\hat{\rho}_{.95})$	23.263*	23.263	3.678	0.633	0.405	0.162
$\hat{\sigma}_{.9}(\hat{\rho}_{.05})$	1.905	1.003	0.765	0.387	0.286	0.136

N	100	500	1000	5000	10000	50000
		$\varrho=0.998$				
λ	0.202	1.001	2.002	10.010	20.020	100.10
$\hat{\varrho}_{.95}$	0.9999*	0.9999*	0.9999	0.9995	0.9991	0.9985
$\hat{\varrho}_{.05}$	0.9925	0.9950	0.9955	0.9968	0.9971	0.9975
$\hat{\sigma}_{.9}(\varrho)$	5.199	2.325	1.644	0.735	0.520	0.233
$\hat{\sigma}_{.9}(\hat{\varrho}_{.95})$	23.263*	23.263*	23.263	1.471	0.775	0.269
$\hat{\sigma}_{.9}(\hat{\varrho}_{.05})$	2.681	1.469	1.095	0.581	0.432	0.208
		$\varrho=0.999$				
λ	0.100	0.500	1.001	5.003	10.005	50.03
$\hat{\varrho}_{.95}$	0.99999*	0.99999*	0.99999*	0.99993	0.99976	0.99933
$\hat{\varrho}_{.05}$	0.99700	0.99710	0.99748	0.99815	0.99840	0.99869
$\sigma_{.9}(\varrho)$	7.35	3.289	2.326	1.040	0.735	0.329
$\sigma_{.9}(\hat{\varrho}_{.95})$	73.566*	73.566*	73.566*	3.932	1.502	0.402
$\sigma_{.9}(\hat{\varrho}_{.05})$	4.244	1.931	1.410	0.765	0.581	0.287

The half confidence interval width $\tilde{\sigma}_p(\varrho) = 1.645\sqrt{(1+\varrho)/N(1-\varrho)}$ at level p for $\frac{\mu}{\sigma_\xi}\cdot\hat{\varrho}_\beta$ means the β level confidence bound of ϱ, $\varrho = e^{-\lambda/N}$, $\lambda = -N\log\varrho$, N is the sample size.

*In the cases marked by * the upper confidence bound for ϱ is equal to 1 and the confidence interval width is ∞ (see section 4).

$$1.645 \sqrt{\frac{1 + \hat{\varrho}_{.05}}{(1 - \hat{\varrho}_{.95})N}} < \varepsilon .$$

In the same way, at given ϱ and ε, one can get the minimal value $\bar{N}(\varrho)$ for which

$$1.645 \sqrt{\frac{1 + \hat{\varrho}_{.95}}{(1 - \hat{\varrho}_{.95})N}} < \varepsilon .$$

e.g. for $\varrho = 0.99 = 1 - \frac{1}{100}$ and $\varepsilon = 0.33$ (when $N = 5000$) one can get $\underline{N}(1 - \frac{1}{100}) = 4320$, $\bar{N}(1 - \frac{1}{100}) = 7680$.

1.6.2. Round-off Errors in Solutions of Ordinary Differential Equations.

When computers first became widely used for solving differential equations, it was observed that some of the commonly used numerical integration formulas, such as Milne's formulas, led to errors in the solution much larger than would be expected from the discretization error alone. Moreover, as the step size was made smaller, these errors for a fixed value of the independent variable x actually became larger than smaller. Let us consider the first order vector differential equation (system of equations) with initial condition

$$(1.6.10) \qquad \underline{y}'(x) = \underline{f}(x, \underline{y}(x)), \quad y(x_o) = \underline{y}(0), \quad \underline{y}^* = (y^1, \ldots, y^k)$$

$$\underline{f}^* = (f^1, \ldots, f^k).$$

Where for simplicity we assume that $\underline{f}$ is sufficiently differentiable. The initial-value problem (1.6.10) possesses a unique solution. To solve numerically equation (1.6.10) let us take a one-step method, with stepsize h, i.e.,

$$(1.6.11) \qquad \underline{y}_{n+1} = \underline{y}_n + h\,\underline{\Phi}(x_n, \underline{y}_n, h), \quad x_n = x_o + nh,$$

where

$$(1.6.12) \qquad \underline{y}(x_n) - \underline{y}_n = \underline{e}_n$$

denotes the discretization error. Using Euler's method, i.e., $\underline{\Phi}(x, \underline{y}) = \underline{f}(x, \underline{y})$ it is easy to prove that

(1.6.13) $$\underline{e}_{n+1} = \underline{e}_n + h[\underline{f}(x_n, \underline{y}(x_n)) - \underline{f}(x_n, \underline{y}_n)] + \frac{h^2}{2}\underline{y}''(\xi),$$

$$\|e_{n+1}\| \le (1 + hL)\|\underline{e}_n\| + \frac{h^2}{2}\cdot K,$$

and this gives, assuming $\|\underline{y}''\| < K$ and $\|\underline{f}(x, \tilde{y}) - \underline{f}(x, \tilde{\tilde{y}})\| \le$ $\le L\ |\ \underline{y} - \tilde{\underline{y}}\ |$, the following estimation

(1.6.14) $$\|\underline{e}_n\| \le \frac{h\cdot K}{L}[e^{L(x_n - x_o)} - 1].$$

The above inequality is the simple consequence of a well known statement and as we are often using (in the continuous time case, see proof b) of Theorem 1 in Section 2.2) let us recall it.

Lemma 1. If the numbers $u_n (n = 0, 1, 2, \ldots)$ satisfy the inequalities

(1.6.15) $$|u_{n+1}| \le A\ |u_n| + B, \qquad n = 0,1, \ldots$$

where A and B are nonnegative constants independent of n, then

(1.6.16) $$|u_n| \le A^n\ |u_o| + \frac{A^n - 1}{A - 1}B\ , \qquad A \ne 1,$$

and $|u_n| \le n\cdot B$, if $A = 1$.

The proof is straightforward by induction. Note that if $A = 1 + \delta <$ $< e^{\delta} (\delta < 0)$ (1.6.16) can be rewritten

(1.6.17) $$|u_n| \le e^{n\delta}\ |u_o| + \frac{e^{n\delta} - 1}{\delta}B.$$

Lemma 2. If $c_1 > 0$, $u(t) \ge 0$, $v(t) \ge 0$ and

$$u(t) \le c_1 + \int_o^t u(s)v(s)ds,$$

then

$$u(t) \le c_1 \exp\{\int_o^t v(s)ds\}.$$

Proof of the Lemma. We have

$$\frac{u(t)v(t)}{c_1 + \int_o^t u(s)v(s)ds} \le v(t).$$

By integration

$$\ln [c_1 + \int_0^t u(s)v(s)ds] - \ln c_1 \le \int_0^t v(s)ds,$$

or

$$u(t) \le c_1 + \int_0^t u(s)v(s)ds \le c_1 \exp \{ \int_0^t v(s)ds\},$$

which gives the desired result. The case $c_1 = 0$ we may get from here by limiting $c_1 \downarrow 0$.

To illustrate the numerical instability problem mentioned in the beginning of this section, let us consider the following method (Milne's formula)

$$(1.6.18) \qquad \underline{y}_{n+1} = \underline{y}_{n-1} + 2\, h\, \underline{f}(x_n, \underline{y}_n)$$

which, depending on $\underline{f}\,(x, \underline{y})$, can lead to extraneous terms in solution because the arithmetic operations are not exact in calculations of initial conditions. And in practice some errors will be introduced, primarily due to inexact starting values. The errors of round-off we shall meet below. Roughly speaking in an unstable method the errors introduced into the calculations grow at an exponential rate.

Now we shall discuss in a more sophisticated form the round-off error propagation in one step methods. Let $\hat{\underline{y}}_n$ denote the numerical approximation of $\underline{y}_n$. The local error, denoted by $\underline{\varepsilon}_n$ at step n is induced by computer round-off (or chopping) and inherent error by inaccuracy of evaluation of function $\underline{\Phi}\,(x_n, \underline{y}_n, b)$.

Instead of equation (1.6.11) we have

$$(1.6.19) \qquad \hat{\underline{y}}_{n+1} = \hat{\underline{y}}_n + h\, \underline{\Phi}\,(x_n, \hat{\underline{y}}_n, h) + \underline{\varepsilon}_{n+1},$$

and the accumulated round-off error $\underline{r}_n = \hat{\underline{y}}_n - \underline{y}_n$ fulfils the equation

$$(1.6.20) \qquad \underline{r}_{n+1} = \underline{r}_n + h[\underline{\Phi}(x_n, \hat{\underline{y}}_n, h) - \underline{\Phi}(x_n, \underline{y}_n, h)] + \underline{\varepsilon}_{n+1}.$$

This means that the accumulated round-off error is not simply the sum of local round-off errors. They may grow-up or decay. It depends on the kind of arithmetic used in computer, the way in which the machine rounds, the order in which the arithmetic operations are performed and on the numerical procedures being used. As over an extended interval the loss of accuracy may be serious, it is desirable to obtain estimates by making some statistical assumptions about the behaviour of local round-off errors $\underline{\varepsilon}_n$.

It is known that by double precision the possible gain in accuracy can be very significant, but we have a loss in performance and efficiency.

A crude bound for the accumulated round-off error $\underline{r}_n$ can be obtained from (1.6.20) and Lemma 1 if we assume that

(1.6.21) $$\|\underline{\varepsilon}_n\| \le \varepsilon, \quad n = 1,2,\ldots$$

Indeed, if $\|\underline{\Phi}(x, \underline{y}, h) - \underline{\Phi}(x, \tilde{\underline{y}}, h)\| \le L^1 \cdot \|y - \tilde{\underline{y}}\|$ we get

(1.6.22)

$$\|\underline{r}_n\| \le \frac{\varepsilon}{h \cdot L^1} [e^{L^1(x_n - x_o)} - 1].$$

Comparing (1.6.22) and (1.6.14) we see, as the accuracy of numerical integration depends upon the discretization error and the accumulated rounding error, that it is impossible to keep both of the errors small. To keep the discretization error small, we will normally choose the stepsize h small. On the other hand, the smaller h is taken, the more integration steps we shall have to perform, and the greater the rounding error is likely to be. An optimum value of the stepsize h exists but in practice it seems difficult to find it.

In order to obtain realistic statements about the behaviour of the propagated round-off errors, we now shall assume that the local round-off errors $\underline{\varepsilon}_n$ are random variables. In the simplest case $\underline{\varepsilon}_n$ is white noise process, i.e., cov $(\underline{\varepsilon}_n, \underline{\varepsilon}_m) = 0$, if $n \neq m$. Further we assume that $E\,\underline{\varepsilon}_n = \underline{0}$, and $E(\underline{\varepsilon}_n \cdot \underline{\varepsilon}_n^*) = B_{\varepsilon_n}$.

Note that in the case $E\,\underline{\varepsilon}_n = \underline{\mu}_n = \mu\,\underline{p}(x_n)$

(1.6.23) $$E\,\underline{r}_n = \frac{\mu}{h}(m(x_n) + O(h)),$$

where $\underline{m}(x)$ is the solution initial value problem

$$\underline{m}'(x) = G(x)\underline{m}(x) + \underline{p}(x), \tag{1.6.24}$$

with the assumption that matrix G(x) is given by

$$\underline{\Phi}(x_n, \underline{y}_n, h) - \underline{\Phi}(x_n, \tilde{\underline{y}}_n, h) = G(x_n)(\underline{y}_n - \tilde{\underline{y}}_n) + \varepsilon\underline{\theta}_n, \tag{1.6.25}$$

$$\varepsilon > 0, \ \|\theta_n\| < 1.$$

Returning to the case when $E\,\underline{\varepsilon}_n = 0$ the behavior of $\underline{r}_n$ in equation (1.6.20) under the condition (1.6.25) depends on $\underline{\varepsilon}_n$ in the following way:

$$\underline{r}_{n+1} = (I + h\,G(x_n))\,\underline{r}_n + \underline{\varepsilon}_{n+1}. \tag{1.6.26}$$

The random sequence $\underline{r}_n$ generally is not stationary because $\underline{r}_0 = 0$, but its steady state stationarity depends on the behaviour of matrix

$$Q_n = I + h \cdot G(x_n) \sim e^{hG(x_n)}. \tag{1.6.26}$$

One can easily find that in case of Gaussian variables $\underline{\varepsilon}_n$ the process $\underline{r}_n$ is Markov's. The covariance matrix

$$B_n = B\,\underline{r}_n\,\underline{r}_n^*, \tag{1.6.27}$$

or if we assume that $\underline{r}_x$ depends continuously on x with the stochastic differential equation

$$d\,\underline{r}_x = G(x)\underline{r}_x\,dx + d\underline{w}_x, \tag{1.6.28}$$

the x-dependent covariance matrix

$$B_x = E\,\underline{r}_x\,\underline{r}_x^*, \quad E\,(d\underline{w}_x\,d\underline{w}_x^*) = B_{w,x}\,dx, \tag{1.6.27'}$$

is the solution of equation (see (2.2.6))

$$B_x' = B_{w,x} + G(x)B_x + B_x\,G_x^*. \tag{1.6.29}$$

From Theorem 1 in Section 2.2 it follows that for large x the process $\underline{r}_x$ is stationary if G(x) = A and in this case $B_x' = 0$ and $B_x = B_0$ is the solution of equation (see (2.2.3))

(1.6.30) $$A\,B_0 + B_0\,A = -\,B_w\,,$$

i.e. $\underline{r}_x$ has a normal distribution with parameters $(0, B_0)$, the roots of the characteristic equation of matrix A must have negative real parts. In the discrete case B_0 is the solution of equation (see (2.1.1))

(1.6.30) $$B_0 = Q\,B_0\,Q^* + B_\varepsilon\,.$$

Note that if h is small $B_\varepsilon \simeq B_w \cdot h$ and from (1.6.30) we see that

$$B_0 \sim \frac{1}{h}\,(B_\varepsilon + O(h)).$$

In agreement with (1.6.22).

The above discussion we can summarize in the following statement (see P. Henrici [1] Section 3.4).

<u>Theorem 1</u>. Suppose the local round-off errors $\underline{\varepsilon}_n$ are Gaussian random variables with parameters $(\underline{0}, B_\varepsilon)$, then the accumulated round-off $\underline{r}_n$, satisfying equations (1.6.26) tends to a stationary, Markov process if the roots of characteristic equation of matrix $G(x) \sim A$ have negative real parts and the covariance matrix B_0 of $\underline{r}_n$ satisfies (1.6.30).

1.6.3. Probability Bounds and Asymptotic Properties of Error Propagation

In this part we investigate the asymptotic behaviour of the accumulated round-off errors in integrating a system of ordinary differential equations by one-step methods in the interval $0 \le x \le b$ on the basis of the related stochastic system of differential equations.

Let be given the following first order vector initial value problem

(1.6.31) $$\underline{y}'(x) = \underline{f}(x, \underline{y}(x)),\ \underline{y}(x_0) = \underline{y}(0),\ x_0 \le x \le b,$$

where $\underline{y}$ and $\underline{f}$ are column vectors. A one-step method is defined by the formula

(1.6.32) $$\underline{y}_{n+1} = \underline{y}_n + h\,\underline{\Phi}(x, \underline{y}_n;\ h),\ h > 0,\ x_n = x_0 + nh,\ \underline{y}_0 = \underline{y}(0),$$

where $\underline{\Phi}(x, \underline{y}; h)$ is called the increment function. We assume that

$$\underline{\Phi}(x_n, \hat{\underline{y}}_n; h) - \underline{\Phi}(x_n, \underline{y}_n; h) = G(x_n)\, \underline{r}_n + h\, \underline{\theta}_n, \tag{1.6.33}$$

where $\hat{\underline{y}}_n$ is the numerical approximation of $\underline{y}_n$. $\underline{r}_n(x) = \underline{r}_n = \hat{\underline{y}}_n - \underline{y}_n$ is the propagated error which fulfils the following equation (see Henrici [1])

$$\underline{r}_{n+1} - \underline{r}_n = h\, G(x_n)\underline{r}_n + \mu(h)\, \underline{p}(x_{n+1}) + B_\varepsilon^{1/2}(x_n)\, \underline{\varepsilon}_{n+1}, \tag{1.6.34}$$

$$\underline{r}_0 = 0,$$

where $\underline{\varepsilon}_n$ is the local error with $E\, \underline{\varepsilon}_n = 0$, $E\, \underline{\varepsilon}_n\, \underline{\varepsilon}_n^* = I$.

The accumulated round-off error, after n step, $\underline{r}_n$ fulfils the above stochastic difference equation (1.6.34) which can be handled as the discretization of the solution of a stochastic Ito type differential equation. This stochastic differential equation we called the related stochastic equation to the system of ordinary differential equation (1.6.31). As a measure of error behaviour we introduce the probability $P\{\max_{0<x<b} |\underline{r}_x|\}$, which can be reduced by the well known "time" scale transformation in diffusional type processes to the calculation of probability for the absolute value of Wiener process to remain less than 1. This results are closely connected with the estimates and asymptotic behaviour of nonexit probabilities of a Wiener process to a moving boundary (see Novikov (1979), (1981)).

Equation (1.6.34) indicates that the propagated error $\underline{r}_n = \underline{r}(x_n)$ can be regarded as a function of x and the solution of the following stochastic differential equation

$$d\underline{r}_x = G(x)\, \underline{r}_x\, dx + \mu\underline{p}(x)\, dx + B_w^{1/2}(x)\, d\underline{w}(x), \quad \underline{r}_0 = 0,$$

where $\underline{w}(x)$ is the standard Wiener process with $E\underline{w}(x) = 0$, $E\, \underline{w}(x)\, \underline{w}^*(x) = I.x$. In this case $\underline{r}_x$ is a Gaussian random process, the solution of the following linear equation

$$\underline{r}_x = \int_0^x [G(u)\, \underline{r}_u + \mu\underline{p}(u)]du + \int_0^x B_w^{1/2}(u)d\underline{w}(u), \quad \underline{r}_0 = 0, \tag{1.6.35}$$

where the second term is a stochastic integral with respect to the Wiener process.

It is possible and more adequate to regard $\underline{w}(x)$ as a wide sense Wiener process (see Liptser, Shiryaev [1], Section 15), but for simplicity we assume that $\underline{w}(x)$ is a Wiener process.

Lemma 3. Let

$$F(x) = \exp\{ \int_0^x G(u)\, du\}, \tag{1.6.36}$$

be the fundamental matrix, i.e., the solution of the differential equation

$$\frac{dF(x)}{dx} = G(x)\, F(x), \quad F(0) = I_{kxk}. \tag{1.6.37}$$

Then (see Appendix B1, th.6)

$$\underline{r}_x = F(x)\{ \int_0^x (F(s))^{-1} \mu\cdot\underline{p}(s)ds + \int_0^x (F(s))^{-1} B_w^{1/2}(s)d\underline{w}(s)\}, \quad \underline{r}_0 = 0. \tag{1.6.38}$$

Lemma 4. Let the elements of $\underline{p}(x)$, $G(x)$, $B_w(x)$ be integrable functions on $0 \le x \le b$, and let $\underline{r}_x$ fulfil the stochastic equation (1.6.35). Then $\underline{m}(x) = E\, \underline{r}_x$ and $B(x) = E\underline{r}_x\underline{r}_x^*$ are the solutions of the differential equations

$$\frac{d\underline{m}(x)}{dx} = \underline{p}(x) + G(x)\, \underline{m}(x), \tag{1.6.39}$$

$$\frac{dB(x)}{dx} = G(x)\, B(x) + B(x)\, G^*(x) + B_w(x). \tag{1.6.40}$$

As the natural measure of the error behaviour we use

$$\underline{P}\{ \max_{0\le x\le b} \| F^{-1}(x)\, \underline{r}_x\| \le k\}, \tag{1.6.41}$$

where matrix $F(x)$ is given by (1.6.36). We propose this bounds in the one dimensional case, instead of the estimation of the mean $\underline{m}(x)$, as it is accepted in the literature (see Henrici [1]).

Let us recall here that in the case $p(x) = 0$ and

$$G(x) = \frac{m'(x)}{m(x)} \tag{1.6.42}$$

with positive continuous function $m(x)$, $x \ge 0$, the following statement holds

Theorem 2.

$$\frac{8}{3\pi} \le P\{|r_x \le km(x),\ 0 \le x \le b\} \exp\{\frac{\pi^2}{8k^2} \int_0^b m^{-2}(x) b_w(x)\, dx\} \le \frac{4}{\pi}. \tag{1.6.43}$$

Proof. The following representation for the probability that a Wiener process w(t) does not exit the interval [-k, k] is known:

$$P\{\sup_{0\le u\le T} |w(u)| \le k\} = \frac{4}{\pi} \sum_{n=0}^{\infty} \frac{(-1)^n}{2\ n+1} \exp\{-(2n+1)^2 \frac{\pi^2 T}{8k^2}\}. \tag{1.6.44}$$

On the other side for the Gaussian random processes r_x, $x \ge 0$, defined by the relation

$$r_x = m(x) \int_0^x m^{-1}(u)\, b_w^{1/2}(u)\, dw(u), \tag{1.6.45}$$

we have

$$P\{\sup_{0\le x\le b} |r_x| \le m(x)\} = P\{\sup_{0\le x\le b} |\int_0^x m^{-1}(u)\, b_w^{1/2}(u)| \le 1\} = \tag{1.6.46}$$

$$= P\{|\tilde{w}(x)| \le 1,\ 0 \le x \le \int_0^b m^{-2}(u) b(u) du\},$$

where $\tilde{w}(x)$ is a new Wiener process obtained by the "time" change

$$u = \int_0^x m^{-2}(s) d(s) ds \tag{1.6.47}$$

in the stochastic integral $\int_0^x m^{-1}(u) b_w^{1/2}(u) dw(u)$ (see Gikhman, Skorokhod [2]).

Taking one and two terms respectively in the alternating series in (1.6.44) one can get the estimates

$$\frac{4}{\pi} [\exp(-\frac{\pi^2}{8k^2} T) - \frac{1}{3} \exp(-\frac{9\pi^2}{8k^2} T)] \le P\{\sup_{0\le t\le T} |(w(u)| \le k\} \le$$

$$\le \frac{4}{\pi} \exp(-\frac{\pi^2}{8k^2} T).$$

This and (1.6.46) imply (1.6.43).

In the k-dimensional case it is known that the following inequalities hold for the standard Wiener process $\underline{w}(t)$ with independent components (see Skorokhod [1])

(1.6.48) $$P\{|\underline{w}(T)| \geq C\} \leq P\{\sup_{0\leq t\leq T} |\underline{w}(t)| \geq C\} \leq 2P\{|\underline{w}(T)| \geq C\},$$

where

(1.6.49) $$P\{|\underline{w}(T)| \geq C\} = \frac{1}{\Gamma(\frac{k}{2})2^{\frac{k-2}{2}}} \int_{\frac{C}{\sqrt{T}}}^{\infty} e^{-\frac{r^2}{2}} r^{k-1}dr \leq$$

$$\leq (\frac{2T}{\pi})^{1/2} \frac{k^{3/2}}{C} e^{-\frac{C^2}{2Tk}}$$

From (1.6.38); assuming $\underline{p}(x) = 0$, we get as a first approximation for the ith component

(1.6.50) $$P\{\sup_{0\leq x\leq b} |F_i^{-1}(x)\underline{r}_x| \leq 1\} = P\{\sup_{0\leq x\leq b} |\int_0^x F_i^{-1}(s)B_w^{1/2}(s)d\underline{w}(s)|\leq 1\} =$$

$$= P\{|\tilde{\underline{w}}(x)| \leq 1, 0 \leq x \leq \int_0^b \|F_i^{-1}(u)B_w^{1/2}(u)\|^2du\},$$

where $\tilde{\underline{w}}(x)$ is a new Wiener process obtained by the "time" change (see McKean [1])

(1.6.51) $$u = \int_0^x \|F_i^{-1}(s)\; B_w^{1/2}(s)\|^2ds,$$

in the stochastic integral $\int_0^x F_i^{-1}(s)B_w^{1/2}(s)d\underline{w}(s)$. Comparing (1.6.48) and (1.6.50) we get the following statement.

<u>Theorem 3</u>. Let $B = \int_0^b \|F^{-1}(u)B_w^{1/2}(u)\|^2du$, then

$$1 - 2\,P\{|\tilde{\underline{w}}(B)| \geq C\} \leq P\{\sup_{0\leq x\leq b} |F_i^{-1}(x)\underline{r}_x| \leq C\} \leq 1 - P\{|\tilde{w}(B)| \geq C\}.$$

Remark. From (1.6.49) it follows that

$$\lim_{\frac{c^2}{T} \to \infty} P\{ \sup_{0 \le t \le T} |\underline{w}(t)| \ge c\} \, e^{(\frac{1}{2} + \varepsilon)\frac{c^2}{T}} = \infty ,$$

$$\lim_{\frac{c^2}{T} \to \infty} P\{ \sup_{0 \le t \le T} |\underline{w}(t)| \ge c\} \, e^{(\frac{1}{2} - \varepsilon)\frac{c^2}{T}} = 0.$$

1.7 The Sunspot Activity

The sunspot activity and the changes in it have been observed long ago. The series is obtainable on a daily basis from 1818, on a monthly basis from 1749 and regular observations on one year basis are obtainable from 1610 (see e.g. Waldmeier [1], H.W. Newton [1]). It is remarkable that the polar light activity, which is highly correlated with the sunspot activity, is registered from 500 B.C. (see Fritz [1], [2] and Slutsky (1935)). The series has no growth in it, but only a systematic oscillation of approximate 11-year periods, which was suggested by Schwabe in 1843. Moreover, in this oscillation, if one examines the graph of the series or the empirical covariance function, the amplitude varies over a forty-to-sixty-year period. This latter oscillation seems not an exact period motion. No theoretical foundation exists to explain the nearly 11 year period (cycle); however, its effects have been felt on the earth in a variety of ways. Another aspect that becomes apparent from an examination of the graph is the non-sinusoidal nature of the waveform. The waves tend to rise quickly and fall slowly and also to remain longer at minimum values than maximum.

The calculations on a daily relative sunspot numbers are based upon counts of spots and group entities of spots on the sun's surface at some time each day. A basic difficulty lies in the fact that the number of spots and groups seen depends on the particular observer and his telescope. Wolf devised a formula, yielding a relative number, where, for a given day, the number of groups, the number of component spots and the efficiency of the observer and his telescope were included.

The random series of sunspot activity was investigated many times by different statisticians. Yule in 1927 first suggested in his pioneering work the autoregressive model to describe this phenomena. Slutsky in 1935 analyzed the data on polar light from 500 B.C. and proposed a 11.103 year period for the sunspot activity. This period was compared by him to the minimum and maximum values of the sunspot activity from 1610 and the fitting was almost exact. With this respect, note that using the number of peaks (maximal values), which is 32, in the time interval of years 1615.5 and 1969.25 one can easily obtain the estimate 11.055 for the period of the sunspot activity.

In Table 1. we collected some of the estimations of the two unknown parameters by different authors to show the sensitive differences between different estimations and the difference between the 11.0 years period of peaks and 11.2 year period in the damped oscillation of covariance function (see Fig. 2.a). We take advantage of continuous time description of the phenomena to explain the differences and underline the role of observation error, which can be measured by the smoothed data.

After subtracting the mean value ($E\xi(t) = 46.92$/year) the number of sunspots $\xi(t)$ satisfies the equation

$$(1.7.1) \qquad d\,\xi'(t) + (a_1\xi'(t) + a_2\xi(t))\,dt = dw(t),$$

which is a second order AR process. The covariance function of the two dimensional process $(\xi(t), \xi'(t))$ has the form

$$(1.7.2) \qquad B(t) = e^{A|t|}\,B(0) =$$

$$= \frac{1}{\omega}\,e^{-\lambda|t|}\begin{pmatrix} \omega\cos\omega t + \lambda\sin\omega|t| & \sin\omega|t| \\ -(\lambda^2+\omega^2)\sin\omega|t| & \omega\cos\omega t - \lambda\sin\omega|t| \end{pmatrix} B(0)$$

where

$$(1.7.3) \qquad A = \begin{pmatrix} 0 & 1 \\ -a_2 & -a_1 \end{pmatrix}, \quad \lambda = \frac{a_1}{2}, \quad \omega = \sqrt{a_2 - \frac{a_1^2}{4}}, \quad \omega^2 = \omega_0^2 - \lambda^2, \quad \omega_0 = \sqrt{a_2},$$

$$B(0) = \sigma_w^2 \begin{pmatrix} \frac{1}{4\lambda(\lambda^2+\omega^2)} & 0 \\ 0 & \frac{1}{4\lambda} \end{pmatrix}, \qquad T = \frac{2\pi}{\omega} \text{ years.}$$

Table 1.

Estimates of parameters in the second order A R model of sunspot activity

	a_1	a_2	a	b	ω	ω_o	$\frac{2\pi}{\omega}$=T years
Schwabe (1843)							11.0
Yule (1927)			-1.5153	0.8025			
Slutsky (1935)						0.5659	11.103
Bartlett (1950)	0.3186	0.3631	-1.4255	0.7272	0.5811		
Anderson (1971)			-1.352	0.655			
Box & Jenkins (1970)			-1.316	0.632			
G. Nemeth (1973)	0.514	0.381			0.5621		11.2
Brillinger (1975)					0.5658		11.1
Kashyap & Rao (1976)			-1.352	0.666			
Proposed	0.07				0.5661		

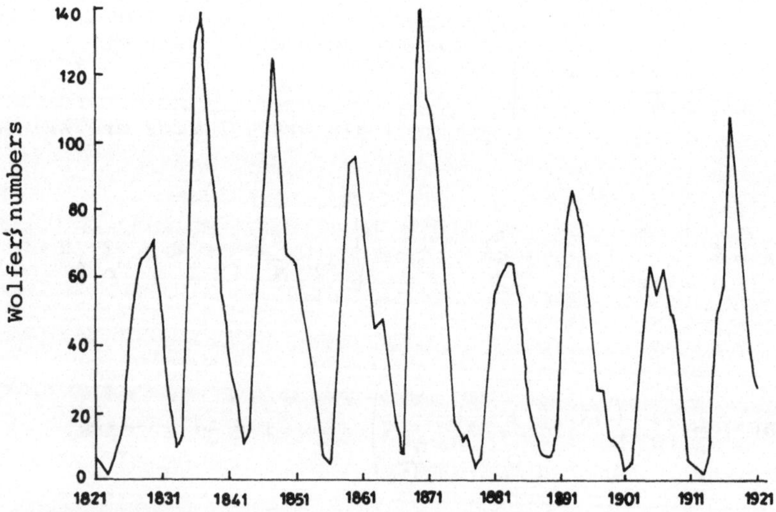

Fig. 1 Annual sunspot activity (Wolfer's numbers (mean sunspot numbers))

Or using the discrete approximation (see Section 4.5) we obtain

(1.7.4) $$\xi_\Delta(n) + a\xi_\Delta(n-1) + b\xi_\Delta(n-2) = \varepsilon(n),$$

where (see (4.5.15) - (4.5.21))

(1.7.5) $$a = -2\,e^{-\lambda\Delta}\cos\omega\Delta, \quad b = e^{-2\lambda},$$

(1.7.6) $$B(n) = \frac{\sigma_w^2}{4\lambda(\lambda^2+\omega^2)}\,e^{-\lambda\Delta_n}\,\frac{\cos(\omega n\Delta-\psi)}{\cos\psi}$$

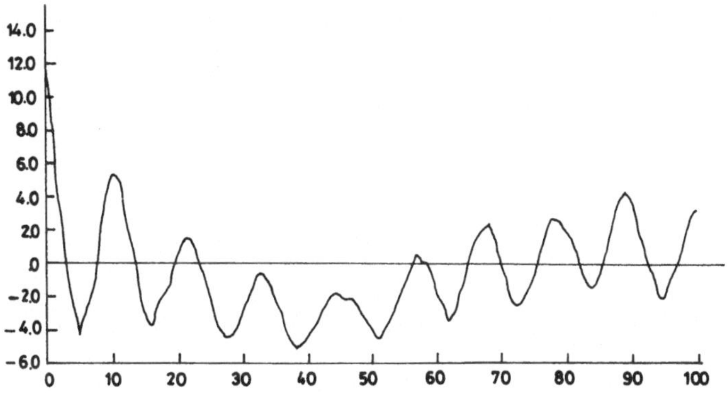

Figure 2.a. The autocovariance estimate for annual mean sunspot numbers for the year 1750-1965.

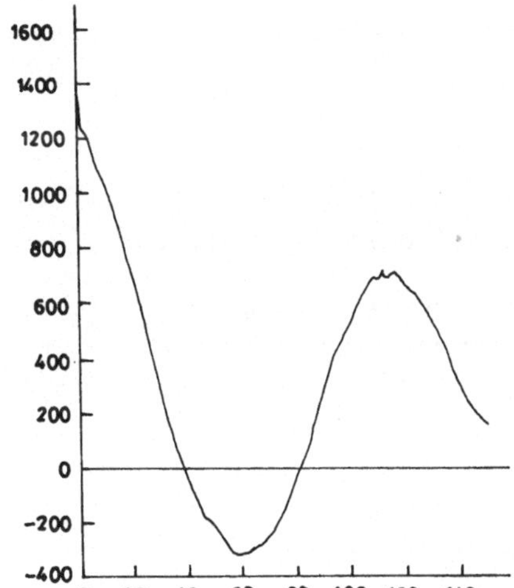

Figure 2.b. The autocovariance estimate for monthly mean sunspot numbers for the years 1750-1965.

As the sample covariance function of a series often is useful for examining the structure of a series in Figures 1 and 2, we present portions of the sample and the sample covariances the mean annual and mean monthly sunspot numbers, respectively.

Figure 3. presents the smoothed periodogram of the series of annual sunspot numbers. Apart from its behaviour at the origin the spectrum consists of a smooth hump-backed curve with a peak at $\frac{\omega}{2\pi} = 0.090$ in agreement with the estimated value of the period T = 11 year. Figures 4.a and 4.b present the smoothed periodogram for m = 2,20, with 2 m + 1 periodogram ordinates averaged. Only the periodograms corresponding to m = 2 suggest a possible peak in the spectrum which corresponds to the eleven-year solar cycle.

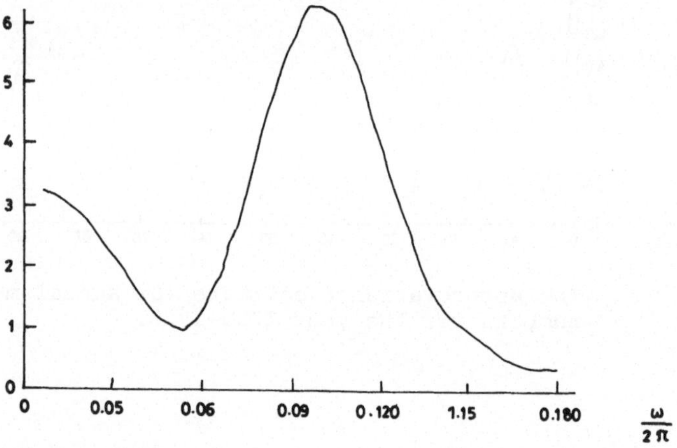

Figure 3. Smoothed periodogram (n = 24) of Wolfer's sunspot numbers.

In order to estimate the unknown parameters in (1.7.1) we take advantage of Section 4.5. The approximate solutions of the likelihood equations give the following estimates (see (4.5.6))

$$(1.7.7)\quad \hat{a}_1 = -\frac{\int_0^T \xi'(t)d\xi'(t)}{\int_0^T (\xi'(t))^2 dt} = -\frac{\frac{1}{2}[\sigma_w^2 T + (\xi'(T))^2 - (\xi'(0))^2]}{\int_0^T (\xi'(t))^2\, dt},$$

$$(1.7.8)\quad \hat{a}_2 = \frac{\int_0^T (\xi'(t))^2\, dt}{\int_0^T (\xi(t))^2\, dt},$$

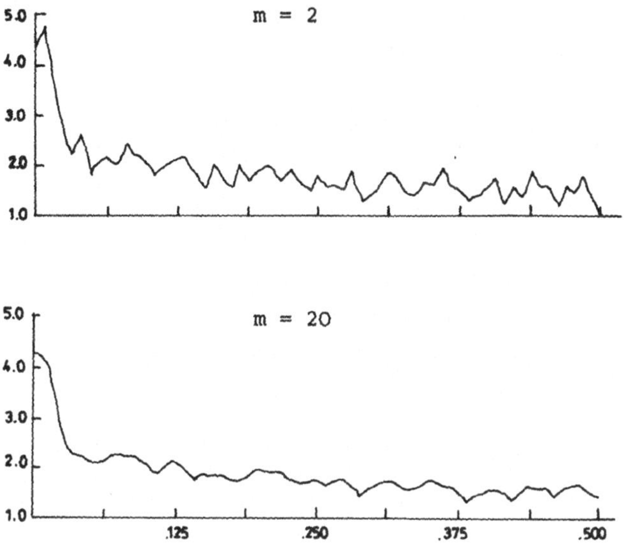

Figure 4.a. $\frac{\lambda}{2\pi}$ Frequency in cycles per month.

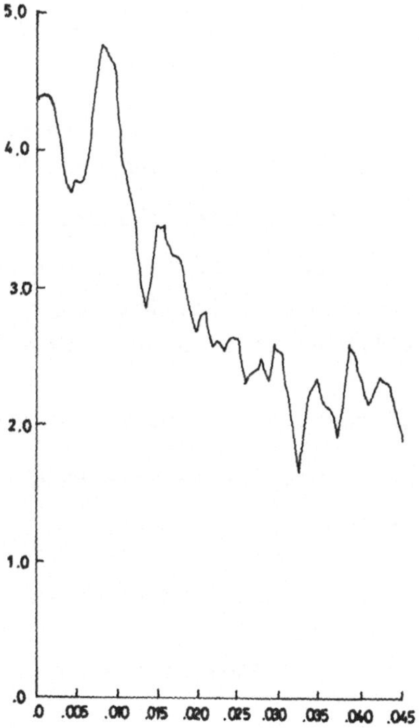

Figure 4.b. $\frac{\lambda}{2\pi}$ Frequency in cycles per month.

with the asymptotic errors on the basis of the ergodic Theorem or Theorem 2 in Section 4.6

$$(1.7.9)\quad D^2(\hat{a}_1) \sim \frac{\sigma_w^2}{\int_0^T (\xi'(t))^2\,dt} \sim \frac{2a_1}{T}, \quad D^2(\hat{a}_2) \sim \frac{\sigma_w^2}{\int_0^T (\xi(t))^2\,dt} \sim \frac{2a_1a_2}{T}.$$

It may be argued that the continuous process model (1.7.1.), which represents continuous sunspot activity, is a more consistent model. The estimates of parameters a_1 and a_2 show that better agreement over the first correlations gives a loss for the later correlations.

The natural parameters describing the process $(\xi(t), \xi'(t))$ are λ and ω_o (see (1.7.3) and (1.7.5)), and if we assume that ω_o is given it remains λ as the unique unknown parameter. ω_o is the frequency of the process $\xi(t)$, while ω ($< \omega_o$) is the frequency of the correlation function (see (1.7.2)).

As λ and ω_o are connected with the energy dissipation (see Sections 1.2 and 1.4), where $Q^{-1} = \frac{2\lambda}{\omega_o}$, it seems that the smaller values of λ are more realistic. This observation means that the less damping in correlogram is more natural, and even more in this case the frequencies ω_o^2 $(= \omega^2 + \lambda^2)$ and ω^2 are closed. The frequency of harmonic oscillator $\omega_o(> \omega)$ has exact meaning in the two dimensional (complex) case, where the oscillation is exactly given. This approach can be used also in the description of linear oscillator in Statistical Mechanics.

As it was mentioned that the period of the process $(T_o = \frac{2\pi}{\omega_o})$ can be estimated by a longer observation time than the correlogramm we recall here the results of Slutsky.

Let us consider the polar lights data of Table 2 after rounding up the epochs by subtracting throughout 1/2 (column 2). Taking the middle term τ_o and T_o as the solutions of the least square estimation

$$\sum_k (\tau_o + kT_o - t_k)^2 = \min,$$

with k being the ordinal numbers we get (see Slutsky, 1935):

$$\hat{\tau}_o = 550.78,$$

$$\hat{T}_o = 11.103.$$

This period $T_o = 11.1$ (for the years 1615 - 1969 we get a rough estimation $T_o = 11.05$) can be accepted as the process frequency. This means $\sqrt{a_2} = \frac{2\pi}{T_o} = \omega_o = 0.5661$. In this case $\lambda = 0.07$, in agreement with the covariances, and assuming a large observation error (see § 1.8).

As σ_w^2 can be exactly calculated (see Theorem 3 in Section 2.2.1) by

$$\frac{1}{N}\sum_{i=1}^{N}(\xi'(t_i) - \xi'(t_{i-1}))^2 \rightarrow \sigma_w^2 \approx 5.08 ,$$

we get for the estimate of a_1 (see G. Németh)

$$\hat{a}_1 = 0.514$$

and an extremely large period T, as

(1.7.11)
$$\hat{\lambda} = 0.257 \qquad \hat{\omega} = \sqrt{\hat{a}_2 - \lambda^2} = 0.5044$$

$$T = \frac{2\pi}{\omega} = 12.46.$$

Using the estimate of Bartlett for λ we get again a very large period T

1.7.12
$$\hat{\lambda} = 0.1593 , \qquad \hat{\omega} = \sqrt{0.3205 - 0.0254} = 0.5432,$$

$$T = \frac{2\pi}{\omega} = 11.57.$$

To explain the differences between the λ values given by different authors we give the confidence limits for λ, using advantage of Remark 1 in section 4.5 that $\hat{\lambda}$ has the distribution given in Table 1 in Chapter 3.

Table 2

Polar lights epochs. τ_o = 550.75 as middle term t_k empirical epoch of polar light (maximal sunspot activity)

$\tau_k = \tau_o + kT_o$

1	2	3	1	2	3
k	t_k	$t_k-\tau_k$	k	t_k	$t_k-\tau_k$
-95	-501.5	2.5	-58	-92.5	0.7
-91	-460.5	-0.9	-55	-61.5	-1.6
-89	-441.5	-4.1	-53	-42.5	-4.8
-81	-348.5	0.0	-48	14.5	-3.4
-69	-215.5	-0.2	-45	50.5	-0.7
-58	-202.5	1.7	-32	194.5	-1.0
-65	-168.5	2.4	-14	397.5	2.2
-64	-160.5	-0.7	-9	451.5	0.6
-59	-101.5	2.8	-4	503.5	-2.9

1	2	3	1	2	3
k	t_k	$t_k-\tau_k$	k	t_k	$t_k-\tau_k$
1	566.5	4.6	51	1117.5	0.5
3	585.5	1.4	59	1203.5	-2.3
6	616.5	-0.9	68	1306.5	0.7
11	676.5	3.6	73	1361.5	0.2
17	742.5	3.0	88	1529.5	1.7
23	807.5	1.4	90	1546.5	-3.5
32	905.5	-0.6	91	1560.5	-0.6
40	992.5	-2.4	93	1580.5	-2.8
49	1098.5	3.7	95	1605.5	0

Let λ = 0.257 then we get:

Lower bounds: $\lambda_{0.1}$ = 0.183, $\lambda_{0.05}$ = 0.161, $\lambda_{0.01}$ = 0.132.

Upper bounds: $\lambda_{0.9}$ = 0.319, $\lambda_{0.95}$ = 0.339, $\lambda_{0.99}$ = 0.374.

The same confidence limits, using the normal approximation (1.7.9) (T = 174 years):

Lower bounds: $\lambda_{0.1} = 0.206$, $\lambda_{0.05} = 0.193$, $\lambda_{0.01} = 0.167$

Upper bounds: $\lambda_{0.9} = 0.306$, $\lambda_{0.95} = 0.319$, $\lambda_{0.99} = 0.345$

1.8. Kalman Filtering with Explicite Solutions (Signal Plus Noise Case)

The problem of optimal linear nonstationary filtering examined by Kalman and Bucy consists of the following. Suppose the process $\theta(s)$, $0 \le s \le t$, is unobservable and one can only observe the values of $\xi(s)$, $0 \le s \le t$, containing imcomplete information on the values on $\theta(t)$. It is required at each moment t to estimate in the optimal way the values of $\theta(t)$ on the basis of the observed process:

$$\xi_0^t = \{\xi(s),\ 0 \le s \le t\}.$$

Taking the conditional expectation

$$m(t) = E(\theta(t)|\mathcal{F}_\xi^t)$$

it is optimal in the mean square sense. We denote the estimation error by

$$\gamma(t) = E(\theta(t) - m(t))^2.$$

The method of Kalman and Bucy yields a closed system of dynamic equations for $m(t)$ and $\gamma(t)$ if $(\theta(t),\xi(t))$ forms a two-dimensional Gaussian random process satisfying the stochastic differential equations

$$d\theta(t) = [a_1(t)\theta(t) + a_2(t)\xi(t)]dt + b_1 dw_1(t) + b_2 dw_2(t),$$

$$d\xi(t) = [A_1(t)\theta(t) + A_2(t)\xi(t)]dt + B_1 dw_1(t) + B_2 dw_2(t).$$

The conditional expectation $m(t) = E(\theta(t)|\mathcal{F}_t^\xi)$ and mean square error of filtering $\gamma(t) = E(\theta(t) - m(t))^2$ satisfy the system of equations

$$dm(t) = [a_1(t)\ m(t) + a_2\xi(t)]dt + \frac{b_1B_1 + b_2B_2 + \gamma(t)A_1(t)}{B_1^2 + B_2^2}[d\xi(t) +$$

$$- (A_1(t)m(t) + A_2(t)\xi(t))dt],$$

$$\dot{\gamma}(t) = 2a_1(t)\gamma(t) + (b_1^2 + b_2^2) - \frac{[b_1B_1 + b_2B_2 + \gamma(t)A_1(t)]^2}{B_1^2 + B_2^2},$$

with the initial conditions

$$\gamma_o = \gamma(0), \quad m(0) = E(\theta(0)|\xi(0)).$$

The above differential equation for $\gamma(t)$ is a Riccati type equation and in the general case no explicite solution exists for it. We shall not deduce here the filtering equations for $m(t)$ and $\gamma(t)$, the reader can find in every textbook on filtering theory. We shall be interested in the cases when for $\gamma(t)$ explicite solution can be given. It turns out (see (9), (14) later) that in the case when the functions a_1, a_2, A_1, A_2, b_1, b_2, B_1, B_2 are constant - in spite of the "general belief" - one can get explicite solutions. This allows us to make some general statements for the estimation theory with rational spectral densities and for estimation of parameters of elementary Gaussian processes with additive noise. In the following we shall use many times the filtering theorem in the multidimensional case. For our goals we formulate it in the Gaussian case (see e.g. in Liptser - Shiryaev [1], Theorem 12.7).

Theorem 1. Let us consider the $(k + \ell)$ dimensional Gaussian random process $(\underline{\theta}(t), \underline{\xi}(t))^* = [(\theta^1(t),\dots, \theta^k(t)), (\xi^1(t),\dots, \xi(t))]$, $0 \le t \le T$, with

$$(1.8.1) \qquad d\underline{\theta}(t) = [\underline{a}_o(t) + a_1(t)\underline{\theta}(t) + a_2(t)\underline{\xi}(t)]dt + $$
$$+ b_1d\underline{w}_1(t) + b_2d\underline{w}_2(t),$$

$$(1.8.2) \qquad d\underline{\xi}(t) = [\underline{A}_o(t) + A_1(t)\underline{\theta}(t) + A_2(t)\underline{\xi}(t)]dt +$$
$$+ B_1d\underline{w}_1(t) + B_2d\underline{w}_2(t),$$

where $\underline{w}_1^* = (w_1^1, w_1^2, \ldots, w_1^k)$ and $\underline{w}_2^* = (w_2^1, \ldots, w_2^\ell)$ are independent Wiener processes. $\underline{\theta}(0)$, $\underline{\xi}(0)$ are Gaussian and independent of the processes $\underline{w}_1(t)$, $\underline{w}_2(t)$, $t \geq 0$. The measurable deterministic functions $a_i(t)$, $A_i(t)$ are square integrable. Let $\underline{m}(t) = E(\underline{\theta}(t)/\mathcal{F}_t^\xi)$, $\gamma(t) = E(\underline{\theta}(t) - \underline{m}(t))(\underline{\theta}(t) - \underline{m}(t))^*$ be the conditional expectation vector and covariance matrix respectively. Then

$$(1.8.3) \quad d\underline{m}(t) = [\underline{a}_0(t) + a_1(t)\underline{m}(t) + a_2(t)\underline{\xi}(t)]dt + [(boB) + \gamma(t)A_1^*(t)]\,(BoB)^{-1}[d\underline{\xi}(t) - (\underline{A}_0(t) + A_1(t)\underline{m}(t) + A_2(t)\underline{\xi}(t))dt],$$

$$(1.8.4) \quad \gamma(t) = a_1(t)\gamma(t) + \gamma(t)a_1^*(t) + bob - [(boB) + \gamma(t)A_1^*(t)] \times (BoB)^{-1}\,[boB + \gamma(t)A_1^*(t)]^*,$$

with the initial conditions

$$\underline{m}(0) = E(\underline{\theta}(0)\,|\,\underline{\xi}(0)), \quad \gamma(0) = (\gamma_{ij}(0)).$$

We are using the notation

$$(1.8.5) \qquad boB = b_1B_1^* + b_2B_2^* \,.$$

As we are interested in stochastic equations with constant coefficients the following lemmas will be very useful, and they can be checked by direct calculations.

Lemma 1. The homogeneous differential equation

$$(1.8.6) \qquad \dot{\gamma}(t) = -A\gamma(t) - B\gamma^2(t), \quad A \geq 0,\ B \geq 0,$$

has the solution

$$(1.8.7) \qquad \gamma(t) = e^{-At}\,[c_0 + \frac{B}{A}(1 - e^{-At})]^{-1}, \quad \gamma(0) = \frac{1}{c_0}\,.$$

The inhomogeneous equation

$$(1.8.8) \qquad \dot{\gamma}(t) = -A\gamma(t) - B\gamma^2(t) + b\,, \quad A \geq 0,\ B \geq 0,\ b \geq 0,$$

has the solution

$$(1.8.9) \qquad \gamma(t) = e^{-\tilde{A}t}\,[c_o + \frac{B}{\tilde{A}}\,(1 - e^{-\tilde{A}t})]^{-1} + c,$$

where c is the root of the following equation, called particular solution,

$$(1.8.10) \qquad Bc^2 + Ac - b = 0, \qquad c = \frac{-A \pm \sqrt{A^2 + 4Bb}}{2B}$$

and

$$(1.8.11) \qquad \gamma(0) = \frac{1}{c_o} + c, \qquad \tilde{A} = A + 2cB$$

In the case when B = 0 we have

$$(1.8.12) \qquad \gamma(t) = \frac{1}{c_o}\,e^{-At} + c\,, \qquad c = \frac{b}{A}\,.$$

Later we shall use these explicite solutions for calculating the Radon-Nikodym derivatives of stationary processes with rational spectral densities (see Section 2.3.4).

Lemma 2. Let a, b, A, B be constant matrices such that the inverse in (1.8.14) exists. Then the matrix solution $\gamma(t)$ of the following Euler-Riccati type equation

$$(1.8.13) \qquad \dot{\gamma}(t) = a\gamma(t) + \gamma(t)a^* + bb^* - \gamma(t)A^*(BB^*)^{-1}A\gamma(t)$$

is given by

$$(1.8.14) \qquad \gamma(t) = e^{\tilde{a}t}\,[\,c_o + \int_0^t e^{\tilde{a}^*u}\,A^*(BB^*)^{-1}A\,e^{\tilde{a}u}\,du]^{-1}\,e^{\tilde{a}^*t} + c,$$

where c is the solution, called a particular or stationary solution, of the following equation:

$$(1.8.15) \qquad ac + ca^* + bb^* - cA^*(BB^*)^{-1}Ac = 0, \; \tilde{a} = a - cA^*(BB^*)^{-1}A,$$

$$\tilde{a}^* = a^* - A^*(BB^*)^{-1}Ac.$$

The proof can be carried out by direct calculations (see Appendix (A1.21)). We note that using standard transformations (see e.g. in

Reid [1]) one can easily see that the above Riccati equation is equivalent to two systems of linear equations with constant coefficients, which are, of course, explicitely solvable.

Although the above lemmas are simple they are useful in many investigations and we should note that even in most textbooks it is tacitly considered that no explicite solutions can be given for Riccati equations (see e.g. Liptser-Shiryaev [2] in Sections 15.3, 16.2, 16.3, 17.1). In this part we shall indicate only some illustrating examples. The first will show that the Kalman-Bucy filtering problem with constant coefficients can be exactly solved even if instead of the Wiener process we take a stationary process with colored spectrum. This solution, taking the appropriate limiting in the spectral densities, gives the answer of a problem of A. Balakrishnan who used the white noise instead of a Wiener process as noise (see Balakrishnan (1978), (1981)).

In the second example we shall deal with the estimation problem of the drift parameter of the elementary Gaussian process in the presence of noise. This problem was discussed by many authors, mostly in the discrete time case, earlier (see e.g. P. Whittle (1953), Dzhaparidze [1], Dzhaparidze (1971), Shaman (1973)).

<u>Example 1</u>. a) Estimation of the unobservable component is the presence of uncorrelated noise (see e.g. in Liptser-Shiryaev [1], Section 15.3.4).

We assume that $\theta(t)$ is a one dimensional first order autoregressive type Gaussian process, i.e., it is the solution of the differential equation

$$(1.8.16) \qquad d\theta(t) = -\alpha\theta(t)dt + \sqrt{c_1}\, dw_1(t),$$

where $w_1(t)$ is a Wiener process.

Let

$$(1.8.17) \qquad \xi(t) = \theta(t) + \varepsilon(t)$$

be the sum of two processes, where

$$(1.8.18) \qquad d\varepsilon(t) = -\beta\varepsilon(t)dt + \sqrt{c_2}\, dw_2(t),$$

and $w_1(t)$, $w_2(t)$, $(0 \le t)$, are independent Wiener processes, independent of $\theta(0)$, $\xi(0)$. The spectral representation of these processes in the following (see Lemmas 5 and 6 in Section 2.1.3).

$$\theta(t) = \sqrt{c_1} \int_{-\infty}^{\infty} \frac{e^{i\lambda t}}{\alpha + i\lambda} \Phi_1(d\lambda) \tag{1.8.16'}$$

and

$$\xi(t) = \sqrt{c_1} \int_{-\infty}^{\infty} \frac{e^{i\lambda t}}{\alpha + i\lambda} \Phi_1(d\lambda) + \sqrt{c_2} \int_{-\infty}^{\infty} \frac{e^{i\lambda t}}{\beta + i\lambda} \Phi_2(d\lambda), \tag{1.8.17'}$$

where $\Phi_1(d\lambda)$, $\Phi_2(d\lambda)$ are orthogonal independent measures, with

$$E\, \Phi j(d\lambda) = 0, \quad E|\Phi\gamma(d\lambda)|^2 = \frac{d\lambda}{2\pi}, \quad \text{and}$$

$$w_j(t) = \int_{-\infty}^{\infty} \frac{e^{i\lambda t}-1}{i\lambda} \Phi_j(d\lambda), \qquad j = 1,2,$$

are Wiener processes.

(1.8.16) - (1.8.17) can be written in the form

$$d\theta(t) = -\alpha\theta(t)dt + \sqrt{c_1}\, dw_1(t), \tag{1.8.19}$$

$$d\xi(t) = d\theta(t) + d\varepsilon(t) = \tag{1.8.20}$$

$$= -(\alpha-\beta)\theta(t)dt - \beta\xi(t)dt + \sqrt{c_1}dw_1(t) + \sqrt{c_2}dw_2(t),$$

where $\xi(t)$ is observable, while $\theta(t)$ is unobservable. The problem of estimating $\theta(t)$ from $\xi(s)$, $0 \le s \le t$, is a conventional problem of estimating the "signal" $\theta(t)$ in additive "noise" $\varepsilon(t)$. There we assume that $\varepsilon(t)$ is also "colored". When $\varepsilon(t)$ is a Wiener process we shall discuss in another respect.

Theorem 1 gives the following filtering equations

$$dm(t) = -\alpha m(t)dt + \frac{c_1 + (\beta-\alpha)\gamma(t)}{c_1 + c_2} [d\xi(t) - ((\beta-\alpha)m(t) - \tag{1.8.21}$$

$$-\beta\xi\,(t))dt],$$

(1.8.22) $$\dot{\gamma}(t) = -2\alpha\gamma(t) + c_1 - \frac{1}{c_1 + c_2}(c_1 + (\beta-\alpha)\gamma(t))^2 =$$

$$= -2[\alpha + \frac{c_1(\beta-\alpha)}{c_1 + c_2}]\gamma(t) - \frac{(\beta-\alpha)^2}{c_1 + c_2}\gamma^2(t) + \frac{c_1 c_2}{c_1 + c_2}.$$

From Lemma 1 one can get

(1.8.23) $$\gamma(t) = e^{-\tilde{A}t}[c_o + \frac{B}{\tilde{A}}(1 - e^{-\tilde{A}t})]^{-1} + c,$$

where

$$A = 2\frac{\alpha c_2 + \beta c_1}{c_1 + c_2}, \quad B = \frac{(\beta-\alpha)^2}{c_1 + c_2}, \quad b = \frac{c_1 c_2}{c_1 + c_2},$$

(1.8.24)

$$c = \frac{-A + \sqrt{A^2 + 4Bb}}{2B} = -\frac{(\alpha c_2 + \beta c_1)}{(\beta-\alpha)^2} +$$

$$+ \frac{\sqrt{(\alpha c_2 + \beta c_1)^2 + (\beta-\alpha)^2 c_1 c_2}}{(\beta-\alpha)^2},$$

$$\tilde{A} = A + 2cB,$$

and

(1.8.25) $$\frac{1}{c_o} = \gamma_o - c, \quad \gamma_o = \frac{c_1 c_2}{2\alpha(\alpha c_2 + \beta c_1)}.$$

For $m(t)$ we get

$$dm(t) = -[\alpha + \frac{c_1(\beta-\alpha)}{c_1 + c_2} + \frac{(\beta-\alpha)^2}{c_1 + c_2}\gamma(t)]\,m(t)dt +$$

$$+ \beta[\frac{c_1}{c_1 + c_1} + \frac{c_1(\beta-\alpha)}{c_1 + c_2}\gamma(t)]\xi(t)dt +$$

$$+ [\frac{c_1}{c_1 + c_2} + \frac{\beta-\alpha}{c_1 + c_2}\gamma(t)]d\xi(t) =$$

$$= -[A/2 + B\gamma(t)]\,m(t)dt + \frac{\beta}{\beta-\alpha}[A/2 + B\gamma(t) - \alpha]\xi(t)dt +$$

$$+ \frac{1}{\beta-\alpha}[A/2 + B\gamma(t) - \alpha]d\xi(t),$$

$$(1.8.26)\quad m(t) = e^{-\int_0^t [A/2 + B\gamma(s)]ds}\{m(0) + \int_0^t e^{\int_0^s [A/2 + B\gamma(u)]du}\{\frac{\beta}{\beta-\alpha}[\frac{A}{2} + B\gamma(s)-\alpha]\xi(s)ds + \frac{1}{\beta-\alpha}[\frac{A}{2} + B\gamma(s)-\alpha]d\xi(s)\}\},$$

where

$$m(0) = \frac{c_1\beta}{\alpha c_2 + \beta c_1}\xi(0).$$

The last term in (1.8.26) can be given (using (1.8.22)) in the following way

$$(1.8.27)\quad \int_0^t \exp[\int_0^s (A/2 + B\gamma(u)\,du]\frac{1}{\beta-\alpha}(A/2 + B\gamma(s)-\alpha)d\xi(s) = \exp\{\int_0^t(A/2 + B\gamma(s))ds\}\cdot\frac{1}{\beta-\alpha}[A/2 + B\gamma(t)-\alpha]\xi(t) - \frac{1}{\beta-\alpha}[A/2 + B\gamma(0) - \alpha]\xi(0) - \int_0^t \exp[\int_0^s(A/2 + B\gamma(u))du]\{\frac{1}{\beta-\alpha}[A/2 + B\gamma(s)-\alpha][A/2 + B\gamma(s)] + \frac{1}{\beta-\alpha}B\dot{\gamma}(s)\}\xi(s)ds =$$

$$= \frac{1}{\beta-\alpha}\exp[\int_0^t(A/2 + B\gamma(s))ds]\cdot[\frac{A}{2} + B\gamma(t)-\alpha]\xi(t) - \frac{1}{\beta-\alpha}[A/2 + B\gamma(0)-\alpha]\xi(0) - \frac{1}{\beta-\alpha}\int_0^t \exp[\int_0^s(A/2 + B\gamma(u))du]\times$$

$$\times\{[A/2 + B\gamma(s)]^2-\alpha[A/2 + B\gamma(s)] + B(-A\gamma(s)-B\gamma^2(s) + b)\}\xi(s)ds.$$

By (1.8.27) we have

$$(1.8.28)\quad m(t) = \exp\{-\int_0^t(A/2 + B\gamma(s)ds\}\{m(0) + \frac{1}{\beta-\alpha}\exp[\int_0^t(A/2 + B\gamma(s)ds]\times$$

$$\times[A/2 + B\gamma(t)-\alpha]\xi(t) - \frac{1}{\beta-\alpha}[A/2 + B\gamma(0)-\alpha]\xi(0) + \frac{1}{\beta-\alpha}\int_0^t \exp[\int_0^s(A/2 + B\gamma(u))du][A\frac{\beta-\alpha}{2} - \beta\alpha - \frac{A^2}{4} - bB + B\gamma(s)(\beta-2A+\alpha) + B^2\gamma^2(s)]\xi(s)ds\},$$

where

$$(1.8.29)\qquad \int_0^t [A/2 + B\gamma(s)]ds = \int_0^t [A/2 + Bc + B(c_o + B/\tilde{A})e^{\tilde{A}s} - B/\tilde{A})^{-1}]ds =$$

$$= (A/2 + Bc)t - \ln \frac{c_o}{c_o + B/\tilde{A}\,(1 - e^{-\tilde{A}t})} .$$

Wishing to estimate $\theta(t)$ from $\xi(s)$, $-T \le s \le t$, where $T > 0$, $t > 0$, we have

$$m(-T) = \frac{c_1\beta}{\alpha c_2 + \beta c_1}\xi(-T) , \quad \gamma(-T) = \frac{\beta c_1 c_2}{2\alpha(\alpha c_2 + \beta c_1)} ,$$

and letting $T \to \infty$ we get that the optimal estimator $\tilde{m}(t)$ can be gotten from (1.8.28) and (1.8.23) as

$$\gamma(t) = [e^{\tilde{A}(t+T)} (c_o + B/\tilde{A}) - B/\tilde{A}]^{-1} + c \to c = \tilde{\gamma}(t), \text{ if } T \to \infty,$$

and

$$m(t) \to \tilde{m}(t) , \text{ if } T \to \infty,$$

where

$$\tilde{m}(t) = \frac{1}{\beta-\alpha}[A/2 + Bc-\alpha]\,\xi(t) + \frac{1}{\beta-\alpha}\int_{-\infty}^{t} e^{-(A/2+Bc)(t-s)}[A\,\frac{\beta-\alpha}{2} -$$

$$- \beta\alpha - \frac{A^2}{4} - bB + Bc(\beta-2A + \alpha) - c^2B^2]\xi(s)ds.$$

In the case when $\alpha = \beta$ we have (see (1.8.12))

$$b = \frac{c_1 c_2}{c_1 + c_2} , \quad c = \frac{b}{A} = \frac{c_1 c_2}{2\alpha(c_1 + c_2)} ,$$

$$\frac{A}{\beta-\alpha} = \frac{2\alpha}{\beta-\alpha} + 2\,\frac{c_1}{c_1 + c_2} , \quad \frac{B}{\beta-\alpha} = \frac{\beta-\alpha}{c_1 + c_2} ,$$

and

$$\tilde{m}(t) = \frac{c_1}{c_1 + c_2}\,\xi(t),$$

as it was expected.

Wishing to consider the "white noise" case let us put $\beta \to \infty$, $c_2 \to \infty$ and

$$\frac{\beta^2}{c_2} \to \beta_o .$$

One can get

$$A \to 2\alpha, \quad B \to \beta_o, \quad c \to \frac{-\alpha + \sqrt{\alpha^2 + b\beta_o}}{\beta_o} = \tilde{c} ,$$

$$\gamma_o \to \frac{c_1}{2\alpha} , \frac{1}{c_o} \to \gamma_o - \tilde{c} = \frac{c_1}{2\alpha} + \frac{\alpha}{\beta_o} - \frac{\sqrt{\alpha^2 + b\beta_o}}{\beta_o} = 1/\tilde{c}_o ,$$

$$2\tilde{A} \to 2\alpha + 2\beta_o \tilde{c}$$

The optimal estimator $\tilde{m}(t)$ and its error $\tilde{\gamma}(t)$ have the following form

$$\tilde{\gamma}(t) = e^{-2\tilde{A}t} [\tilde{c}_o + \frac{\beta_o}{2\tilde{A}} (1 - e^{-2\tilde{A}t})]^{-1} + \tilde{c} ,$$

and using (1.8.28), (1.8.29)

$$\tilde{m}(t) = e^{-(\alpha + \beta_o\tilde{c})t} \left[\frac{\tilde{c}_o}{\tilde{c}_o + \frac{\beta_o}{2\tilde{A}} - \frac{\beta_o}{2\tilde{A}} e^{-2\tilde{A}t}}\right] \{m(0) + \tag{1.8.30}$$

$$+ \int_0^t e^{(\alpha+\beta_o\tilde{c})s} \left[\frac{\tilde{c}_o}{\tilde{c}_o + \frac{\beta_o}{2\tilde{A}} - \frac{\beta_o}{2\tilde{A}} e^{-2\tilde{A}t}}\right]^{-1} [\beta_o\gamma(s)\xi(s)ds +$$

$$+ \frac{\beta_o}{\beta-\alpha} \gamma(s)d\xi(s)\Big] =$$

$$= e^{-(\alpha+\beta_o\tilde{c})t} \left[\frac{\tilde{c}_o}{\tilde{c}_o + \frac{\beta_o}{2\tilde{A}} - \frac{\beta_o}{2\tilde{A}} e^{-2\tilde{A}t}}\right] \{m(0) +$$

$$+ \beta_o \int_0^t e^{(\alpha+ \beta_o\tilde{c})s} \left[e^{-2\tilde{A}s}\left[\tilde{c}_o + \frac{\beta_o}{2\tilde{A}} - \frac{\beta_o}{2\tilde{A}} e^{-2\tilde{A}s}\right]^{-1} +\right.$$

$$\left. + \tilde{c}\right]\left[\frac{\tilde{c}_o}{\tilde{c}_o + \frac{\beta_o}{2\tilde{A}} - \frac{\beta_o}{2\tilde{A}} e^{-2\tilde{A}s}}\right]^{-1} \xi(s)ds\}.$$

b) Estimation of the unobservable component in the presence of correlated noise. Let us assume that

$$(1.8.31)\qquad d\theta(t) = -\alpha\theta(t)dt + \sqrt{c_1}dw(t),$$

$$(1.8.32)\qquad d\xi(t) = d\theta(t) + d\varepsilon(t) = -\alpha\theta(t)dt - \beta\varepsilon(t)dt + \sqrt{c_1}dw_1(t) + \sqrt{c_2}dw_2(t) = -(\alpha-\beta)\theta(t)dt - \beta\xi(t)dt + (\sqrt{c_1} + \sqrt{c_2})dw(t).$$

By applying the Kalman-Bucy equation we find

$$(1.8.33)\qquad dm(t) = -\alpha m(t)dt + \frac{\sqrt{c_1}(\sqrt{c_1} + \sqrt{c_2}) - \gamma(t)(\alpha-\beta)}{(\sqrt{c_1} + \sqrt{c_2})^2}[d\xi(t) + ((\alpha-\beta)m(t) + \beta\xi(t))dt] = -[\frac{A}{2} + B\gamma(t)]m(t)dt + - \frac{\beta}{\beta-\alpha}[A/_2 + B\gamma(t)-\alpha]\xi(t)dt + \frac{1}{\beta-\alpha}[A/_2 + B\gamma(t)-\alpha]d\xi(t),$$

$$(1.8.34)\qquad \dot{\gamma}(t) = -2\alpha\gamma(t) + c_1 - [\sqrt{c_1}(\sqrt{c_1} + \sqrt{c_2}) + (\beta-\alpha)\gamma(t)]^2\frac{1}{(\sqrt{c_1} + \sqrt{c_2})^2} = -A\gamma(t) - B\gamma^2(t),$$

where

$$A = 2\alpha + \frac{2\sqrt{c_1}(\beta-\alpha)}{\sqrt{c_1} + \sqrt{c_2}}, \qquad B = \frac{(\beta-\alpha)^2}{(\sqrt{c_1} + \sqrt{c_2})^2}$$

with the initial conditions

$$(1.8.35)\qquad m(0) = \xi(0)\frac{E\xi(0)\theta(0)}{E\xi^2(0)}, \quad \gamma_0 = E\theta^2(0) - \frac{(E\theta(0)\xi(0))^2}{E\xi^2(0)}.$$

In order to find the values m(0) and $\gamma(0)$ we shall take advantage of the elementary Gaussian processes that

$$(1.8.36)\qquad B(0) = \begin{pmatrix} b_{11} & b_{12} \\ b_{12} & b_{22} \end{pmatrix} = \begin{pmatrix} E\,\theta^2(0) & E\,\theta(0)\xi(0) \\ E\theta(0)\xi(0) & E\,\xi^2(0) \end{pmatrix},$$

is the solution of the equation (see (2.2.3))

$$(1.8.37)\qquad AB(0) + B(0)\,A^* = -B_w$$

where

$$A = \begin{pmatrix} -\alpha & 0 \\ \beta-\alpha & -\beta \end{pmatrix}, \quad B_w = \begin{pmatrix} c_1 & \sqrt{c_1}(\sqrt{c_1} + \sqrt{c_2}) \\ \sqrt{c_1}(\sqrt{c_1} + \sqrt{c_2}) & c_1+c_2 \end{pmatrix}.$$

From (37) one can get

$$b_{11} = \frac{c_1}{2\alpha}, \quad b_{12} = \frac{1}{\alpha+\beta}[\sqrt{c_1}(\sqrt{c_1} + \sqrt{c_2}) + \frac{c_1}{2\alpha}(\beta - \alpha)],$$

(1.8.38)

$$b_{22} = \frac{1}{\beta}[c_1 + c_2 + \frac{\beta-\alpha}{\alpha+\beta}(\sqrt{c_1}(\sqrt{c_1} + \sqrt{c_2}) + \frac{c_1(\beta-\alpha)}{2\alpha})].$$

Thus we get

$$(1.8.39)\quad m(0) = \xi(0)\frac{b_{12}}{b_{22}} = \xi(0)\frac{\beta}{\alpha+\beta}\,\frac{\sqrt{c_1}(\sqrt{c_1} + \sqrt{c_2}) + \frac{c_1(\beta-\alpha)}{2\alpha}}{c_1+c_2 + \frac{\beta-\alpha}{\alpha+\beta}[\sqrt{c_1}(\sqrt{c_1}+\sqrt{c_2}) + \frac{\beta-\alpha}{2}c_1]},$$

$$(1.8.40)\quad \gamma(0) = b_{11} - \frac{(b_{12})^2}{b_{22}} =$$

$$= \frac{c_1}{2\alpha} - \frac{\beta}{(\alpha+\beta)^2}\,\frac{\left[\sqrt{c_1}(\sqrt{c_1}+\sqrt{c_2}) + \frac{c_1(\beta-\alpha)}{2\alpha}\right]^2}{c_1+c_2+\frac{\beta-\alpha}{\alpha+\beta}[\sqrt{c_1}(\sqrt{c_1}+\sqrt{c_2})+\frac{\beta-\alpha}{2}c_1]}.$$

Using Lemma 1. we have the error and optimal estimation

$$(1.8.41)\quad \gamma(t) = e^{-At}\left[\frac{1}{\gamma_0} + \frac{B}{A}(1-e^{-At})\right]^{-1},$$

and

$$(1.8.42)\quad m(t) = e^{-\int_0^t [A/2+B\gamma(s)]ds}\{m(0) +$$

$$+ \int_0^t e^{\int_0^s [A/2+B\gamma(u)]du}[\frac{\beta}{\beta-\alpha}(\frac{A}{2} + B\gamma(s)-\alpha)\xi(s)ds +$$

$$+ \frac{1}{\beta-\alpha}[\frac{A}{2} + B\gamma(s)-\alpha]]d\xi(s)\},$$

where (see (29))

$$(1.8.43)\quad \int_0^t [A/2 + B\gamma(s)]ds = \frac{A}{2}t - \ln\frac{1/\gamma_0}{1/\gamma_0 + \frac{B}{A} - \frac{B}{A}e^{-At}},$$

and (compare with (27))

$$(1.8.44)\qquad \frac{1}{\beta-\alpha}\int_0^t e^{\int_0^s(\frac{A}{2}+B\gamma(u))du}[\tfrac{A}{2}+B\gamma(s)-\alpha]d\xi(s)=$$

$$=\frac{1}{\beta-\alpha}\{e^{\int_0^t[A/2+B\gamma(s)]ds}[\tfrac{A}{2}+B\gamma(t)-\alpha]\xi(t)-$$

$$-[\tfrac{A}{2}+B\gamma(0)-\alpha]\xi(0)-\int_0^t e^{\int_0^s[\frac{A}{2}+B\gamma(u)]du}\{[\tfrac{A}{2}+B\gamma(s)-\alpha][\tfrac{A}{2}+$$

$$+B\gamma(s)]+B(-A\gamma(s)-B\gamma^2(s))\}\xi(s)ds\}.$$

In the special case when $\alpha = \beta$

$$\gamma(t) = \gamma_0 e^{-2\alpha t},$$

$$m(0) = \xi(0)\,\frac{\sqrt{c_1}(\sqrt{c_1}+\sqrt{c_2})}{2(c_1+c_2)},\qquad \gamma(0) = \frac{c_1}{4\alpha}\,\frac{(\sqrt{c_1}-\sqrt{c_2})^2}{c_1+c_2},$$

$$m(t) = e^{-\alpha t}[m(0)+\alpha\frac{\sqrt{c_1}}{\sqrt{c_1}+\sqrt{c_2}}\int_0^t e^{\alpha s}\xi(s)ds+$$

$$+\frac{\sqrt{c_1}}{\sqrt{c_1}+\sqrt{c_2}}\int_0^t e^{\alpha s}d\xi(s)]=$$

$$= e^{-\alpha t}[m(0)-\frac{c_1}{\sqrt{c_1}+\sqrt{c_2}}\xi(0)]+\frac{c_1}{\sqrt{c_1}+\sqrt{c_2}}\xi(t).$$

The "white noise" case can be considered taking the limit $\beta \to \infty$, $c_2 \to \infty$ and

$$(1.8.45)\qquad \frac{\beta^2}{c_2} \to \beta_0.$$

We have

$$A \to 2(\alpha+\sqrt{c_1\beta_0}),\qquad B \to \beta_0,$$

$$\gamma(0) \to \frac{c_1}{2\alpha},\quad m(0) \to 0.$$

Furthermore,

$$\gamma(t) = \bar{e}^{(2\alpha+ 2\sqrt{c_1\beta_o})t}\left[\frac{c_1}{2\alpha} + \frac{\beta_o}{2\alpha+2\sqrt{c_1\beta_o}}(1-\bar{e}^{(2\alpha+2\sqrt{c_1\beta_o})t})\right]^{-1}$$

$$(1.8.46)\ m(t) = \bar{e}^{\int_o^t[\alpha+\sqrt{c_1\beta_o} + \beta_o\gamma(s)]ds} \{\int_o^t \exp\ [\int_o^s \alpha+\sqrt{c_1\beta_o} +$$

$$+ \gamma(u)\beta_o]du]\ (\sqrt{c_1\beta_o} + \beta_o\gamma(s))\xi(s)ds\}.$$

Example 2. Parameter estimation in the presence of "noise." Using the same notations as in Example 1. we are interested in the estimation of α in equation (16) if the observable process is $\xi(t)$, given by (17). First, we assume that $\varepsilon(t) = \sqrt{c_2}\ w_2(t)$, i.e., $\varepsilon(t)$ is a Wiener process and later we discuss the case when $\varepsilon(t)$ is also an elementary Gaussian process.

This problem, in the discrete time case, was considered in the author's dissertation (1962). And was many times discussed by A. Balakrishnan (see [3], and (1978)) in the general case.

a) Let $\theta(t)$, $\xi(t)$ be given by the equations

$$(1.8.47)\qquad d\theta(t) = -\ \alpha\theta(t)dt + \sqrt{c_1}dw_1(t),$$

$$(1.8.48)\qquad d\xi(t) = d\theta(t) + d\varepsilon(t) = -\ \alpha\theta(t)dt + \sqrt{c_1}dw_1(t) + \sqrt{c_2}\ dw_2(t),$$

where $w_1(t)$, $w_2(t)$ are independent Wiener processes. In this case

$m(t) = E(\theta(t)\,|\,\mathcal{F}_t^{\xi})$ and $\gamma(t) = E(\theta(t) - m(t))^2$ are the solutions of the following equations

$$(1.8.49)\qquad dm(t) = -\ \alpha m(t)dt + \frac{c_1-\alpha\gamma(t)}{c_1+c_2}\ (d\xi(t) + \alpha m(t)dt) =$$

$$= -\ \frac{\alpha}{c_1+c_2}\ [c_2 + \alpha\gamma(t)]m(t)dt + \frac{c_1-\alpha\gamma(t)}{c_1+c_2}\ d\xi(t),$$

$$(1.8.50)\qquad \dot{\gamma}(t) = -2\alpha\gamma(t) - \frac{(c_1-\alpha\gamma(t))^2}{c_1+c_2} + c_1 =$$

$$= -\frac{2\alpha c_2}{c_1+c_2}\gamma(t) - \frac{\alpha^2}{c_1+c_2}\gamma^2(t) + \frac{c_1c_2}{c_1+c_2},$$

with the initial conditions

$$(1.8.51)\qquad m(0) = E(\theta(0)|\xi(0)\ ,\quad \gamma(0) = E(\theta(0)-m(0))^2.$$

Using Lemma 1. one can get

$$(1.8.52)\qquad \gamma(t) = e^{-\tilde{A}t}[c_o + \frac{B}{\tilde{A}}(1-e^{-\tilde{A}t})]^{-1} + c$$

where

$$B = \frac{\alpha^2}{c_1+c_2},\quad c = \frac{1}{\alpha}[c_2 + \sqrt{c_2^2 + c_1c_2}]\ ,$$

$$\tilde{A} = \frac{2\alpha c_2}{c_1+c_2} + \frac{2\alpha}{c_1+c_2}[c_2 + \sqrt{c_2^2 + c_1c_2}],\quad \frac{1}{c_o} = \gamma(0) - c.$$

Note, that using the representation

$$(1.8.52)\qquad dm(t) = -\alpha m(t)dt + \frac{c_1-\alpha\gamma(t)}{c_1+c_2}d\tilde{w}(t),$$

where

$$\tilde{w}(t) = \xi(t) + \alpha\int_o^t m(s)ds$$

is a Wiener process one can estimate α estimating the "diffusion" parameter in (1.8.52). We take advantage of Girsanov's theorem (see Theorem 3. in Section 2.3.4) in the form

$$(1.8.53)\qquad \frac{dP_\xi}{dP_w}(\xi) = \exp\{\int_o^t \alpha(s,\xi)d\xi - \frac{1}{2}\int_o^t \alpha^2(s,\xi)ds\},$$

where

$$(1.8.54)\qquad \alpha(t,\xi) = E(-\alpha\theta(t)|\mathcal{F}_t^\xi) = -\alpha E(\theta(t)|\mathcal{F}_t^\xi) = -\alpha m(t).$$

We have from (1.8.49)

$$(1.8.55)\qquad m(t) = e^{-\frac{\alpha}{c_1+c_2}\int_o^t (c_2+\alpha\gamma(s))ds}\{m(0)+\int_o^t e^{\frac{\alpha}{c_1+c_2}\int_o^s (c_1+\alpha\gamma(u))du}\, \times$$

$$\times\, \frac{(c_1-\alpha\gamma(s))}{c_1+c_2}\, d\xi(s)\} =$$

$$= e^{-\frac{\alpha}{c_1+c}\int_o^t [c_2+\alpha\gamma(s)]ds}\{m(0) +$$

$$+ e^{\frac{\alpha}{c_1+c_2}\int_o^t [c_2+\alpha\gamma(s)]ds}\, \frac{c_1-\alpha\gamma(t)}{c_1+c_2}\,\xi(t) + \frac{c_1-\alpha\gamma(0)}{c_1+c_2}\,\xi(0) -$$

$$- \int_o^t e^{\frac{\alpha}{c_1+c_2}\int_o^s (c_2+\alpha\gamma(u)]du}\, [\frac{\alpha}{c_1+c_2}(c_2+\alpha\gamma(s))\, \frac{c_1-\alpha\gamma(s)}{c_1+c_2} +$$

$$- \frac{\alpha}{c_1+c_2}\,\dot{\gamma}(s)]\xi(s)ds\} = e^{-\frac{\alpha}{c_1+c_2}\int_o^t [c_2+\alpha\gamma(s)]ds}\{m(0) +$$

$$+ e^{\frac{\alpha}{c_1+c_2}\int_o^t [c_2+\alpha\gamma(s)]ds}\, \frac{c_1-\alpha\gamma(t)}{c_1+c_2}\,\xi(t) - \frac{c_1-\alpha\gamma(0)}{c_1+c_2}\,\xi(0) +$$

$$- \int_o^t e^{\frac{\alpha}{c_1+c_2}\int_o^s [c_2+\alpha\gamma(u)]du}\, \frac{\alpha^2}{(c_1+c_2)}\,\gamma(s)\xi(s)ds\}.$$

By (1.8.29) and (1.8.55) one can give the final expression of $m(t)$, as

$$(1.8.56)\qquad \int_o^t [c_2+\alpha\gamma(s)]ds = (c_2+\alpha c)t - \frac{\alpha}{B}\,\ln \frac{c_o}{c_o+\frac{B}{\tilde{A}}-\frac{B}{\tilde{A}}\,e^{-\tilde{A}t}}\,.$$

If $c_2 \to 0$ one can get the maximum likelihood estimator of α without noise, which will be discussed in Section 3.3. Note that in this case no finite system of sufficient statistics exists.

b) Let $\theta(t)$, $\xi(t)$ be given by equations (1.8.19) and (1.8.20) we can use the relations (53), (54) with the difference that $m(t)$ and $\gamma(t)$ are given by (23) and (28).

CHAPTER 2

ELEMENTARY GAUSSIAN PROCESSES

2.1. Processes With Discrete Time

2.1.1. Main Theorems. A k-dimensional stochastic process $\underline{\xi}^*(t)=(\xi^1(t), \ldots, \xi^k(t))$ is called an elementary Gaussian if it is stationary, Markov and Gaussian. In continuous time case it is assumed that the process is one of diffusion type.

It is assumed that the process $\underline{\xi}(t)$ is nondegenerate and it is linearly regular (or purely non-deterministic). Nondegeneracy means that the components of $\underline{\xi}(t)$ are pointwise linearly independent. A stationary process $\underline{\xi}(t)$ is called deterministic (or linearly singular) if the least squares prediction of $\underline{\xi}(t+s)$ is always correct with probability 1, i.e.,

$$\underline{\xi}(s+t) = A(t)\underline{\xi}(s)$$

where $A(t)$ is deterministic. $\underline{\xi}(t)$ is called linearly regular if it has no deterministic component. In this case the Wold's expansion holds (discrete time)

$$\underline{\xi}(t) = \sum_{k=0}^{\infty} A_k \underline{\varepsilon}(t-k)$$

where $E\underline{\varepsilon}(t) = 0$, $\mathrm{cov}(\underline{\varepsilon}(t),\underline{\varepsilon}(t))$ exists, $\mathrm{cov}(\underline{\varepsilon}(t),\underline{\varepsilon}(t-k)) = 0$, $k > 0$. Such an $\underline{\varepsilon}(t)$ process will be called a white noise process. In the sequel we shall denote the time by n in discrete time case.

The connection between stochastic difference equations and elementary processes can be characterized. Let $\underline{\varepsilon}(n)$ be a k dimensional Gaussian white noise process with the parameters

$$E\underline{\varepsilon}(n) = 0,\ \mathrm{cov}(\underline{\varepsilon}(n),\underline{\varepsilon}(n)) = E(\underline{\varepsilon}(n)\underline{\varepsilon}^*(n)) = B_\varepsilon,$$

where rank $B_\varepsilon \geq 1$. Let Q be a non-singular kxk matrix. Let us assume that equation

$$B(0) = QB(0)Q^* + B_\varepsilon \tag{2.1.1}$$

has a non-singular, symmetric, positive definite solution B(0). Let $\underline{\xi}(0)$ be a Gaussian vector with the parameters $E\underline{\xi}(0) = \underline{0}$, $\mathrm{cov}(\underline{\xi}(0), \underline{\xi}(0)) = B(0)$, and independent of $\underline{\varepsilon}(n)$, $n > 0$, or $F^{\varepsilon}_{[1,n]}=\sigma(\underline{\varepsilon}(1),\ldots, \underline{\varepsilon}(n))$.

Theorem 1. Let $\underline{\xi}(n)$ be defined by the following difference equation (with non-singular Q).

(2.1.2) $\underline{\xi}(n) = Q\underline{\xi}(n-1) + \underline{\varepsilon}(n), \; n = 1,2,\ldots,$

where $\underline{\varepsilon}(n)$ is independent of F^{ξ}_{n-1} and it is a Gaussian white noise, with $E\underline{\varepsilon}(n) = \underline{0}$, $\text{cov}(\underline{\varepsilon}(n),\underline{\varepsilon}(n)) = B_{\varepsilon}$. Then $\underline{\xi}(n)$ is a nondegenerate elementary Gaussian process with $E\underline{\xi}(n) = 0$ and covariance

(2.1.3) $B(\ell) = \text{cov}(\underline{\xi}(n+\ell),\underline{\xi}(n)) = Q^{\ell} \cdot B(0), \; \ell \geq 0,$

where $B(0)$ is prescribed by (2.1.1).

Proof. The normality of $\underline{\xi}(n)$ follows by induction from linearity of (2.1.2). By repeated application of (2.1.2) we get

$$\underline{\xi}(n+\ell) = \underline{\varepsilon}(n+\ell) + Q\underline{\varepsilon}(n+\ell-1)+\ldots +Q^{n+\ell-1}\underline{\varepsilon}(1) + Q^{n+\ell}\underline{\xi}(0),$$

and this gives (multiplying by $\underline{\xi}^*(n)$ and taking expectation)

$$\text{cov}(\underline{\xi}(n+\ell),\underline{\xi}(n)) = Q^{\ell}[B_{\varepsilon}+QB\,Q^* +\ldots+ Q^{n-1}B_{\varepsilon}Q^{*n-1}+ Q^{n}B(0)Q^{*n}] =$$

$$= Q^{\ell}[B_{\varepsilon} + QB_{\varepsilon}Q^{*}+\ldots +Q^{n-1}(B_{\varepsilon}+ QB(0)\,Q^{*})\,Q^{*n-1}]$$

$$= Q^{\ell}[B_{\varepsilon} + QB_{\varepsilon}Q^{*}+\ldots +Q^{n-1}\;B(0)\;Q^{*n-1}] =$$

$$= Q^{\ell}\;B(0),$$

which proves the stationarity.

The markovity immdiately follows from the normality and (2.1.2) as

$$E(\underline{\xi}(n)\,|\,F^{\xi}_{n-1}) = Q\underline{\xi}(n-1),$$

$$\text{cov}(\underline{\xi}(n) - Q\underline{\xi}(n-1),\; \underline{\xi}(n) - Q\underline{\xi}(n-1)\,|\,F^{\xi}_{n-1}) = B_{\varepsilon}.$$

This proves the theorem.

Now we prove that equation (2.1.1) has a solution.

Lemma 1. Let $\underline{\xi}(n)$ be Gaussian and satisfy equation (2.1.2), $\underline{\xi}(0) = \underline{0}$. Let

$$B_{\varepsilon} + QB_{\varepsilon}\,Q^{*}+\ldots+ Q^{k-1}\,B_{\varepsilon}\,Q^{*k-1} = A\,A^{*},$$

$$A = [B_{\varepsilon}^{1/2}QB_{\varepsilon}^{1/2}\ldots\;(Q^{k-1}\;B_{\varepsilon}^{1/2})],$$

If A has rank k and B_{ε} is positive semidefinite, then the matrix $B_n = E\underline{\xi}(n)\underline{\xi}(n)$ at $n \geq k$ is positive definite.

Proof. From (2.1.2)

$$B_{n+1} = \text{cov}(\underline{\xi}(n+1),\underline{\xi}(n+1)) = QB_n\,Q^{*} + B_{\varepsilon},\; B_{o} = 0,$$

and

$$B_n = B_\varepsilon + QB_\varepsilon Q^* + \ldots + Q^{n-1} B_\varepsilon Q^{*(n-1)}.$$

At $n = k$ we have $B_k = A A^*$ and therefore, for $n > k$,

$$B_n = A A^* + \sum_{j=k}^{n-1} Q^j B_\varepsilon Q^{*j}.$$

This shows that B_n is non-singular because rank $A = k$ (by assumption) implies that rank $A A^* = k$.

<u>Lemma 2. In addition, the conditions in Lemma 1, let Q have eigenvalues λ_i in the unit circle, i.e., $|\lambda_i| < 1$. Then $\lim_{n\to\infty} B_n = B^0$ exists and does not depend on B_0. B^0 is the unique solution (in the class of symmetric positive definite matrices) of the matrix equation (2.1.1) and</u> $SpB^0 < \infty$.

<u>Proof.</u> According to Lemma 1, B_n is positive definite if $n \geq k$ and we have

$$\underline{x}^* B_n \underline{x} > 0$$

for any nonzero vector $\underline{x}$. Further, $\underline{x}^* B_n \underline{x}$ is monotone non-decreasing in n $(n \geq k)$ as

$$\underline{x}^* B_{n+1} \underline{x} = \underline{x}^* B_\varepsilon \underline{x} + \ldots + \underline{x}^* Q^{k-1} B_\varepsilon Q^{*(k-1)} \underline{x} + \ldots + \underline{x}^* Q^n B_\varepsilon Q^{*n} \underline{x} =$$

$$= \underline{x}^* A A^* \underline{x} + \underline{x}^* Q^k B_\varepsilon Q^{*k} \underline{x} + \ldots + \underline{x}^* Q^n B_\varepsilon Q^{*n} \underline{x} \geq \underline{x}^* B_n \underline{x},$$

which proves the positive definiteness of the matrix B^0. That

$$\sup_n \underline{x}^* B_n \underline{x} < \infty$$

follows from the Jordan representation of Q, (see Appendix A, (A2.5)), which proves the existence of B^0 and that $SpB^0 < \infty$.

The uniqueness of B^0, in the class of the positive definite symmetric matrices, follows from the following consideration. Let B^1 and B^2 two such solutions and let

$$B_n^i = Q B_{n-1}^i Q^* + B, \quad B_0^i = B^i, \quad i = 1,2.$$

Then from what we proved above

$$\lim_{n\to\infty} B_n^i = B^0 = B^i, \quad i = 1,2.$$

Lemma 3. The series

$$(2.1.4) \qquad \sum_{n=0}^{\infty} Q^n \underline{\varepsilon}(n).$$

where $\underline{\varepsilon}(n)$ is a white noise process with $E\underline{\varepsilon}(n) = 0, \mathrm{cov}(\underline{\varepsilon}(n), \underline{\varepsilon}(n)) = B_\varepsilon$, converges in mean square and with probability 1 if and only if $|\lambda_i| < 1$ where λ_i are the eigenvalues of the matrix Q.

Proof. The convergence of (2.1.4) holds if and only if the series of covariance matrices $\sum_{n=0}^{\infty} Q^n B_\varepsilon Q^{*n}$ is convergent (Kolmogorov's three series theorem for independent vector variables). The above series is convergent if and only if $Q^n B_\varepsilon Q^{*n} \to 0$, $n \to \infty$. By Jordan representation this is true if and only if $|\lambda_i| < 1$, $(i=1,2,\ldots,k)$ (see (A2.5)).

In the preceding we examined the one sided process ($n \geq 0$), but stationarity in $-\infty < n < \infty$ is more interesting. The converse of Theorem 1 has to be discussed too.

Theorem 2. The process $\underline{\xi}(n)$ is an elementary Gaussian process if and only if it is the solution of the following stochastic difference equation

$$(2.1.2') \qquad \underline{\xi}(n) = Q\underline{\xi}(n-1) + \underline{\varepsilon}(n), \quad n = 0, \pm 1, \pm 2, \ldots,$$

in the following sense:

a) Let $\underline{\varepsilon}(n)$ be a k-dimensional Gaussian white noise (independent, identically distributed sequence of random vectors) with covariance matrix B_ε ($B_\varepsilon \neq 0$, $E\underline{\varepsilon}(n) = 0$) and let Q be a non-singular $k \times k$ matrix with eigenvalues in the unit circle. Then equation (2.1.2') has a unique regular stationary solution which is a Gaussian Markov process where $\mathrm{cov}(\underline{\xi}(n), \underline{\xi}(n)) = B(0)$ is the solution of (2.1.1).

b) Let $\underline{\xi}(n)$ $(n=0, \pm 1, \pm 2, \ldots)$ be a nondegenerate, linearly regular k dimensional elementary Gaussian process with $E\underline{\xi}(n) = \underline{0}$, $\mathrm{cov}(\underline{\xi}(n), \underline{\xi}(n)) = B(0)$, $B(0) \neq 0$. Then there exists a nonsingular matrix $Q_{k \times k}$, with eigenvalues in the unit circle and a sequence of independent, identically distributed Gaussian vectors $\underline{\varepsilon}(n)$ such that equation (2.1.2') holds. $E\underline{\varepsilon}(n) = \underline{0}$ and B_ε is uniquely determined by (2.1.1).

Proof. a) By Lemma 3 the series $\sum_{i=0}^{\infty} Q^i \underline{\varepsilon}(n-i)$ is convergent as the eigenvalues of Q are inside the unit circle. By iteration of (2.1.2') we get

(2.1.2") $\underline{\xi}(n) = \underline{\varepsilon}(n) + Q\underline{\varepsilon}(n-1) + \dots + Q^{\ell-1}\,\underline{\varepsilon}(n-\ell+1) + Q^{\ell}\,\underline{\xi}(n-\ell)$.

Using Theorem 1 the process $\underline{\xi}(n)$ is stationary so $||Q^{\ell}\,\underline{\xi}(n-\ell)|| \to 0$ (see (A2.5)), as $\ell \to \infty$, and by the well known theorem on normal convergence (see Appendix B, Lemma 4) the infinite series

$$\sum_{i=0}^{\infty} Q^{i}\,\underline{\varepsilon}(n-i)$$

is a Gaussian vector variable, which satisfies (2.1.2). The Markov property immediately follows from (2.1.2′) as we proved it in Theorem 1. The regularity follows from (2.1.2").

The uniqueness of the solution follows from the following. Let $\underline{\xi}_1(n)$ and $\underline{\xi}_2(n)$ two such solutions, then by (2.1.2")

$$E(\underline{\xi}_1(n) - \underline{\xi}_2(n))(\underline{\xi}_1(n) - \underline{\xi}_2(n))^* = E\,Q^{\ell}[\underline{\xi}_1(n-\ell) - \underline{\xi}_2(n-\ell)]$$

$$[(\underline{\xi}_1(n-\ell) - \underline{\xi}_2(n-\ell))\quad Q^{\ell}] = Q^{\ell}\,\tilde{B}(0)\,Q^{*\ell} \to 0$$

as $\ell \to \infty$.

b) By the theorem on normal correlation (see Appendix B, Lemma 4) and Markov property

$$E(\underline{\xi}(n)\,|\,F^{\xi}_{[-\infty,n-1]}) = Q_n\,\underline{\xi}(n-1),$$

where $Q_n = B(1)B^{+}(0)$. From this it follows that the vectors

$$\underline{\varepsilon}(n) = \underline{\xi}(n) - Q\underline{\xi}(n-1)$$

are independent Gaussian. Indeed, because of the Markov property of $\underline{\xi}(n)$ for $m < n$

$$E(\underline{\xi}(n) - E(\underline{\xi}(n)|\underline{\xi}(n-1))|\underline{\xi}(m), \underline{\xi}(m-1)) = E(\underline{\xi}(n)|\underline{\xi}(m)) - E(\underline{\xi}(n)|\underline{\xi}(m)) = \underline{0},$$

and

$$E\,\underline{\varepsilon}(n)\,\underline{\varepsilon}^*(m) = E(\underline{\xi}(n) - Q_n\underline{\xi}(n-1))(\underline{\xi}(m) - Q_m\,\underline{\xi}(m-1))^* =$$

$$= E[E(\underline{\xi}(n) - Q_n\underline{\xi}(n-1))(\underline{\xi}(m) - Q_m\,\underline{\xi}(m-1))|\underline{\xi}(m), \underline{\xi}(m-1)] = 0.$$

From stationarity of the process $\underline{\xi}(n)$ we get that $Q_n = Q$ and the distribution of $\underline{\varepsilon}(n)$ does not depend on n. This proves the theorem.

<u>Remark 1.</u> From the proof of Tehorem 2 one may see that the best mean square extrapolation of $\underline{\xi}(n+1)$ based on the realization $\underline{\xi}(n), \underline{\xi}(n-1), \dots,$ is $Q\underline{\xi}(n)$ as

$$E\,(\underline{\xi}(n-1)\,|\,F^{\xi}_{n}) = Q\,\underline{\xi}(n).$$

Further, the covariance matrix of the error equals B_{ε} as

$$E(\underline{\xi}(n+1) - Q\,\underline{\xi}(n))(\underline{\xi}(n+1) - Q\,\underline{\xi}(n))^* = B_\varepsilon .$$

Remark 2. The random vector $\underline{\xi}(n)$ is $F^{\varepsilon}_{[-\infty,n]}$ measurable and therefore $\underline{\varepsilon}(n+1)$ is independent of $F^{\xi}_{[-\infty,n]}$. The proof of Theorem 2 shows that $F^{\xi}_{[-\infty,n]} = F^{\varepsilon}_{[-\infty,n]}$. The stochastic process $\underline{\varepsilon}(n)$ is said to be the innovation process.

Remark 3. From (2.1.2") we get the following relations

$$E(\underline{\xi}(n)\,|\,F^{\xi}_{n-\ell}) = Q^{\ell}\underline{\xi}(n-\ell),\ \ell > 0,$$

(2.1.5) $B(\ell) = E\,(\underline{\xi}(n+\ell)\ \underline{\xi}^*(n)) = Q^{\ell}\,B(0),\ B(-\ell) = B(0)\,Q^{*\ell},\ \ell > 0,$

and

(2.1.1′) $B(0) = B_\varepsilon + QB_\varepsilon\,Q^* + \ldots = \sum_{i=0}^{\infty} Q^i\,B_\varepsilon\,Q^{*i}.$

Remark 4. Multiplying (2.1.2′) by $\underline{\varepsilon}(n)$ and taking expectation we get

$$E\underline{\xi}(n)\ \underline{\varepsilon}^*(n) = B_\varepsilon,$$

and multiplying by $\underline{\xi}(n)$, taking into account (2.1.5)

$$B(0) = E\,\underline{\xi}(n)\ \underline{\xi}^*(n) = QE\,\underline{\xi}(n-1)\ \underline{\xi}^*(n) + E\,\underline{\varepsilon}(n)\ \underline{\xi}^*(n) =$$
$$= QB(0)\,Q^* + B_\varepsilon,$$

and we find equation (2.1.1).

2.1.2. The Structure of Degenerate and Deterministic Processes.

As we mentioned earlier, the k-dimensional stochastic process $\underline{\xi}(t)$ will be called degenerate if there are constants $c_1,\ldots,c_k$ not all 0, such that

(2.1.6) $$\sum_{j=1}^{k} c_j\ \xi^j(t) = 0,$$

with probability 1, for all t. Equation (2.1.6) is true if and only if, with b_{ij} denoting ij-th component of $B(0)$, $B(0) = (b_{ij})$,

$$E\,|\sum_{j=1}^{k} c_j\ \xi^j(t)\,|^2 = \sum_{j,\ell=1}^{k} b_{j\ell}\,c_j\,\bar{c}_\ell = 0,$$

i.e., if and only if the correlation matrix $B(0)$ is singular. The only degenerate one-dimensional process is $\xi(t) = 0$ (with probability 1).

As with a non-singular transformation of $\underline{\xi}(t)$ the symmetric non-negative definite matrix $B(0)$ is in diagonal form, with only 0's and 1's down the main diagonal, say 0 to the ℓ-th place and 1 thereafter we get that

$$\xi^j(t) \equiv 0,\ \text{if}\ j \le \ell.$$

So the following lemma is true.

Lemma 4. Every degenerate elementary Gaussian process is the direct product of processes of type $\xi^j(t) = 0$ and of a nondegenerate elementary process.

Theorem 3. Let $\underline{\xi}(t)$ be a deterministic elementary Gaussian process, with correlation function $B(t)$.

a) If the parameter t is restricted to the integers, there is a nonsingular matrix Q such that

$$(2.1.7)\qquad \begin{aligned} &(1)\ \underline{\xi}(n) = Q\underline{\xi}(n-1),\ \underline{\xi}(n) = Q^n\,\underline{\xi}(0),\\ &(2)\ B(n) = Q^n\,B(0),\ n = 0, \pm 1, \pm 2, \ldots,\\ &(3)\ B(0) = Q\,B(0)\,Q^*. \end{aligned}$$

If the process $\underline{\xi}(n)$ is nondegenerate, Q is uniquely determined.

b) If the time t is continuous and $\underline{\xi}(t)$ is continuous, there is a matrix A such that

$$(2.1.7')\qquad \begin{aligned} &(1')\ \frac{d\underline{\xi}(t)}{dt} = A\,\underline{\xi}(t),\ \underline{\xi}(t) = e^{At}\,\underline{\xi}(0),\\ &(2')\ B(t) = e^{A|t|}B(0),\ -\infty < t < \infty,\\ &(3')\ AB(0) + B(0)\,A^* = 0. \end{aligned}$$

If the process is nondegenerate, A is uniquely determined.

c) Conversely if B(0) is any symmetric non-negative definite matrix, and if Q(A) is any matrix satisfying in (2.1.7) (3) ((3')), where Q is non - singular, there is a deterministic elementary Gaussian process with correlation function (2.1.7) (2) ((2')) and satisfying in (2.1.7) (1) ((1')).

Proof. If the given process can be expressed as a direct product, it will be sufficient to prove a) and b) for each factor. From this and Lemma 4 we see that it will be sufficient to prove a), b) for nondegenerate processes.

a) If $\underline{\xi}(n)$ is deterministic, from definition

$$\underline{\xi}(n+1) = Q\underline{\xi}(n),$$

and (1) is true for all n if Q is non-singular. Using (1)

$$B(n) = E\,(\underline{\xi}(n)\ \underline{\xi}^*(0)) = E\,(Q^n\underline{\xi}(0)\underline{\xi}^*(0)) = Q^n\,B(0),$$

and

$$B(0) = E\,(\underline{\xi}(1)\ \underline{\xi}^*(2)) = Q\,B(0)\,Q^*,$$

where B(0) is non-singular. Then Q is uniquely determined by (2) with n = 1, and Q cannot be singular because of (3).

<u>Corollary.</u> We can choose a process $\tilde{\underline{\xi}}(n)$ equivalent to $\underline{\xi}(n)$ in which $B(0) = I = QQ^*$, so Q^* is orthogonal and has the normal form: all elements of Q are 0 except for two dimensional rotation matrices (complex eigenvalue) or 1's or - 1's down the main diagonal. It is now obvious that the process $\underline{\xi}(n)$ is the direct product of processes of types

(A) $\xi(n) = \xi(0),\ E\xi(0) = 0,\ E(\xi(0))^2 = \sigma^2 > 0,$

(A') $\xi(n) = (-1)^n\,\xi(0),\ E\xi(0) = 0,\ E(\xi(0))^2 = \sigma^2 > 0.$

(B) $\underline{\xi}(n) = (\xi^1(n),\ \xi^2(n))$ is two-dimensional

$$E\xi^j(0) = 0,\ E(\xi^j(0))^2 = \sigma^2 > 0,\ E\xi^1(0)\,\xi^2(0) = 0,\ j = 1,2,$$

$$\xi^1(n) = \xi^1(0)\cos\theta n - \xi^2(0)\sin\theta n,$$

$$\xi^2(n) = \xi^1(0)\sin\theta n + \xi^2(0)\cos\theta n,$$

$$B(n) = e^{i\theta n}\,B(0) = \sigma^2\begin{pmatrix}\cos n\theta & \sin n\theta\\ -\sin n\theta & \cos n\theta\end{pmatrix}$$

b) If t is continuous and $\underline{\xi}(t)$ is deterministic we get

(1") $\underline{\xi}(s+t) = A(t)\,\underline{\xi}(s),$

and further

(2") $B(t) = E(\underline{\xi}(s+t)\underline{\xi}^*(s)) = A(t)B(0),$

(3") $B(0) = E(\underline{\xi}(t)\underline{\xi}^*(t)) = A(t)\,B(0)\,A\,(t),$

where

(2.1.8) $A(s+t) = A(s)\,A(t).$

By the continuity assumption on process $\underline{\xi}(t)$

$$\lim_{t\to 0} A(t)B(0) = B(0)$$

which implies

(2.1.9) $$\lim_{t\to 0} A(t) = I.$$

It is well known that any solution of (2.1.8) with (2.1.9) can be written in the form

$$A(t) = e^{tA}.$$

Matrix A(t) is uniquely determined by (2") since B(0) is non-singular. Expanding the right side of (3") in powers of t we get (3).

If $B(0) = I$ we get $A(t)\,A^*(t) = I$ and $A + A^* = 0$, and A is in the real canonical form of skew symmetric matrices: its elements vanish except for possible two rowed matrices

$$\begin{pmatrix} 0 & \theta \\ -\theta & 0 \end{pmatrix}$$

down the main diagonal. From this it is clear that the nondegenerate process is a direct product of factors of type (B) and (A).

c) Let $\underline{\xi}(0)$ a Gaussian variable $E\underline{\xi}(0) = 0$, $E\underline{\xi}(0)\underline{\xi}(0)=B(0)$. Let $\underline{\xi}(n) = Q^n\underline{\xi}(0)$. It is easy to prove that $\underline{\xi}(n)$ is an elementary deterministic Gaussian process. Theorem 3 is now completely proved. As a consequence of Theorems 2 and 3 we get the following:

Theorem 4. Every nondegenerate elementary process $\underline{\xi}(n)$ is the direct product of processes of regular type and deterministic type and the representation

$$\underline{\xi}(n) = \sum_{i=0}^{\infty} Q^i \varepsilon(n-i) + Q^n \underline{\tilde{\varepsilon}}, \tag{2.1.10}$$

where $\underline{\varepsilon}(n)$, $\underline{\tilde{\varepsilon}}$, $n=0, \pm 1, \ldots$, are mutually independent Gaussian variables, exhibits the decomposition into factor processes. The transition matrix Q is uniquely determined. Conversely if Q is a matrix with all characteristic values of modulus less than 1 or of modules 1 then there exists an elementary nondegenerate Gaussian process with the given Q, B(0), B_ε.

Making a suitable change of variables, if necessary, it can be supposed that Q has the form

$$\begin{pmatrix} Q_1 & 0 \\ 0 & Q_2 \end{pmatrix},$$

where the characteristic values of Q_1 have moduli less than 1, and those of Q_2 have modulus 1. It is easy to prove that $B_{\tilde{\varepsilon}}$ and B_ε can be written in the same terms

$$B_{\tilde{\varepsilon}} = \begin{pmatrix} 0 & 0 \\ 0 & B_{\tilde{\varepsilon}}^2 \end{pmatrix}, \qquad B_\varepsilon = \begin{pmatrix} B_\varepsilon^1 & 0 \\ 0 & 0 \end{pmatrix},$$

which proves the theorem.

2.1.3. The Spectral Representation of processes, Autoregressive and Moving Average Type Processes.

The spectral representation of stationary processes plays an important role in the theory of stochastic processes. We assume that the reader is acquainted with the elements of spectral theory. For details see Gikhman-Skorokhod [1], Rozanov [1], Shiryaev [2]. Let $\underline{\xi}(n)$, $(E\underline{\xi}(n) = 0)$, be a multidimensional wide sense stationary process, then it permits the following representation

$$(2.1.11) \qquad \underline{\xi}(n) = \int_{-\pi}^{\pi} e^{i\lambda n}\, \underline{\Phi}_{\xi}(d\lambda),$$

where $\underline{\Phi}_{\xi}(d\lambda)$ is a random orthogonal measure with $E\underline{\Phi}_{\xi}(d\lambda) = \underline{0}$, and

$E\, \underline{\Phi}_{\xi}(d\lambda)\, \overline{\underline{\Phi}_{\xi}(d\lambda)^*} = F_{\xi}(d\lambda)$. The matrix function $F_{\xi}(d\lambda)$ is said to be the spectral measure and it is non-negative definite. $F_{\xi}(\lambda)$ is called spectral distribution. If $F_{\xi}(\cdot)$ is absolutely continuous (with respect to the Lebesque measure)

$$\frac{F_{\xi}(d\lambda)}{d\lambda} = \frac{dF_{\xi}}{d\lambda} = f_{\xi}(\lambda),$$

is called the spectral density matrix of the process $\underline{\xi}(t)$. In the case of Gaussian processes $\underline{\Phi}$ is a Gaussian random orthogonal spectral measure. The Bochner-Hincsin-Cramer theorem states that the covariance function has the following representation

$$(2.1.12) \qquad B(n) = E\, \underline{\xi}(m+n)\, \underline{\xi}^*(m) = \int_{-\infty}^{\infty} e^{i\lambda n} F_{\xi}(d\lambda).$$

For stochastic processes with continuous time parameter (assuming that the stationary process $\underline{\xi}(t)$ is continuous) a similar representation holds

$$(2.1.11') \qquad \underline{\xi}(t) = \int_{-\infty}^{\infty} e^{i\lambda t}\, \underline{\Phi}_{\xi}(d\lambda)$$

where $\underline{\Phi}_{\xi}(d\lambda)$ is a random orthogonal measure with $E\, \underline{\Phi}_{\xi}(d\lambda) = \underline{0}$ and $E\, \underline{\Phi}(d\lambda)\, \underline{\Phi}^*(d\lambda) = F_{\xi}(d\lambda)$. In case of absolute continuity the spectral density matrix $f_{\xi}(\lambda) = \frac{dF_{\xi}}{d\lambda}$ exists, and

$$(2.1.12') \qquad B(t) = E\, \underline{\xi}(t+s)\, \underline{\xi}^*(s) = \int_{-\infty}^{\infty} e^{i\lambda t}\, F_{\xi}(d\lambda) = \int_{-\infty}^{\infty} e^{i\lambda t} f_{\xi}(\lambda)\, d\lambda.$$

In the following we recall some fundamental properties of random orthogonal measures. Let $\underline{\Phi}(d\lambda)$ be a random orthogonal measure with $E\, \underline{\Phi}(d\lambda) = \underline{0}$ and $E\, \underline{\Phi}(d\lambda)\, \overline{\underline{\Phi}^*(d\lambda)} = \frac{d\lambda}{2\pi} I$. It is well known that for any matrix function $\psi(\lambda)$ with

$$(2.1.13) \qquad \int \|\psi(\lambda)\|^2 d\lambda < \infty$$

the stochastic integral

$$(2.1.14)\qquad \int\psi(\lambda)\ \underline{\Phi}^*(d\lambda)$$

exists. It is the limit in mean square of the integral of simple functions $\psi_n(\lambda)$ $(n=1,2,\ldots,)$, for which

$$\int||\psi(\lambda)-\psi_n||^2\, d\lambda \to 0,\ \text{as}\ n\to\infty.$$

In this case $\underline{\zeta}_n = \int\psi_n(\lambda)\ \underline{\Phi}(d\lambda)$ and

$$E(\zeta_n - \int\psi(\lambda)\ \underline{\Phi}(d\lambda))\overline{(\underline{\zeta}_n - \int\psi(\lambda)\ \underline{\Phi}(d\lambda))^*} \to 0.$$

Integral (2.1.14) has the following two important properties

$$(2.1.15)\qquad E\int\psi(\lambda)\ \underline{\Phi}\ (d\lambda) = 0$$

$$(2.1.16)\qquad E\ \int\psi_1(\lambda)\ \underline{\Phi}(d\lambda)\ \overline{(\int\psi_2(\lambda)\ \underline{\Phi}(d\lambda))^*} = \frac{1}{2\pi}\int\psi_1(\lambda)\ \psi_2(\lambda)\ d\lambda.$$

Note that (2.1.11) and (2.1.15) give (2.1.12) and (2.1.13).

Lemma 5. If $\underline{\Phi}(d\lambda)$ is a Gaussian random measure with $E\underline{\Phi}(d\lambda)=\underline{0}$,

$$E\ \underline{\Phi}(d\lambda)\ \underline{\Phi}\ (d\lambda) = \frac{d\lambda}{2\pi}\ I,\ -\infty<\lambda<\infty,\quad \text{then}$$

$$(2.1.17)\qquad \underline{w}(t) = \int_{-\infty}^{\infty}\frac{e^{i\lambda t}-1}{i\lambda}\ \underline{\Phi}(d\lambda)$$

is a Wiener process.

Proof. $\underline{w}(t)$ is Gaussian, being the limit of Gaussian random variables (Appendix B, Lemma 4) and $\underline{w}(0) = \underline{0}$. By (2.1.15) $E\underline{w}(t) = \underline{0}$. Further, let $s_2 < s_2 < t_1 < t_2$, $\Delta = (t_1,t_2)$, $\Delta' = (s_1,s_2)$, then using (2.1.16) we get

$$E(\underline{w}(t_2)-\underline{w}(t_1))(\underline{w}(s_2)-\underline{w}(s_1))^* = \frac{1}{2\pi}\int_{-\infty}^{\infty}\frac{e^{i\lambda t_2}-e^{i\lambda t_1}}{i\lambda}\ \frac{e^{-i\lambda s_2}-e^{-i\lambda s_1}}{-i\lambda}\ I\ d\lambda.$$

Let

$$\chi_\Delta(t) = \begin{cases} 1, & t\in\Delta,\\ 0, & t\,\overline{\in}\,\Delta,\end{cases}$$

then from the Parseval's equality

$$\frac{1}{2\pi}\int_{-\infty}^{\infty}(e^{i\lambda t_2}-e^{i\lambda t_1})(e^{-i\lambda s_2}-e^{-i\lambda s_1})\ I\frac{d\lambda}{\lambda^2} = \int_{-\infty}^{\infty}\chi_\Delta(t)\ \chi_{\Delta'}(t)I dt = 0.$$

In the same way we may get

$$E(\underline{w}(t_2) - \underline{w}(t_1)(\underline{w}(t_2)-\underline{w}(t_1))^* = \int_{-\infty}^{\infty} |\chi_\Delta(t)|^2 I dt = (t_2-t_1)I,$$

and

$$E\underline{w}(t)\ w^*(s) = E(\underline{w}(t)-\underline{w}(s)+\underline{w}(s))\underline{w}^*(s) = sI, \quad s<t,$$

which proves the lemma.

Lemma 6. Let the function $g(z)$ be a spectral characteristic (or in other words transfer function), i.e., $\int_{-\infty}^{\infty}|g(i\lambda)|^2 d\lambda < \infty$, with respect to the Gaussian random spectral measure $\Phi(d\lambda)$, $E\Phi(d\lambda) = 0$, $E|\Phi(d\lambda)|^2 = \frac{d\lambda}{2\pi}$. Let the stochastic process $\zeta(t)$ be defined by

$$\zeta(t) = \int_{-\infty}^{\infty} e^{i\lambda t} g(i\lambda)\ \Phi(d\lambda). \tag{2.1.18}$$

Then (with probability 1)

$$\int_0^t \zeta(s)\ ds < \infty, \qquad t < \infty, \tag{2.1.19}$$

$$\int_0^t \zeta(s)\ ds = \int_{-\infty}^{\infty} \frac{e^{i\lambda t}-1}{i\lambda}\ g(i\lambda)\Phi(d\lambda). \tag{2.1.20}$$

Proof. From

$$\int_0^t E|\zeta(s)|ds \le \int_0^t |E\zeta^2(s)|^{1/2}\ ds \le (t\int_0^t E\ \zeta^2(s)ds)^{1/2} =$$

$$= t(\frac{1}{2\pi}\int_{-\infty}^{\infty}|g(i\lambda)|^2 d\lambda)^{1/2} < \infty,$$

and from Fubini's theorem we get that $\int_0^t \zeta(s)ds$ exists. Using the representation (2.1.18) we get

$$\int_0^t \zeta(s)ds = \int_0^t\int_{-\infty}^{\infty} e^{i\lambda s}\ g(i\lambda)\ \Phi(d\lambda)\ ds,$$

and changing the order of integration, which is permitted, we get

$$\int_0^t \zeta(s)\ ds = \int_{-\infty}^{\infty}\int_0^t e^{i\lambda s}\ ds\ g(i\lambda)\Phi(d\lambda) = \int_{-\infty}^{\infty} \frac{e^{i\lambda t}-1}{i\lambda}\ g(i\lambda)\Phi(d\lambda), \tag{2.1.21}$$

proving (2.1.20). In order to prove (2.1.21) let $\psi(\lambda)$ be square integrable, i.e. $\int_{-\infty}^{\infty}|\psi(\lambda)|^2\ d\lambda < \infty$, then from Fubini's theorem and (2.1.16)

$$E\int_0^t\int_{-\infty}^{\infty} e^{i\lambda s} g(i\lambda)\ \Phi(d\lambda)\ ds\ \overline{\int\psi(\lambda)\ \Phi(d\lambda)} = \frac{1}{2\pi}\int_0^t\int_{-\infty}^{\infty} e^{i\lambda s}\ g(i\lambda)\ \overline{\psi}(\lambda)\ d\lambda ds =$$

$$= \frac{1}{2\pi}\int_{-\infty}^{\infty}(\int_0^t e^{i\lambda s}ds)\ g(i\lambda)\overline{\psi}(\lambda)\ d\lambda = E\int_{-\infty}^{\infty}(\int_0^t e^{i\lambda s}ds) g(i\lambda)\Phi(d\lambda)\overline{\int_{-\infty}^{\infty}\psi(\lambda)\Phi(d\lambda)},$$

which proves (2.1.21).

Note that Lemma 6 remains valid in the case when $g(z)$ is a matrix function and $\underline{\eta}(t)$ is a vector process.

Processes defined by a spectral characteristic (transfer function) are called linear transforms, i.e., if $\underline{\xi}(t)$ has the random orthogonal spectral measure $\underline{\Phi}_\xi(d\lambda)$ and

$$\underline{\zeta}(t) = \int e^{i\lambda t} g(i\lambda)\ \underline{\Phi}_\xi(d\lambda),$$

with spectral characteristic $g(z)$ than process $\underline{\zeta}(t)$ is called a linear transform of $\underline{\xi}(t)$. It is known that $\underline{\zeta}(t)$ is a linear transform of $\underline{\xi}(t)$ with spectral characteristic $g(z)$ if and only if

(2.1.22) $$F_\zeta(d\lambda) = g(i\lambda)\ F_\xi(d\lambda)\ \overline{g^*(i\lambda)},$$

$$F_{\xi\zeta}(d\lambda) = E\ \underline{\Phi}_\zeta(d\lambda)\ \overline{\underline{\Phi}^*_\xi(d\lambda)} = g(i\lambda)\ F_\xi(d\lambda).$$

Let

(2.1.23) $$\underline{\xi}(n) = \int_{-\pi}^{\pi} e^{i\lambda n}(I-e^{-i\lambda}Q)^{-1}\ \underline{\Phi}(d\lambda),$$

where $\underline{\Phi}$ is Gaussian orthogonal random vector measure with

$$E\ \underline{\Phi}\ (d\lambda) = \underline{0},\qquad E\ \underline{\Phi}\ (d\lambda)\ \underline{\Phi}^*(d\lambda) = \frac{d\lambda}{2\pi}\ I,$$

and all the eigenvalues of Q are inside of the unit circle. Then we get by (2.1.15), (2.1.16) and (2.1.12′) that

$$E\underline{\xi}(n) = 0,\ E\underline{\xi}(n)\_\xi(n) = \frac{1}{2\pi}\int_{-\pi}^{\pi}(I-e^{-i\lambda}Q)^{-1}(I-e^{i\lambda}Q)^{*-1}d\lambda,$$

which means that the spectral density of $\underline{\xi}(n)$ exists and is equal to

(2.1.24) $$f_\xi(\lambda) = \frac{1}{2\pi}\ (I-e^{-i\lambda}Q)^{-1}(I-e^{i\lambda}Q)^{*-1}.$$

From the measure $\underline{\Phi}(d\lambda)$ constructing the process

(2.1.25) $\underline{\varepsilon}(n) = B_\varepsilon^{1/2} \int_{-\pi}^{\pi} e^{i\lambda n} \underline{\Phi}(d\lambda)$,

it is clear that

$$E\ \underline{\varepsilon}(n) = \underline{0}, \quad E\ \underline{\varepsilon}(n)\ \underline{\varepsilon}^*(m) = \delta(n,m) B_\varepsilon$$

where $\delta(n,m)$ is the Kronecker function; i.e., $\underline{\varepsilon}(n)$ is a Gaussian white noise process.
From (2.1.23) we have

$$\underline{\xi}(n) - Q\underline{\xi}(n-1) = \int_{-\pi}^{\pi} e^{i\lambda n} (I - e^{-i\lambda} Q)(I - e^{-i\lambda} Q)^{-1} \underline{\Phi}(d\lambda) =$$

$$= \int_{-\pi}^{\pi} e^{i\lambda n} \underline{\Phi}(d\lambda) = B_\varepsilon^{-1/2} \underline{\varepsilon}(n),$$

and $\underline{\xi}(n)$ is an elementary Gaussian process satisfying equation (2.1.2′). This shows that in the general case the spectral density of $\underline{\xi}(n)$ has the form

(2.1.24′) $f_\xi(\lambda) = \frac{1}{2\pi} (I - e^{-i\lambda} Q)^{-1} B_\varepsilon (I - e^{i\lambda} Q)^{*-1}$,

which is a "rational" function in $e^{i\lambda}$. It is easy to verify by direct calculations that (see (2.1.5))

$$B_\xi(n) = \int_{-\pi}^{\pi} e^{i\lambda n} f_\xi(\lambda) d\lambda = \frac{1}{2\pi} \int_{-\pi}^{\pi} e^{i\lambda n} [I + e^{-i\lambda} Q + \ldots] B_\varepsilon [I + e^{i\lambda} Q + \ldots] d\lambda =$$

$$= Q^n [B_\varepsilon + Q B_\varepsilon Q + \ldots] = Q^n \cdot B(o).$$

Further more, we have

(2.1.26) $F_\xi(\lambda) = \frac{1}{2\pi} \sum_{n=-\infty}^{\infty} Q^n B(o) \frac{e^{-i\lambda n} - 1}{-i\lambda}$

Let $\xi(n)$ be a one-dimensional Gaussian and stationary process. $\xi(n)$ is said to be an autoregressive type (AR) process of order p if it satisfies the following difference equation

(2.1.27) $\xi(n) + [a_1 \xi(n-1) + \ldots + a_p \xi(n-p)] = \varepsilon(n)$,

where $\varepsilon(n)$ is a Gaussian white noise, $E\varepsilon(n) = 0$, $E\ \varepsilon^2(n) = \sigma_\varepsilon^2$. Using the following notations

(2.1.28) $\zeta^j(n) = \xi(n+j)$, $j = 1,2,\ldots,p$; $n = 0, \pm 1, \pm 2, \ldots$,

$$(2.1.29)\quad Q = \begin{pmatrix} 0 & 1 & 0 & \cdots & 0 \\ 0 & 0 & 1 & \cdots & 0 \\ \cdot & & & & \\ \cdot & & & & \\ \cdot & & & & \\ 0 & 0 & 0 & \cdots & 1 \\ -a_p & -a_{p-1} & \cdot & \cdots & -a_1 \end{pmatrix}, \quad B_\varepsilon = \begin{pmatrix} 0 & 0 & \cdots & 0 \\ 0 & 0 & \cdots & 0 \\ \cdot & & & \\ \cdot & & & \\ \cdot & & & \\ 0 & 0 & \cdots & \sigma_\varepsilon^2 \end{pmatrix}$$

$$\underline{\varepsilon}(n) = \begin{pmatrix} 0 \\ 0 \\ \cdot \\ \cdot \\ \cdot \\ 0 \\ \varepsilon(n) \end{pmatrix},$$

we get that the process $\underline{\zeta}^*(n) = (\zeta^1(n),\dots,\zeta^p(n)) = (\xi(n+1),\dots,\xi(n+p))$ is an elementary p-dimensional Gaussian process:

$$\underline{\zeta}(n) = Q\,\underline{\zeta}(n-1) + \underline{\varepsilon}(n).$$

As $\xi(n)$ is the first component of the elementary process $\underline{\xi}(n)$ we introduce the following definition.

<u>Definition 1.</u> A one-dimensional stationary Gaussian process $\xi^1(n)$ will be called a component process of a k-dimensional elementary Gaussian process if there are k-1 stationary Gaussian processes $\xi^2(n),\dots,\xi^k(n)$ such that the process $\underline{\xi}^*(n) = (\xi^1(n),\dots,\xi^k(n))$ is an elementary Gaussian process.

<u>Definition 2.</u> We call a one-dimensional stationary Gaussian process $\xi(n)$ with discrete time an autoregressive moving average (ARMA) process if it satisfies the following equation

$$(2.1.30)\qquad \xi(n) + \sum_{i=1}^{p} a_i \xi(n-i) = \sum_{i=1}^{q} b_i \varepsilon(n-i),$$

where $\varepsilon(n)$ is a sequence of i.i.d. Gaussian random variables, and $\varepsilon(n)$ is independent of F_{n-1}^{ξ}.
In case $b_i = 0 (i \geq 1)$ the process is an autoregressive one and in

case $a_i = 0$ ($i \geq 1$) the process is a moving average one (MA).

Theorem 5. Equation (2.1.30) has a unique regular stationary solution if and only if the roots λ_i of the characteristic polinomial $P(z) = z^p + \sum_{i=1}^{p} a_i z^{p-i}$ are inside the unit circle ($|\lambda_i| < 1$). In this case $\xi(n)$ is the first component of a $k = \max(p,q+1)$ dimensional elementary Gaussian process.

Proof. Let us assume that $\xi^1(n) = \xi(n)$ and consider the system of equations (assuming $p \geq 1$, case $p = 0$ is trivial),

$$\xi^i(n) = \xi^{i+1}(n-1) + c_{i-1}\varepsilon(n), \qquad 1 \leq i < p,$$

$$\xi^p(n) = -\sum_{i=1}^{p} a_{p+1-i}\,\xi^i(n-1) + \sum_{i=p+1}^{q+1} b_{i-1}\xi^i(n-1) + c_{p-1}\varepsilon(n),$$

(2.1.31)

$$\xi^{p+1}(n) = \varepsilon(n),$$

$$\xi^i(n) = \xi^{i-1}(n-1), \quad p+1 < i \leq q+1.$$

Naturally in the case $q < p$ the suitable terms and equations are omitted. If the constants c_j ($j=0,\ldots,(p-1)$) satisfy the equations

$$c_o = b_o,$$

$$c_1 + a_1 c_o = b_1,$$

$$c_2 + a_1 c_1 + a_2 c_o = b_2,$$

(2.1.32) $\vdots$

$$c_{p-1} + a_1 c_{p-2} + \ldots + a_{p-1} c_o = b_{p-1},$$

then the system (2.1.31) is equivalent to the equation (2.1.30). Simple calculation shows that the characteristic polynomial $P_1(z)$ of (2.1.31) is equal to $P(z)$ if $q < p$, and $z^{q+1}P(z)$ otherwise. So the system (2.1.31) of stochastic difference equations has a unique stationary solution, which is a k-dimensional Gaussian Markov process and its first component will be the unique stationary solution of the equation (2.1.30). Q.E.D.

It is easy to see that (2.1.31), in case $q \neq 0$, is not the same as representation (2.1.28).
Note that the solution of equation (2.1.30) can be obtained in a constructive way similarly to that which we get for the first order autoregressive process (see Theorem 2).

$$(2.1.33) \qquad \xi(n) = \sum_{k=0}^{\infty} c_k \varepsilon(n-k).$$

Indeed, if the coefficients c_k satisfy the infinite recursive system of equations

$$c_o = b_o,$$
$$c_1 + a_1 c_o = b_1,$$
$$\vdots$$
$$c_k + \sum_{i=1}^{p} a_i c_{k-i-1} = b_k, \quad p \leq k ,$$

where the first p equations coincide with system (2.1.32), and $\sum_{k=0}^{\infty} |c_k|^2 < \infty$, then the process (2.1.33) is a correctly defined stationary Gaussian process satisfying (2.1.30).

As $b_k = 0$ for $k > q$ and the roots of the characteristic polynomial $P(z)$ are inside the unit circle system (2.1.34) has a unique solution with the desired property.

We recall that in Theorem 2 we get that the k-dimensional stationary, regular, Gaussian Markov process $\underline{\xi}(n)$ has the representation $\underline{\xi}(n) = \sum_{i=0}^{\infty} Q^i \underline{\varepsilon}(n-i)$. As the matrix Q satisfies its own characteristic equation with coefficients a_i, i.e., (see Appendix A, (A1.2), (A1.11'))

$$Q^k + \sum_{i=1}^{k} a_i Q^{k-i} = 0,$$

all the elements of $\{Q^n\}$ satisfy a recursive system of equations similar to (2.1.34) therefore the components of $\underline{\xi}(n)$ are sums of ARMA processes. Notice that if $\xi(n) = \sum_{j=1}^{\ell} d_j \xi^j(n)$, ($d_j$ are constants), where

$$(2.1.35) \quad \xi^j(n) = - \sum_{i=1}^{p} a_i \xi(n-i) + \sum_{i=0}^{q} b^j(i) \varepsilon^j(n-i), \quad j = 1,2,\dots,\ell,$$

and $\{\underline{\varepsilon}(n)\}$ is a sequence of i.i.d. Gaussian vectors, then $\xi(n)$ is an ARMA process. This establishes the converse of Theorem 5.

Theorem 6. Any component of a regular elementary Gaussian process is an ARMA process.

In terms of the spectral density function Theorems 5 and 6 can be expressed in the following way. Let $\xi(n)$ be a one-dimensional Gaussian stationary process with the spectral representation

$$(2.1.36) \qquad \xi(n) = \int_{-\pi}^{\pi} e^{i\lambda n} \frac{Q(e^{i\lambda})}{P(e^{i\lambda})} \Phi(d\lambda)$$

where $\Phi(d\lambda)$ is a Gaussian orthogonal measure with

$$E\Phi(d\lambda) = 0, \quad E|\Phi(d\lambda)|^2 = \frac{d\lambda}{2\pi}, \text{ and}$$

$$P(z) = z^p + a_1 z^{p-1} + \ldots + a_p, \quad Q(z) = b_o z^q + b_1 z^{q-1} + \ldots + b_q.$$

Assume that all the roots of the polynomial $P(z)$ are inside the unit circle. It follows from (2.1.36) (in the same way as we proved (2.1.24) and (2.1.25)) that $\xi(n)$ has the spectral density

$$(2.1.37) \qquad f_\xi(\lambda) = \frac{1}{2\pi} \frac{|Q(e^{i\lambda})|^2}{|P(e^{i\lambda})|^2},$$

and

$$(2.1.38) \quad \xi(n) + a_1\xi(n-1) + \ldots + a_p\xi(n-p) = b_p\varepsilon(n) + \ldots + b_q\varepsilon(n-q),$$

where $\varepsilon(n)$ is a Gaussian white noise $E\varepsilon(n) = 0$, $E\varepsilon^2(n) = 1$,

$$(2.1.39) \qquad \varepsilon(n) = \int_{-\pi}^{\pi} e^{i\lambda n} \Phi(d\lambda)$$

Representation (2.1.38) coincides with (2.1.30), i.e., $\xi(n)$ is an ARMA process.

The representation of $\xi(n)$ as the first component of max(p,q+1) dimensional elementary Gaussian process in the form (2.1.31) we shall call representation (A).

Let

$$g_1(z) = \frac{1}{z} g_2(z) + c_o,$$

$$g_2(z) = \frac{1}{z} g_3(z) + c_1,$$

$$\vdots$$

(2.1.40)
$$g_{p-1}(z) = \frac{1}{z} g_p(z) + c_{p-2},$$

$$g_p(z) = -\sum_{i=0}^{p-1} a_{p-i} g_{i+1}(z) + \sum_{i=p+1}^{q+1} b_{i-1} z^{-(i-1)} + c_{p-1},$$

$$g_{p+1}(z) = 1,$$

$$g_{p+2}(z) = \frac{1}{z},$$

$$\vdots$$

$$g_{q+1}(z) = \frac{1}{z^{q-p}},$$

where the constants c_j $(j=0,\ldots,p-1)$ are given in (2.1.32), and the functions $g_k(z)$ $(k=1,\ldots,p)$ are the ratios of two polynomials, which can be calculated, e.g.,

$$g_1(z) = \frac{Q(z)}{P(z)} = \frac{\sum_{i=0}^{q} b_i z^{q-i}}{z^p + \sum_{i=1}^{p} a_i z^{p-i}} .$$

Let $g_j(z)$ be the spectral characteristic of the process $\xi^j(n)$ with respect to $\Phi(d\lambda)$, i.e.,

(2.1.41) $$\xi^j(n) = \int_{-\pi}^{\pi} e^{i\lambda n} g_j(e^{i\lambda}) \Phi(d\lambda),$$

then the random vector process $(\xi^1(n),\ldots,\xi^{q+1}(n)) = \underline{\xi}^*(n)$ is an elementary Gaussian process

(2.1.42) $$\underline{\xi}(n) = Q\,\underline{\xi}(n-1) + \underline{\varepsilon}(n)$$

where

$$(2.1.42') \qquad Q = \begin{pmatrix} 0 & 1 \dots 0 & 0 \dots 0 \\ \vdots & \vdots & \vdots \quad \vdots \\ 0 & 0 \dots 1 & 0 \dots 0 \\ -a_p & -a_{p-1} \; -a_1 & b_p \dots b_q \\ 0 & 0 \dots 0 & 0 \dots 0 \\ 0 & 0 \dots 0 & 1 \dots 0 \\ \vdots & & \\ 0 & 0 \dots 0 & 0 \dots 1 \; 0 \end{pmatrix}, \quad \underline{\varepsilon}(n)=\varepsilon(n) \begin{pmatrix} c_o \\ \vdots \\ c_{p-1} \\ 1 \\ 0 \\ \vdots \\ 0 \end{pmatrix}$$

On the basis of Theorems 5, 6, and by representation (2.1.40) we proved the following theorem.

Theorem 7. *The regular stationary Gaussian process* $\xi(n)$ *is a component process of an elementary Gaussian process if and only if its spectral density function is rational in* $e^{i\lambda}$, *i.e.,*

$$f_\xi(\lambda) = \frac{1}{2\pi} \frac{|Q\ (e^{i\lambda})|^2}{|P\ (e^{i\lambda})|^2},$$

and the roots of polynomial $P(z)$ *has modulus less than 1.*

Remark 1. The process $\xi(n)$ is an autoregressive type stationary process if and only if

$$f_\xi(\lambda) = \frac{1}{2\pi} \frac{c}{|P\ (e^{i\lambda})|^2}, \qquad c > o ,$$

where $P(z)$ has roots in the unit circle.
There exists a more simple and straightforward representation, called representation (B). Let the spectral representation of $\xi(n)$ be (2.1.36), i.e.,

$$\xi(n) = \int_{-\pi}^{\pi} e^{i\lambda n} \frac{Q(e^{i\lambda})}{P(e^{i\lambda})} \Phi(d\lambda),$$

where $\Phi(d\lambda)$ is a random Gaussian orthogonal spectral measure with $E\Phi(d\lambda) = 0$, $E|\Phi(d\lambda)|^2 = \frac{d\lambda}{2\pi}$. For simplicity let $p > q$ and let us denote the partial rational decompositions in the following way (the roots of $P(z)$ are λ_i, and for simplicity they are distinct)

$$\frac{Q(z)}{P(z)} = \frac{Q_1}{z-\lambda_1} + \dots + \frac{Q_p}{z-\lambda_p},$$

(2.1.43)
$$\frac{zQ(z)}{P(z)} = \frac{Q_1'}{z-\lambda_1} + \dots + \frac{Q_p'}{z-\lambda_p},$$

$$\vdots$$

$$\frac{z^{p-q-1}Q(z)}{P(z)} = \frac{Q_1^{(p-q-1)}}{z-\lambda_1} + \dots + \frac{Q_p^{(p-q-1)}}{z-\lambda_p}.$$

Then we get

$$\xi^1(n)=\xi(n)=\int_{-\pi}^{\pi} e^{i\lambda n}\left(\frac{Q_1}{e^{i\lambda}-\lambda_1}+\dots+\frac{Q_p}{e^{i\lambda}-\lambda_p}\right)\Phi(d\lambda)=Q_1\ \zeta_1(n)+\dots+$$

$$+Q_p\zeta_p(n),$$

$$\xi^2(n)=\xi^1(n+1)=\int_{-\pi}^{\pi} e^{i\lambda n}\frac{e^{i\lambda}Q(e^{i\lambda})}{P(e^{i\lambda})}\Phi(d\lambda)=Q_1'\zeta_1(n)+\dots+Q_p'\ \zeta_p(n),$$

$$\vdots$$

$$\xi^{p-q}(n)=\xi^1(n+p-q-1)=\int_{-\pi}^{\pi} e^{i\lambda(n+p-q-1)}\frac{Q(e^{i\lambda})}{P(e^{i\lambda})}\Phi(d\lambda)=Q_1^{(p-q-1)}\zeta_1(n)+\dots$$

$$+Q_p^{(p-q-1)}\ \zeta_p(n),$$

$$\xi^{p-q+1}(n)=\zeta_{p-q+1}(n),$$

$$\vdots$$

$$\xi^p(n)=\zeta_p(n),$$

where the $\zeta_i(n)$ are first order autoregressive processes $(i=1,2,\dots,p)$

$$\zeta_i(n) = \lambda_i\zeta_i(n-1) + \varepsilon(n), \qquad \text{(with the same } \varepsilon(n)\text{)}.$$

Solving equations (2.1.44) with respect to $\zeta_1(n),\dots,\zeta_p(n)$ we get representation (B):

$$\xi^1(n) = \xi(n),$$

$$\xi^2(n) = \xi^1(n+1),$$

$$\vdots$$

$$\xi^{p-q}(n) = \sum_{i=1}^{p} Q_i^{(p-q-1)} \zeta_i(n) = \sum_{i=1}^{p} \lambda_i Q_i^{(p-q-1)} \zeta_i(n-1) +$$

$$+ \varepsilon(n) \sum_{i=1}^{p} Q_i^{(p-q-1)} = \sum_{i=1}^{p} d_i \xi^i(n-1) +$$

(2.1.45) $$+ \varepsilon(n) \sum_{i=1}^{p} Q_i^{(p-q-1)},$$

$$\xi^{p-q+1}(n) = \lambda_{p-q+1} \xi^{p-q+1}(n-1) + \varepsilon(n),$$

$$\vdots$$

$$\xi^p(n) = \lambda_p \xi^p(n-1) + \varepsilon(n).$$

From representation (B) we see that the components $\xi^1(n),\dots,\xi^{p-q}(n)$ are observable, while $\xi^{p-q+1}(n),\dots, \xi^p(n)$ are unobservable.

We say that the stationary process $\eta(n)$ is reversed (in time) with respect to the process $\xi(n)$ if the covariance function $B_\eta(t) = B_\xi(-t)$. The process $\xi(t)$ which is reversed with respect to itself is called symmetric. These definitions are the same in the multidimensional case. If $\underline{\eta}$ is reversed with respect to $\underline{\xi}(t)$ then for the spectral density matrix we get

(2.1.46) $$f_\eta(\lambda) = \overline{f_\xi(\lambda)}.$$

It is obvious that the one dimensional elementary process is symmetric. In the multidimensional case as $\underline{\eta}(n)$ is also elementary Gaussian process

(2.1.47) $$\underline{\eta}(n) = \underline{\eta}(n-1) + C\,\underline{\varepsilon}(n), \qquad E\underline{\varepsilon}(n) = o,\ \operatorname{cov}(\varepsilon(n),\underline{\varepsilon}(n)) = I,$$

where

$$\underline{\xi}(n) = Q\,\underline{\xi}(n-1) + \underline{\varepsilon}(n).$$

From (2.1.24) and (2.1.46) we get

(2.1.48) $$(I-e^{-i\lambda}\tilde{Q})^{-1}CC^*(I-e^{i\lambda}\tilde{Q})^{*-1} = (I-e^{i\lambda}Q)^{-1}(I-e^{-i\lambda}Q)^{*-1},$$

and (by the comparison of coefficients on both sides)

(2.1.49) $$\begin{aligned} Q^* &= \tilde{Q}\, C\, C^* \\ I + Q\, Q^* &= CC^* + \tilde{Q}\, C\, C^*\, \tilde{Q}^*. \end{aligned}$$

$\underline{\xi}(n)$ is symmetric if the matrix Q is symmetric.

2.2 Processes With Continuous Time

2.2.1. Main Theorems. The real observations are realized by discrete time, and even more in digital computer processing, as is the rule, everything is discretized in the analog-to-digital (A-D) conversion process. Nevertheless, the time-continuous model has its advantages. In many cases it is much better to work with the time-continuous model, as we shall see it later. Some phenomena can be described more adequately in that way; in other cases the results have more simple form, if at all they can be got. The exact correspondence between discrete and continuous time cases will be formulated. As the exact analogon of an i.i.d. sequence of Gaussian random variables (white noise) does not exist in the continuous case, as the derivative of the Wiener process does not exist, we follow the construction of Ito integrals for our special case of linear systems with constant coefficients.

Let $(\underline{w}(t), F_t)$ be a k-dimensional standard Wiener process with local parameters $E\underline{w}(t) = \underline{0}$ (the drift is $\underline{0}$), $E(d\underline{w}(t)d\underline{w}(t)^*) = I.dt$ (with unit diffusion parameter). Let us consider the linear stochastic differential equation with the non-singular matrix A_{kxk} and, may be singular, positive semidefinite matrix $B_w \neq 0$

(2.2.1) $\quad d\,\underline{\xi}(t) = A\,\underline{\xi}(t)dt + B_w^{1/2}d\underline{w}(t),$

or in integral form

(2.2.1′) $\quad \underline{\xi}(t) = \underline{\xi}(t_o) + A\int_{t_o}^{t} \underline{\xi}(s)\,ds + B_w^{1/2}(\underline{w}(t) - \underline{w}(t_o)),$

where $\underline{\xi}(t_0)$ is normally distributed and independent of $F^w_{[t_0,t]}, t > t_0$. We prove the following statement.

Theorem 1. The continuous k-dimensional random process $\underline{\xi}(t)$ is an elementary (i.e., stationary Markov) Gaussian process if and only if it is the solution of the stochastic differential equation (2.2.1) in the following sense.

a) If $\underline{\xi}(t)$ is a continuous, $E\underline{\xi}(t) = \underline{0}$, elementary Gaussian process then there exists a unique matrix A_{kxk} with eigenvalues in the left halfplane and a Wiener process $(\underline{w}(t), F^{\xi}_t) = 0$, $E(\underline{w}(t)\underline{w}^*(t)) = B_w \cdot t$, such that (2.2.1) holds and

(2.2.2) $\quad B(t) = E\,\underline{\xi}(s+t)\,\underline{\xi}^*(s) = e^{At}B(0), \quad t > 0,$

where

(2.2.3) $$AB(0) + B(0)\,A^* = -B_w,$$

with the solution

(2.2.4) $$B(0) = \int_0^\infty e^{As}\, B_w\, e^{A^*s}\, ds.$$

b) Let $A_{k\times k}$ be a non-singular matrix with eigenvalues in the left halfplane, and B_w non-negative definite, then the only stationary regular solution of (2.2.1) with continuous $\underline{\xi}(t)$, $-\infty < t < \infty$, is an elementary Gaussian process. Its co-variance matrix function has the form (2.2.2) with $B(0)$ satisfying (2.2.3). In the case when $\underline{\xi}(t)$ is defined for $t \geq 0$, $\underline{\xi}(0)$ and F_t^w are independent, $\underline{\xi}(0)$ is normally distributed with parameters $\underline{0}$, $B(0)$.

Proof of b). The proof is substantially the same as the proof in the discrete time case. First we prove that the solution of (2.2.1) has the following form

(2.2.5) $$\underline{\xi}(t) = \int_{-\infty}^{t} e^{A(t-s)}\, d\underline{w}(s), \quad -\infty < t < \infty,$$

or

(2.2.5') $$\underline{\xi}(t) = e^{A(t-t_o)}\,\underline{\xi}(t_o) + \int_{t_o}^{t} e^{A(t-s)}\, d\underline{w}(s),$$

and is unique. That (2.2.5') or (2.2.5) is a solution of (2.2.1) follows immdiately from (A2.10) (see Appendix A) and the differential of (2.2.5') equals to

$$d\underline{\xi}(t) = [Ae^{A(t-t_o)}\,\underline{\xi}(t_o) + Ae^{At}\int_{t_o}^{t} e^{-As}\, d\underline{w}(s)]dt + e^{At}\, e^{-At}\, d\underline{w}(t) =$$

$$= A\,\underline{\xi}(t)\, dt + d\underline{w}(t)$$

In a similar way we get from (2.2.5)

$$A\int_0^t \underline{\xi}(s)ds = A\int_0^t\int_{-\infty}^{s} e^{A(s-n)} d\underline{w}(n)ds = A\int_0^t\int_{-\infty}^{o} e^{A(s-n)}\, d\underline{w}(n)\, ds +$$

$$+ A\int_o^t\int_o^s e^{A(s-n)}\, d\underline{w}(n)\, ds = [e^{At}-I]\,\underline{\xi}(0) + \int_o^t [e^{A(t-s)}-I]d\underline{w}(s) =$$

$$= \underline{\xi}(t) - \underline{\xi}(0) + [\underline{w}(0) - \underline{w}(t)],$$

which proves that (2.2.5) is a solution (compare the proof with Theorem 6 in Appendix B). To prove the uniqueness let $\underline{\eta}(t)$ be another solution of (2.2.1′) (assuming $\underline{\eta}(t_0) = \underline{\xi}(t_0)$), then

$$\underline{\Delta}(t) = \eta(t) - \underline{\xi}(t) = \int_{t_o}^{t} A\Delta(s)ds,$$

and

$$\sum_{i=1}^{k} |\Delta^i(t)| \leq \int_{t_o}^{t} \sum_{i,j=1}^{k} |a_{ij}| \sum_{i=1}^{k} |\Delta^i(s)|ds.$$

Using the well known lemma (see e.g., Lemma 2. Ch.1 Section 6.) that if for some
$c \geq 0$

$$u(t) \leq v(t) + c \int_{o}^{t} u(s)ds, \quad t \geq 0,$$

then

$$u(t) \leq v(t) + c \int_{o}^{t} e^{c(t-s)} v(s)ds,$$

we obtain (in our case $v(s) \equiv 0$)

$$\Delta(t) \equiv 0$$

with probability 1, which proves the uniqueness.

Now let $B_t = E\underline{\xi}(t)\underline{\xi}^*(t)$ then we state that

$$\text{(2.2.6)} \qquad \frac{dB_t}{dt} = AB_t + B_t A^* + B_w,$$

and

$$\text{(2.2.7)} \qquad B(t,s) = E\, \underline{\xi}(t)\, \underline{\xi}^*(s) = \begin{cases} e^{A(t-s)} B_s & , t \geq s, \\ B_t\, e^{A^*(s-t)} & , t \leq s. \end{cases}$$

Indeed, from (2.2.5′) and (2.1.15) - (2.1.16), using that $\underline{\xi}(t_0)$ and $\underline{w}(t)\,(t \geq t_0)$ are independent

$$B_t = E[e^{A(t-t_o)}\underline{\xi}(t_o) + \int_{t_o}^{t} e^{A(t-s)}d\underline{w}(s)][e^{A(t-t_o)} + \int_{t_o}^{t} e^{A(t-s)}d\underline{w}(s)]^* =$$

$$= e^{At}[B_{t_o} + \int_{t_o}^{t} e^{-As} B_w e^{-A^*s} ds]e^{A^*t},$$

and by derivation we get (2.2.6).

Further (let t>s)

$$E\underline{\xi}(t)\underline{\xi}^*(s) = e^{At}[B_o + E\int_o^t e^{-An}\, d\underline{w}(n) \ (\int_o^t \chi_{[u\le s]} e^{-An} d\underline{w}(n))^*]e^{-A^*s} =$$

$$= e^{A(t-s)}\ e^{As}[B_o + \int_o^s e^{-An}\ B_w\ e^{-A^*n}\ dn]e^{-A^*s} = e^{A(t-s)}\ B_s,$$

which proves (2.2.7).

Representations (2.2.5) and (2.2.5') give that $\underline{\xi}(t)$ is a Gaussian (see Lemma 4 in Appendix B) and Markov. That it is a diffusion type process follows from normality, as all the moments of $\underline{\xi}(t)$ $(E(\xi^i(t))^k < \infty,\ k = 1,2,\dots,)$ exist. Particularly

$$(2.2.8)\quad E(\underline{\xi}(t)\,|\,\underline{\xi}(s),\ \underline{\xi}(u)) = E\ (\underline{\xi}(t)\,|\,\underline{\xi}(s)), \qquad u < s < t.$$

From the normality we have (Lemma 4, Appendix B).

$$(2.2.9)\quad E(\underline{\xi}(t)\,|\,\underline{\xi}(s)) = R(t,s)\ \underline{\xi}(s), \qquad R(t,s) = B(t,s)\ B_s^+.$$

Equations (2.2.8) and (2.2.9) show that, $u < s < t$,

$$E(\underline{\xi}(t) - R(t,s)\underline{\xi}(s)\,|\,\underline{\xi}(s),\ \underline{\xi}(u)) = \underline{0},$$

and

$$E(\underline{\xi}(t)\underline{\xi}^*(u)B_u - R(t,s)\ \underline{\xi}(s)\ \underline{\xi}^*(u)\ B_u^+) = 0.$$

The last equation gives

$$(2.2.10)\quad R(t,u) = R(t,s)\ R(s,u).$$

The stationarity of solution (2.2.5) immediately follows by direct calculation and so matrix $B_t = B(0)$. That $B(0)$ is the solution of (2.2.3) follows from (2.2.6) ($B_t' \equiv 0$), and as A has eigenvalues in the left halfplane (2.2.4) is the only solution. From (2.2.10) we get

$$B(t,s) = e^{A(t-s)}\ B(0),$$

which proves (2.2.2).

It remains to prove that $\underline{\xi}(0)$ and $\underline{w}(t)$ $(t\ge 0)$ are independent, i.e., $E\underline{\xi}(0)\underline{w}(t) = 0 (t\ge 0)$. Before doing this we prove that $\underline{\xi}(t)$ has the spectral density function

$$(2.2.11)\qquad f_\xi(\lambda) = \frac{1}{2\pi}(-i\lambda I + A)^{-1}B_w(i\lambda I + A)^{*-1}.$$

Indeed, let

$$(2.2.12)\quad \underline{\xi}(t) = \int_{-\infty}^{\infty} e^{i\ t}(-i\lambda I + A)^{-1} B_w^{1/2} \underline{\phi}(d\lambda) = \int_{-\infty}^{\infty} e^{i\lambda t} \underline{\phi}_{\xi}(d\lambda),$$

where $\underline{\phi}$ is a Gaussian orthogonal vector measure with

$$E\underline{\phi}(d\lambda) = \underline{0},\quad E(\underline{\phi}(d\lambda)\ \underline{\phi}^*(d\lambda)) = \frac{d\lambda}{2\pi} I.$$

Then using (2.1.15), 2.1.16) we obtain

$$(2.2.13)\quad E\underline{\xi}(t)\underline{\xi}^*(t) = B(0) = \frac{1}{2\pi}\int_{-\infty}^{\infty} (-i\lambda I + A)^{-1} B_w (i\lambda I + A)^{*-1}\ d\lambda,$$

which proves (2.2.11).

One can prove by direct calculations or by using Lemma 5 in 12.1 that the Wiener process $\underline{w}(t)$ in (2.2.1′) has the following representation (see (2.2.12), in case when B_w is positive semidefinite $B_w^{-1/2} = (B_w^{1/2})^+$)

$$(2.2.14)\quad \underline{w}(t) = \int_{-\infty}^{\infty} \frac{e^{i\lambda t} - 1}{i\lambda} (i\lambda I - A) B_w^+\ \underline{\phi}_{\xi}(d\lambda) = \int_{-\infty}^{\infty} \frac{e^{i\lambda t} - 1}{i\lambda} \underline{\phi}(d\lambda),\ t \geq 0.$$

For $T < 0$ let us introduce the Wiener process

$$(2.2.15)\quad \underline{w}_T(t) = \int_{-\infty}^{\infty} \frac{e^{i\lambda t} - e^{i\lambda T}}{i\lambda}\ \underline{\phi}(d\lambda),$$

then

$$\underline{\xi}(t) = \underline{\xi}(T) + A\int_T^0 \underline{\xi}(n)dn + A\int_0^T \underline{\xi}(n)dn + B_w^{1/2} d\underline{w}_T(t),$$

and

$$E\underline{w}(t)w_T^*(0) = E\int_{-\infty}^{\infty} \frac{e^{i\lambda t} - 1}{i\lambda}\ \underline{\phi}(d\lambda) \int_{-\infty}^{\infty} \frac{1 - e^{i\lambda T}}{i\lambda}\ \underline{\phi}(d\lambda) = 0.$$

Further

$$E\ \underline{\xi}(0)\ \underline{w}^*(t) = E\ \underline{\xi}(T)\ \underline{w}^*(t) + A\int_T^0 E\underline{\xi}(n)\ \underline{w}^*(t)\ dn.$$

Solving the last equation with respect to $E\underline{\xi}(T)\underline{w}^*(t)$ (for $T < 0$) we get

$$E\underline{\xi}(T)\ \underline{w}^*(t) = e^{At} E\underline{\xi}(0)\underline{w}^*(t),$$

or

$$(2.2.16)\quad E\underline{\xi}(0)\underline{w}^*(t) = e^{-At}\ E\underline{\xi}(T)\underline{w}^*(t).$$

As $E\underline{\xi}(T)\underline{w}\ (t)$ is bounded and A has eigenvalues in the left halfplane from (2.2.16)

$$\lim_{T\to-\infty}\ E\underline{\xi}(0)\ \underline{w}^*(t) = 0$$

The proof of part b) is realized.

<u>Proof of a).</u> From the Gaussian-Markov property of the process $\underline{\xi}(t)$ we get (2.2.8), (2.2.9) and (2.2.10). Further, from stationarity we obtain $R(t,s) = R(t-s)$ and

$$R(t-n) = R(t-s)\; R(s-n), \tag{2.2.17}$$

where the matrix function $R(v)$ is continuous as $\underline{\xi}(t)$ is continuous (in mean square too). The unique solution of the functional equation (2.2.17) under the initial condition $R(0) = I$ is the matrix function

$$R(t-n) = e^{A(t-n)}, \tag{2.2.18}$$

with a constant matrix A, where

$$A = \lim_{t\to o} \frac{R(t)-I}{t} .$$

Multiplying (2.2.9) by $\underline{\xi}^*(s)$ and taking expectation we get

$$E(E(\underline{\xi}(t)\,|\,\underline{\xi}(s))\underline{\xi}^*(s)) = E(E(\underline{\xi}(t)\underline{\xi}^*(s)\,|\,\underline{\xi}(s))) = B(t-s) = \tag{2.2.19}$$

$$= e^{A(t-s)}B(0), \quad t > s ,$$

which gives (2.2.2). Now let

$$\underline{w}(t) = \underline{\xi}(t) - \underline{\xi}(0) - \int_o^t A\underline{\xi}(s)\, ds. \tag{2.2.20}$$

Then direct calculations show that $E\underline{w}(t) = 0$, and

$$E\underline{w}(t)\;\underline{w}^*(t) = E(\underline{\xi}(t)-\underline{\xi}(0)-A\int_o^t\underline{\xi}(s)ds)(\underline{\xi}(t)-\underline{\xi}(0)-A\int_o^t\underline{\xi}(s)ds)^* = \tag{2.2.21}$$

$$= B(0)-B(0)\; e^{A^*t} - \int_o^t AB(0)e^{A^*(t-s)}ds - e^{At}B(0)+B(0)+$$

$$+ \int_o^t A\; e^{A(t-s)}\; B(0)\; ds - \int_o^t e^{A(t-s)}\; B(0)\; A^* ds +$$

$$+ \int_o^t \Big[\int_o^s A\; e^{A(s-n)}\; B(0)\; A^* dn + \int_s^t A\; B(0)\; e^{A^*(n-s)}\; A^* dn\Big]ds =$$

$$= -[B(0)\; A^* + A\; B(0)] \cdot t.$$

From (2.2.3) and (2.2.21) we obtain that

$$E\underline{w}(t)\;\underline{w}^*(t) = B_w t,$$

with a positive semidefinite B_w, and in the same way

$$E\underline{w}(t+s)\underline{w}^*(t) = B_w s, \quad s > 0 ,$$

which proves that $\underline{w}(t)$ is a Wiener process.
As $e^{At} \to 0$, if $t \to \infty$, this involves the condition on eigenvalues of matrix A. To prove relation (2.2.3), i.e., that B_w is non-negative definite denote the following. The matrix

$$(2.2.22) \qquad E(\underline{\xi}(t+dt)-\underline{\xi}(t))(\underline{\xi}(t+dt)-\underline{\xi}(t))^* =$$

$$=E(A\int_t^{t+dt} \underline{\xi}(s)ds)(A\int_t^{t+dt} \underline{\xi}(s)ds)^* + 2E(\underline{w}(t+dt)-\underline{w}(t))\cdot$$

$$\cdot(A\int_t^{t+dt} \underline{\xi}(s)ds)^* + E(\underline{w}(t+dt)-\underline{w}(t))(\underline{w}(t+dt)-\underline{w}(t))^*$$

is positive semidefinite. As the first two terms on the right side has order $\mathcal{O}(dt)^2$ and the left side equals to (see (2.2.7))

$$B(0)-B(0)e^{A^*dt} - e^{Adt}B(0) + B(0) = - B(0)A + AB(0)\, dt + \sigma(dt)^2 ,$$

we get that

$$B_w = - AB(0) - B(0)A^*$$

is positive semidefinite. This proves the theorem.

<u>Remark 1.</u> The proof of Theorem 1 shows that a Gaussian process is elementary if and only if its covariance matrix has the form (2.2.7), or the spectral density function form (2.2.11), i.e.,

$$E\underline{\xi}(t+s)\,\underline{\xi}^*(s) = e^{At}B(0), \qquad t \geq 0.$$

<u>Remark 2.</u> From (2.2.9) we have that the Gaussian random variables

$$\underline{\xi}\left(\frac{it}{n}\right) - e^{A\frac{1}{n}}\,\underline{\xi}\left(\frac{(i-1)t}{n}\right), \qquad i = 1,2,\dots,n,$$

are independent and for their sum

$$(2.2.23) \qquad \lim_{n\to\infty} \sum_{i=1}^{n} \left[\underline{\xi}\left(\frac{it}{n}\right) - e^{A\frac{1}{n}}\,\underline{\xi}\left(\frac{(i-1)t}{n}\right)\right] = \underline{w}(t), \quad \underline{w}(0) = \underline{0},$$

we get a Gaussian variable (Lemma 4, Appendix B) with independent increments. But

$$(2.2.24) \qquad \sum_{i=1}^{n} \left[\underline{\xi}\left(\frac{it}{n}\right) - e^{A\frac{1}{n}}\,\underline{\xi}\left(\frac{(i-1)t}{n}\right)\right] = \underline{\xi}(t)-\underline{\xi}(0)-A\sum_{i=1}^{n} \frac{1}{n}\,\underline{\xi}\left(\frac{it}{n}\right) + \sigma\left(\frac{1}{n}\right),$$

and the right side tends to (as $n\to\infty$)

$$(2.2.25)\qquad \underline{\xi}(t) - \underline{\xi}(0) - A\int_0^t \underline{\xi}(s)\,ds.$$

(2.2.24) and (2.2.25) give the representation of process $\underline{\xi}(t)$ in the form (2.2.1′).

Remark 3. Using the spectral representation of $\underline{\xi}(t)$

$$\underline{\xi}(t) = \int_{-\infty}^{\infty} e^{i\lambda t}(i\lambda I - A)^{-1}\,\underline{\Phi}(d\lambda)$$

we get

$$(2.2.21')\qquad \underline{\xi}(t)-\underline{\xi}(0)-A\int_0^t \underline{\xi}(s)ds = \int_{-\infty}^{\infty}(e^{i\lambda t}-1)(i\lambda I - A)^{-1}\,\underline{\Phi}(d\lambda) +$$

$$+ A\int_{-\infty}^{\infty}\frac{e^{i\lambda t}-1}{i\lambda}(i\lambda I-A)^{-1}\underline{\Phi}(d\lambda) =$$

$$= \int_{-\infty}^{\infty}\frac{e^{i\lambda t}-1}{i\lambda}(i\lambda I-A)(i\lambda I-A)^{-1}\underline{\Phi}(d\lambda) = \int_{-\infty}^{\infty}\frac{e^{i\lambda t}-1}{i\lambda}\,\underline{\Phi}(d\lambda),$$

which on the basis of Lemma 5 in §2.1 is a Wiener process. From (2.2.21′) and (2.2.21) it is easy to obtain that $-AB(0) - B(0)A$ is positive semidefinite.

The following theorem explains the connection between the discrete and continuous time elementary Gaussian processes.

Theorem 2. The continuous process $\underline{\xi}(t)$, $-\infty < t < \infty$, is an elementary Gaussian process if and only if for any $\delta > 0$ the discrete time process $\underline{\xi}(n\delta)$ is elementary Gaussian.

Proof. Necessity is trivial. For the proof of sufficiency let us first notice that the joint distribution of random vectors $\underline{\xi}(k_1\delta),\ldots$ $\underline{\xi}(k_n\delta)$ for every $\delta > 0$ and every finite sequence $k_1,\ldots,k_n$ of integers is Gaussian. Hence, by the continuity of the process $\underline{\xi}(t)$ and Lemma 4 in Appendix B it is Gaussian. Stationarity is obvious. To prove markovity we note that on the basis of Remark 1 it is sufficient to prove that for $t_2 > t_1$,

$$(2.2.26)\qquad E(\underline{\xi}(t_2)\,|\,\underline{\xi}(t_1)) = e^{A(t_2-t_1)}\underline{\xi}(t_1).$$

From the Gaussian property it follows that there exists a matrix function $R(t_2-t_1)$ for which $E(\xi(t_2)\,|\,\underline{\xi}(t_1)) = R(t_2-t_1)\underline{\xi}(t_1)$. As the process $\underline{\xi}(n\delta)$ is Markov for any $\delta > 0$ we get

$$(2.2.27) \qquad R(m\delta) + R(n\delta) = R((m+n)\delta).$$

Because of the continuity of process $\underline{\xi}(t)$ $R(t_2-t_1)$ is also continuous, and as (2.2.27) coincides with (2.2.17) and it satisfies the initial condition $R(0) = I$, we get (2.2.26).

Let $\underline{w}_1(t)$ and $\underline{w}_2(t)$ be two Wiener processes with $E\underline{w}_k(t)=\underline{0}$, $B_{w_1} \neq B_{w_2}$, where $E\underline{w}_k(t)\underline{w}_k(t) = t \cdot B_{w_k}$, (k=1,2). On the basis of Lemma 5, Appendix B, the two processes are distinguishable with probability 1 having a realization on $[0,T]$; as

$$(2.2.28) \qquad \lim_{n\to\infty} \sum_{i=1}^{2^n} (\underline{w}_k(t_i)-\underline{w}_k(t_{i-1}))(\underline{w}_k(t_i)-w_k(t_{i-1}))^* = B_{w_k} T,$$

$$k = 1,2, \quad t_i = \frac{Ti}{2^n}.$$

Two elementary Gaussian processes can be separated if their matrices of diffusion B_{w_1} and B_{w_2} are different. Moreover, later (in §2.3) we shall see that there is no possibility to differentiate them (with probability 1) on a finite interval if only the drift (transition) matrices A_1 and A_2 are not equal.

<u>Theorem 3.</u> <u>Let $\underline{\xi}(t)$ be a k-dimensional elementary Gaussian process with parameters A and B_w (see (2.2.1')). Then with probability 1</u>

$$(2.2.29) \qquad \lim_{n\to\infty} \sum_{i=1}^{2^n} (\underline{\xi}(t_i)-\underline{\xi}(t_{i-1}))(\underline{\xi}(t_i)-\underline{\xi}(t_{i-1}))^* = B_w T.$$

<u>Proof.</u> On the basis of (2.2.1') we can write that

$$(2.2.30) \qquad \sum_{i=1}^{2^n} (\underline{\xi}(t_i)-\underline{\xi}(t_{i-1}))(\underline{\xi}(t_i)-\underline{\xi}(t_{i-1}))^* =$$

$$= \sum_{i=1}^{2^n} (\underline{w}(t_i) - \underline{w}(t_{i-1}))(\underline{w}(t_i)-\underline{w}(t_{i-1}))^* +$$

$$+ \sum_{i=1}^{2^n} \int_{t_{i-1}}^{t_i} A\underline{\xi}(t)dt\,[\underline{w}(t_i)-\underline{w}(t_{i-1})]^* + \sum_{i=1}^{2^n}$$

$$[\underline{w}(t_i) - \underline{w}(t_{i-1})] \int_{t_{i-1}}^{t_i} \underline{\xi}^*(t)\, A^*dt + \sum_{i=1}^{2^n} \int_{t_{i-1}}^{t_i} A\underline{\xi}(t)dt \int_{t_{i-1}}^{t_i} \underline{\xi}^*(t)A^*dt.$$

As with probability 1 the vector functions $\int_0^t A\underline{\xi}(s)\,ds$ and $\int_0^t \underline{\xi}^*(s)A^*ds$ have bounded variation the last three terms tend to 0 with probability 1 if $n \to \infty$. This proves the theorem.

2.2.2. Stationary Gaussian Processes With Rational Spectral Density Functions. The one dimensional Gaussian process $\xi(t)$ is said to be a p-order autoregressive (AR) stationary process if it satisfies the following equation

(2.2.31) $$d\xi^{(p-1)}(t) + [a_1\xi^{(p-1)}(t) + \ldots + a_p\xi(t)]dt = dw(t),$$

where

$$P(z) = z^p + a_1 z^{p-1} + \ldots + a_p$$

has roots in the left halplane. By the notions

$$\xi^1(t) = \xi(t),\ \xi^2(t) = \frac{d\,\xi^1(t)}{dt},\ \ldots,\ \xi^p(t) = \frac{d\xi^{p-1}(t)}{dt},$$

$$A = \begin{pmatrix} 0 & 1 & 0 & \ldots & 0 \\ 0 & 0 & 1 & & 0 \\ \cdot & & & & \\ \cdot & & & & \\ \cdot & & & & \\ -a_p & -a_{p-1} & -a_{p-2} & \ldots & -a_1 \end{pmatrix},\quad B_w = \begin{pmatrix} 0 & 0 & \ldots & 0 \\ 0 & 0 & \ldots & 0 \\ \cdot & & & \\ \cdot & & & \\ \cdot & & & \\ 0 & 0 & \ldots & \sigma_w^2 \end{pmatrix},$$

$$\underline{\xi}(t) = \begin{pmatrix} \xi^1(t) \\ \xi^2(t) \\ \cdot \\ \cdot \\ \cdot \\ \xi^p(t) \end{pmatrix},$$

$$B^{-1}(0) = b_{ij}^{-1}, \quad b_{ij}^{-1} = \begin{cases} 0 & i \neq j \pmod 2,\ a_o = 1 \\ \frac{2}{\sigma_w^2} \sum_{\ell} (-1)^{\ell} a_{i-\ell}\, a_{j+1+\ell} & \end{cases}$$

from (2.2.31) we obtain

$$d\underline{\xi}(t) = A\,\underline{\xi}(t)dt + d\underline{w}(t),$$

and (2.2.31) and (2.2.32) are equivalent. The spectral representation and spectral density function of $\xi(t)$ (see (2.2.11) has the following form

$$(2.2.32) \quad \xi(t) = \int_{-\infty}^{\infty} e^{i\lambda t}\, \Phi_{\xi}(d\lambda) = \int_{-\infty}^{\infty} \frac{e^{i\lambda t}}{P(i\lambda)}\, \Phi(d\lambda),$$

$$(2.2.33) \quad f_{\xi}(\lambda) = \frac{\sigma_w^2}{2\pi}\ \frac{1}{|P(i\lambda)|^2}\ .$$

We define an autoregressive moving average (ARMA) process as a one dimensional stationary process with rational spectral density and spectral representation

$$(2.2.34) \quad f_{\xi}(\lambda) = \frac{1}{2\pi}\ \frac{|Q(i\lambda)|^2}{|P(i\lambda)|^2}, \qquad \xi(t) = \int_{-\infty}^{\infty} e^{i\lambda t}\, \frac{Q(i\lambda)}{P(i\lambda)}\, \Phi(d\lambda),$$

where (the coefficients are real)

$$P(z) = z^p + a_1 z^{p-1} + \ldots + a_p, \quad Q(z) = b_o z^q + b_1 z^{q-1} + \ldots + b_q,$$

$$q < p,$$

and $P(z)$, $Q(z)$ have different roots, the roots of $P(z)$ have negative real parts,

$$E\Phi(d\lambda) = 0, \quad E|\Phi(d\lambda)|^2 = \frac{d\lambda}{2\pi}\ .$$

It is said that the function $a(t)$ is "finite" if all of its derivatives exist, $a(t)$ is not 0 only in a bounded interval and its Fourier transform

$$\tilde{a}(\lambda) = \int_{-\infty}^{\infty} e^{i\lambda t}\, a(t)\, dt$$

exists. It $a(t)$ is "finite"

(2.2.35) $$\int_{-\infty}^{\infty} \lambda^{2\ell} |\tilde{a}(\lambda)|^2 f_\xi(\lambda)\, d\lambda < \infty,$$

for any $\ell \geq 0$, where $f_\xi(\lambda)$ is given by (2.2.34). We prove the following lemma.

Lemma 1. If $\xi(t)$ is Gaussian with spectral density (2.2.34), with $q < p$, and the roots of $P(z)$ are on the left halfplane, then there exists a Wiener process $w(t)$ such that for any "finite" function $a(t)$

(2.2.36) $$\int_{-\infty}^{\infty} [P(\tfrac{d}{dt})\ a(t-s)]\ \xi(s)\, ds = \int_{-\infty}^{\infty} [Q(\tfrac{d}{dt})\ a(t-s)]dw(s).$$

Proof. Let

(2.2.37) $$\eta(t) = \int_{-\infty}^{\infty} e^{i\lambda t}\ \tilde{a}(\lambda)\ \Phi_\xi(d\lambda) = \int_{-\infty}^{\infty} e^{i\lambda t}\tilde{a}(\lambda)\ \frac{Q(i\lambda)}{P(i\lambda)}\ \Phi(d\lambda) =$$

$$= \int_{-\infty}^{\infty} a(t-s)\ \xi(s)\ ds,$$

where $\Phi(d\lambda)$ is a Gaussian spectral measure with $E\Phi(d\lambda) = 0$, and $E|\Phi(d\lambda)|^2 = \frac{d\lambda}{2\pi}$. As (2.2.35) holds all the derivatives of $\eta(t)$, $\eta^{(\ell)}(t)$ ($\ell = 0,1,2,\ldots$) exist. From (2.1.22) we obtain

(2.2.38) $$P(\tfrac{d}{dt})\ \eta(t) = \int_{-\infty}^{\infty} [P(\tfrac{d}{dt})\ a(t-s)]\xi(s)\ ds =$$

$$= \int_{-\infty}^{\infty} e^{i\lambda t}\ P(i\lambda)\ \tilde{a}(\lambda)\ \Phi(d\lambda) = \int_{-\infty}^{\infty} e^{i\lambda t} P(i\lambda)$$

$$\tilde{a}(\lambda)\ \frac{Q(i\lambda)}{P(i\lambda)}\ \Phi(d\lambda) = \int_{-\infty}^{\infty} e^{i\lambda t}\ Q(i\lambda)\ \tilde{a}(\lambda)\ \Phi(d\lambda).$$

Using Lemma 5 in §2.1 one can obtain

(2.2.39) $$\int_{-\infty}^{\infty} e^{i\lambda t}\ Q(i\lambda)\ \tilde{a}(\lambda)\ \Phi(d\lambda) = \int_{-\infty}^{\infty} [Q(\tfrac{d}{dt})\ a(t-s)]dw(s),$$

where

$$w(t) = \int_{-\infty}^{\infty} \frac{e^{i\lambda t}-1}{i\lambda}\ \Phi(d\lambda).$$

Equations (2.2.38) and (2.2.39) show the validity of the lemma.

As in the discrete time case (see Theorem 5 in §2.1 and (2.1.39)) it seems important to give the definition of an ARMA process by an expression of stochastic differential equations. With respect to this we prove two auxiliary statements, in some sense the improvement of Lemma 1.

Lemma 2. (Representation B.) The regular stationary Gaussian ARMA process $\xi(t)$, permitting the spectral representation (2.2.34) is the first component of the p-dimensional stationary Gaussian process $\xi^*(t)$ satisfying the following linear system of stochastic equations (assuming for simplicity that $P(z)$ has distinct roots λ_i, with $Re\lambda_i < 0$)

(2.2.40)
$$\frac{d\,\xi^1(t)}{dt} = \xi^2(t),$$
$$\vdots$$
$$\frac{d^{p-q-1}\,\xi^1(t)}{dt^{p-q-1}} = \xi^{p-q}(t),$$
$$d\,\xi^{p-q}(t) = \sum_{i=1}^{p} d_i\,\xi^i(t)dt + b_o\,dw(t),$$
$$d\,\xi^{p-q+1}(t) = \lambda_{p-q+1}\,\xi^{p-q+1}(t)\,dt + dw(t),$$
$$\vdots$$
$$d\,\xi^p(t) = \lambda_p\xi^p(t)\,dt + dw(t),$$

where

(2.2.41)
$$\xi^1(t) = \xi(t) = \int_{-\infty}^{\infty} e^{i\lambda t}\left[\frac{Q_1}{i\lambda-\lambda_1} + \dots + \frac{Q_p}{i\lambda-\lambda_p}\right]\Phi(d\lambda) =$$
$$= \int_{-\infty}^{t}\left[Q_1\,e^{\lambda_1(t-s)} + \dots + Q_p\,e^{\lambda_p(t-s)}\right]dw(s) =$$
$$= Q_1\,\zeta_2(t) + \dots + Q_p\,\zeta_p(t),$$

$$\xi^2(t) = \frac{d\,\xi^1(t)}{dt} = \int_{-\infty}^{\infty} e^{i\lambda t}\,\frac{i\lambda Q(i\lambda)}{P(i\lambda)}\,\Phi(d\lambda) =$$
$$= \int_{-\infty}^{\infty} e^{i\lambda t}\left[\frac{Q_1'}{i\lambda-\lambda_1} + \dots + \frac{Q_p'}{i\lambda-\lambda_1}\right]\Phi(d\lambda) =$$
$$= Q_1'\,\zeta_1(t) + \dots + Q_p'\,\zeta_p(t),$$

$\vdots$

$$\xi^{p-q}(t) = \frac{d\, \xi^{p-q-1}(t)}{dt} = \int_{-\infty}^{\infty} e^{i\lambda t} \, \frac{(i\lambda)^{p-q-1} Q(i\lambda)}{P(i\lambda)} \, \Phi(d\lambda) =$$

$$= Q_1^{(p-q-1)} \, \zeta_1(t) + \ldots + Q_p^{(p-q-1)} \, \zeta_p(t),$$

$$\xi^{p-q+1}(t) = \zeta_{p-q+1}(t),$$

$\vdots$

$$\xi^{p}(t) = \zeta_p(t),$$

<u>and</u>

(2.2.42) $$d\, \zeta_j(t) = \lambda_j \, \zeta_j(t)\, dt + dw(t),$$

$$E\, \zeta_j(0)\, w\,(t) = 0, \quad t \geq 0, \; j = 1,2,\ldots,p.$$

<u>Proof.</u> Immediately follows from (2.2.41) and Lemma 5 in §2.1. Indeed, let

$$g(i\lambda) = \frac{1}{i\lambda - a}, \quad \text{Re } a < 0,$$

then obviously (see Remark 3 in 12.2.1).

(2.2.43) $$\zeta(t) - \zeta(0) - a \int_0^t \zeta(s)\, ds = \int_{-\infty}^{\infty} [e^{i\lambda t} - 1] \frac{1}{i\lambda - a} \, \Phi(d\lambda) -$$

$$- a \int_{-\infty}^{\infty} \frac{e^{i\lambda t} - 1}{i\lambda} \, \frac{1}{i\lambda - a} \, \Phi(d\lambda) = \int_{-\infty}^{\infty} \frac{(e^{i\lambda t} - 1)(i\lambda - a)}{i\lambda} \cdot \frac{1}{i\lambda - a} \Phi(d\lambda) =$$

$$= \int_{-\infty}^{\infty} \frac{e^{i\lambda t} - 1}{i\lambda} \, \Phi(d\lambda) = w(t),$$

i.e.,

(2.2.43′) $$d\zeta(t) = a\, \zeta(t)\, dt + dw(t), \quad \zeta(t) = \int_{-\infty}^{t} e^{a(t-s)} \, dw(s),$$

with the Wiener process

(2.2.44) $$w(t) = \int_{-\infty}^{\infty} \frac{e^{i\lambda t} - 1}{i\lambda} \, \Phi(d\lambda).$$

(2.2.43′) proves (2.2.41) and so (2.2.40) is also proved.

Lemma 3. (Representation A.) The stationary Gaussian ARMA process $\xi(t) = \xi^1(t)$ permitting the spectral representation given by (2.2.34), is the first component of the p-dimensional stationary Gaussian process $\underline{\xi}^*(t) = (\xi^1(t), \dots, \xi^p(t))$ satisfying the linear stochastic equations

(2.2.45) $$d\xi^j(t) = \xi^{j+1}(t)\,dt + \beta_j\,dw(t), \quad j=1,2,\dots,p-1,$$

$$d\,\xi^p(t) = -\sum_{k=0}^{p-1} a_{p-k}\,\xi^{k+1}(t)dt + \beta_p\,dw(t),$$

with the Wiener process (2.2.44).

The coefficients $\beta_1, \beta_2, \dots, \beta_p$ are given by

(2.2.46) $$\beta_p = -[a_1\,\beta_{p-1}+\dots+a_{p-1}\,\beta_1] + b_q,$$

$$\beta_{p-j} = -\sum_{i=1}^{p-j-1} \beta_{p-j-i}\,a_i + b_{q-j}, \quad j = 1,2,\dots, p-1,$$

where

$$b_{-1} = \dots = b_{-(p-q-1)} = 0, \quad q \leq p-1, \ \beta_1 = b_{q-p-1}.$$

The components $\xi^j(t)$ are given by

(2.2.47) $$\xi^j(t) = \int_{-\infty}^{\infty} e^{i\lambda t}\,w_j(i\lambda)\,\Phi(d\lambda), \quad j = 1,2,\dots,p,$$

where

(2.2.48) $$w_j(z) = \frac{1}{z}\,[w_{j+1}(z) + \beta_j], \quad j = 1,2,\dots, p-1,$$

$$w_p(z) = \frac{1}{z}\Big[-\sum_{k=0}^{p-1} a_{p-k}\,w_{k+1}(z) + \beta_p\Big],$$

and $E\,\xi^j(0)\,w(t) = 0$, $(t \geq 0,\ j = 1,2,\dots,p)$.

Proof. Evidently

$$\xi^j(t) - \xi^j(0) = \int_{-\infty}^{\infty} [e^{i\lambda t}-1]\,w_j(i\lambda)\,\Phi(d\lambda), \quad j = 1,2,\dots,p-1,$$

and from (2.2.48)

(2.2.49) $$\xi^j(t) - \xi^j(0) = \int_{-\infty}^{\infty} \frac{e^{i\lambda t}-1}{i\lambda}\,w_{j+1}(i\lambda)\Phi(d\lambda) + \beta_j \int_{-\infty}^{\infty} \frac{e^{i\lambda t}-1}{i\lambda}\Phi(d\lambda).$$

From Lemma 6 in §2.1 we obtain

$$(2.2.50)\quad \int_{-\infty}^{\infty} \frac{e^{i\lambda t}-1}{i\lambda} w_{j+1}(i\lambda)\ \Phi(d\lambda) = \int_0^t \int_{-\infty}^{\infty} e^{i\lambda s} w_{j+1}(i\lambda)\Phi(d\lambda)\ ds =$$

$$= \int_0^t \xi^{j+1}(s)ds,$$

and from Lemma 5 in §2.1

$$w(t) = \int_{-\infty}^{\infty} \frac{e^{i\lambda t}-1}{i\lambda}\ \Phi(d\lambda)$$

is a Wiener process. (2.2.50) and (2.2.49) give for $t > s$

$$\xi^j(t) - \xi^j(s) = \int_s^t \xi^{j+1}(n)dn + \beta_j \left[w(t)-w(s)\right],\quad j = 1,2,\ldots, p-1,$$

and we get (2.2.45). The last equation in (2.2.45) may be proved in the same manner.
The fact that $\xi^j(0)$ and $w(t)$ are independent can be proved as in Theorem 1.

Note that

$$(2.2.51)\quad w_p(z) = \frac{1}{z}\Big[-\sum_{k=0}^{p-1} a_{p-k}\Big\{\frac{1}{z^{p-k-1}}\quad w_p(z) + \sum_{i=1}^{p-k-1} \beta_{p-i} z^{i-1}\Big\} + \beta_p\Big],$$

$$w_p(z) = \frac{\beta_p z^{p-1} - \sum_{k=0}^{p-1}\sum_{i=1}^{p-k-1} a_{p-k}\ \beta_{p-i}\ z^{k+i-1}}{P(z)} =$$

$$= \frac{\beta_p z^{p-1} - \sum_{j=0}^{p-2} z^j \sum_{k=0}^{j} a_{p-k}\ \beta_{p+k-j-1}}{P(z)},$$

$$(2.2.52)\quad w_1(z) = \frac{\dfrac{\beta_p z^{p-1} - \sum_{j=0}^{p-2} z^j \sum_{k=0}^{j} a_{p-k}\ \beta_{p+k-j-1} + \sum_{i=1}^{p-1} \beta_{p-i}\ z^{i-1}\ (z^p + \sum_{\ell=1}^{p} a_\ell z^{p-\ell})}{P(z)}}{} = \frac{Q(z)}{P(z)},$$

which proves the lemma.

From Lemma 3 (or 2) we get the following theorem (see Theorem 7 in the discrete time case).

<u>Theorem 4. The regular stationary Gaussian process</u> $\xi(t)$ <u>is an ARMA type if and only if it is a component of a p-dimensional elementary Gaussian process.</u>

<u>Remark 1.</u> The covariance function $B_\xi(t)$ of process $\xi(t)$ with spectral representation (2.2.34) satisfies the following differential equation

$$(2.2.53) \qquad B_\xi^{(p)}(t) + \sum_{i=1}^{p} a_i \, B_\xi^{(p-i)}(t) = 0, \qquad t \geq 0,$$

$$B_\xi^{(p)}(t) + \sum_{i=1}^{p} (-1)^i \, B_\xi^{(p-i)}(t) = 0, \quad t \leq 0,$$

(compare with (A2.14).

<u>Remark 2.</u> In case $q = 0$, $Q(z) = b_0$, one gets an autoregressive process with observable components $\xi(t), \xi'(t), \dots, \xi^{(p-1)}(t)$ and one stochastic equation

$$(2.2.54) \qquad d\xi^{(p-1)}(t) + \sum_{i=1}^{p-1} a_{p-i} \, \xi^{(i)}(t) \, dt = dw(t),$$

with the solution

$$(2.2.55) \qquad \xi(t) = \int_{-\infty}^{\infty} \frac{e^{i\lambda t}}{P(i\lambda)} \, \Phi(d\lambda) = \sum_{i=1}^{p} c_i \int_{-\infty}^{t} e^{\lambda_i (t-s)} \, dw(s).$$

If $\xi(t)$ is autoregressive $\xi(n\delta)$ $(\delta>0)$ generally is an ARMA process.

<u>Remark 3.</u> If $q \geq 1$ the components $\xi^{p-q+1}(t), \dots, \xi^p(t)$ in (2.2.40) are unobservable, and they are parts of equation for $\xi^{p-q}(t)$. $\xi^{p-q}(t)$ is an Ito type process, i.e., depends on functionals of $\xi(s)$, $0 \leq s \leq t$. The coefficients d_i in (2.2.40) which are calculated by inverting the relation

$$\underline{\xi}(t) = D \, \underline{\xi}(t)$$

may be complex.
In the following we consider two simple examples:

<u>Example 1.</u> Let the spectral density have the form

$$(2.2.56) \qquad f_\xi(\lambda) = \frac{\sigma_w^2}{2\pi} \, \frac{1}{(\lambda^2+\theta^2)^2} = \frac{\sigma_w^2}{2\pi} \, \frac{1}{(i\lambda+\theta)^2(-i\lambda+\theta)^2}$$

Then $\xi(t)$ is an autoregressive process

$$(2.2.57)\qquad \frac{d\xi(t)}{dt} = \xi'(t),$$

$$d\xi'(t) + [2\theta\xi'(t) + \theta^2\xi(t)]\,dt = dw\,(t),$$

with the representation

$$\underline{\xi}(t) = \int_{-\infty}^{t} e^{A(t-s)}\,d\underline{w}(s), \quad A = \begin{pmatrix} 0 & 1 \\ -\theta^2 & -2\theta \end{pmatrix}, \quad B_w = \begin{pmatrix} 0 & 0 \\ 0 & \sigma_w^2 \end{pmatrix},$$

$$\xi(t) = \int_{-\infty}^{t} [e^{-\theta(t-s)} + \theta(t-s)\,e^{-\theta(t-s)}]\,dw(s).$$

Example 2. Now let the spectral density have the form

$$(2.2.58)\qquad f_\xi(\lambda) = \frac{\sigma_w^2}{2\pi}\,\frac{\lambda^2+2\theta^2}{(\lambda^4+4\theta^4)} = \frac{\sigma_w^2}{2\pi}\cdot\frac{(\lambda+\theta\sqrt{2}i)(\lambda-\theta\sqrt{2}\,i)}{(\lambda-\theta\sqrt{2}\,e^{i\frac{\pi}{4}})(\lambda+\theta\sqrt{2}\,e^{i\frac{\pi}{4}})(\lambda-\theta\sqrt{2}\,e^{-i\frac{\pi}{4}})(\lambda+\theta\sqrt{2}\,e^{-i\frac{\pi}{4}})},$$

then

$$\frac{Q(z)}{P(z)} = \frac{z+\theta\sqrt{2}}{(z+\theta-i\theta)(z+\theta+i\theta)} = \frac{Q_1}{(z+\theta-i\theta)} + \frac{Q_2}{(z+\theta+i\theta)},$$

where

$$Q_1 = \frac{1}{2} + i\,\frac{1-\sqrt{2}}{2}, \qquad Q_2 = \frac{1}{2} - i\,\frac{1-\sqrt{2}}{2}.$$

Further, by Representation B (Lemma 2) we obtain that

$$\xi(t) = \xi^1(t) = \int_{-\infty}^{t} Q_1\,e^{(i\theta-\theta)(t-s)}\,dw(s) + \int_{-\infty}^{t} Q_2\,e^{-(i\theta+\theta)(t-s)}\,dw(s) =$$

$$= Q_1\zeta_1(t) + Q_2\zeta_2(t),$$

$$\xi^2(t) = \zeta_2(t),$$

and

$$(2.2.59)\quad d\xi^1(t) = [(i\theta-\theta)\ \xi^1(t) + 2Q_2\theta\xi^2(t)]dt + \sigma_w dw(t),$$

$$d\xi^2(t) = -\ (i\theta+\theta)\ \xi^2(t)\ dt + \sigma_w\ dw\ (t),$$

$$A = \begin{pmatrix} i\theta-\theta & 2Q_2\theta \\ 0 & -(i\theta+\theta) \end{pmatrix}, \qquad B_w = \sigma_w^2 \begin{pmatrix} 1 & 1 \\ 1 & 1 \end{pmatrix}.$$

The reader may find that $f_\xi(\lambda)$ is $f_{11}(\lambda)$ in the following matrix (2.2.11))

$$(i\lambda I-A)^{-1}\ B_w\overline{(i\lambda I-A)}^{*-1} = \begin{pmatrix} f_{11}(\lambda) & f_{12}(\lambda) \\ f_{21}(\lambda) & f_{22}(\lambda) \end{pmatrix}$$

By Representation A (Lemma 3) we get

$$\beta_1 = 1, \quad \beta_2 = b_1 - a_1 = \theta\sqrt{2} - 2\theta$$

and

$$d\ \xi^1(t) = \xi^2(t)\ dt + \sigma_w\ dw(t),$$

$$(2.2.60)\quad d\ \xi^2(t) = (-2\ \theta^2\xi^1(t) - 2\theta\xi^2(t))\ dt + \theta(\sqrt{2}-2)\ \sigma_w dw(t).$$

2.3 Density Functions and Sufficient Statistics

2.3.1. The Discrete Time Case. Let $\underline{\xi}(n)$ be a k-dimensional elementary Gaussian process ($n=0,\pm1,\ldots$), for which (2.1.2) holds, i.e.,

$$\underline{\xi}(n) = Q\,\underline{\xi}(n-1) + \underline{\varepsilon}(n),$$

where $B(0)$, Q, B_ε are satisfying (2.1.2), and $E\underline{\xi}(n) = 0$. The sample $\underline{\xi}(0), \underline{\xi}(1), \ldots, \underline{\xi}(N)$ of size $N+1$ is normally distributed, which we denote by $\sim N(.,.)$. To prove this let us use the following relations

$$\underline{\xi}(0) = \underline{\xi}(0) \quad \sim \quad N(\underline{0}, B(0)),$$

$$\underline{\xi}(1) - Q\underline{\xi}(0) = \underline{\varepsilon}(1) \quad \sim \quad N(\underline{0}, B_\varepsilon), \tag{2.3.1}$$

$$\vdots$$

$$\underline{\xi}(N) - Q\underline{\xi}(N-1) = \underline{\varepsilon}(N) \quad \sim \quad N(\underline{0}, B_\varepsilon),$$

where the random vector variables on the right side are independent and Gaussian. The density function of the variables $\underline{\xi}(0), \underline{\varepsilon}(1), \ldots \underline{\varepsilon}(N)$ is the following

$$f_{\xi(0),\varepsilon(1),\ldots,\varepsilon(N)}(\underline{x}_o, \underline{y}_1,\ldots,\underline{y}_N) = (2\pi)^{-\frac{k(N+1)}{2}} |B(0)|^{-1/2} |B_\varepsilon|^{-\frac{N}{2}} \exp\{$$

$$-\frac{1}{2}\underline{x}_o^* B^{-1}(0)\,\underline{x}_o - \frac{1}{2}\sum_{i=1}^{N} \underline{y}_i^* B_\varepsilon \underline{y}_i\},$$

and from (2.3.1) the random variables $\underline{\xi}(0),\ldots,\underline{\xi}(N)$ have the density function

$$f_{\xi(0),\ldots,\xi(N)}(\underline{x}_o,\ldots,\underline{x}_N) = (2\pi)^{-\frac{k(N+1)}{2}} |B(0)|^{-1/2} |B_\varepsilon|^{-\frac{N}{2}} \exp\{ \tag{2.3.2}$$

$$-\frac{1}{2}\underline{x}_o^* B^{-1}(0)\,\underline{x}_o - \frac{1}{2}\sum_{i=1}^{N} (\underline{x}_i - Q\underline{x}_{i-1})^* B_\varepsilon (\underline{x}_i - Q\underline{x}_{i-1})\},$$

as the Jacobian of the linear transformation (2.3.1) equals to 1. (2.3.2) may be rewritten into the following form

(2.3.2') $$f(\underline{x}_o,\dots,\underline{x}_N) = (2\pi)^{-\frac{k(N+1)}{2}} |B(0)|^{-1/2} |B_\varepsilon|^{-\frac{N}{2}} \exp\Big\{$$

$$-\frac{1}{2}\underline{x}_o^* (B^{-1}(0) + Q^*B_\varepsilon^{-1}Q)\underline{x}_o - \frac{1}{2}\sum_{i=1}^{N-1} \underline{x}_i^* (B_\varepsilon^{-1}+Q^*B_\varepsilon^{-1}Q)\underline{x}_i +$$

$$+\frac{1}{2}\sum_{i=1}^{N}\left[\underline{x}_i^* Q^*B_\varepsilon^{-1}\underline{x}_i + \underline{x}_i^*B_\varepsilon^{-1}Q\,\underline{x}_{i-1}\right] - \frac{1}{2}\underline{x}_N^* B_\varepsilon^{-1}\underline{x}_N\Big\}.$$

The system of sufficient statistics can be easily calculated for some special matrices.

Example 1. Let $\xi(n)$ be one-dimensional elementary Gaussian satisfying equation

$$\xi(n+1) = \rho\xi(n) + \varepsilon(n+1),$$

where

$$E\varepsilon(n) = E\xi(n) = 0, \qquad \sigma_\varepsilon^2 = (1-\rho^2)\sigma_\xi^2.$$

Let

$$\zeta(n) = \xi(n) + m$$

then

(2.3.3) $$f_{\zeta(0),\dots,\zeta(N)}(x_o,\dots,x_N) = (2\pi)^{-\frac{N+1}{2}} \sigma_\xi^{-(N+1)} (1-\rho^2)^{-\frac{N}{2}} \exp\Big\{$$

$$-\frac{1}{2(1-\rho^2)\sigma_\xi^2}\left[(x_o-m)^2(1-\rho^2) + \sum_{i=1}^{N}(x_i-\rho x_{i-1}-m(1-\rho))^2\right]\Big\} =$$

$$= (2\pi)^{-\frac{N+1}{2}} (1-\rho^2)^{1/2} \sigma_\varepsilon^{-(N+1)} \exp\Big\{-\frac{1}{2\sigma_\varepsilon^2}\Big[(x_o-m)^2(1-\rho^2) +$$

$$+\sum_{i=1}^{N}(x_i-\rho x_{i-1}-m(1-\rho))^2\Big]\Big\}.$$

The inverse of the covariance matrix R_{N+1}^{-1} of variables $\zeta(0),\dots,\zeta(N)$ has the form

$$R_{N+1}^{-1} = \frac{1}{\sigma_\varepsilon^2} \begin{pmatrix} 1 & -\rho & 0 & 0 \ldots 0 & 0 \\ -\rho & 1+\rho^2 & -\rho & 0 \ldots 0 & 0 \\ 0 & -\rho & 1+\rho^2 & -\rho \ldots 0 & 0 \\ \cdot & & & & \\ \cdot & & & & \\ \cdot & & & & \\ 0 & 0 \ldots & & 1+\rho^2 & -\rho \\ 0 & 0 \ldots & & -\rho & 1 \end{pmatrix}$$

$$R_{N+1} = \sigma_\xi^2 \begin{pmatrix} 1 & \rho & \rho^2 \ldots & \rho^N \\ \rho & 1 & \rho & \rho^{N-1} \\ \cdot & & & \\ \cdot & & & \\ \cdot & & & \\ \rho^N & \rho^{N-1} & \ldots & 1 \end{pmatrix}$$

The system of sufficient statistics of the unknown parameters $(m, \rho, \sigma_\varepsilon^2)$ or (m, ρ, σ_ξ^2) is

$$\left\{\zeta(0) + \zeta(N),\ \sum_{1}^{N-1} \zeta(i),\ \zeta^2(0) + \zeta^2(N),\ \sum_{i=1}^{N-1} \zeta^2(i),\ \sum_{i=1}^{N} \zeta(i)\zeta(i-1)\right\}.$$

<u>Example 2.</u> The density function of the finite realization $\xi(0), \ldots, \xi(N)$ of a stationary Gaussian autoregressive (AR) process satisfying (2.1.27)

$$\xi(n) + a_1\xi(n-1) + \ldots + a_p\xi(n-p) = \varepsilon(n), \quad n = 0, \pm 1, \ldots,$$

can be determined in the following way (assuming $E\xi(n) = E\varepsilon(n) = 0$). The covariance matrix R_{N+1} of the random variables $\xi(0), \xi(1), \ldots, \xi(N)$ is symmetric with respect to both diagonals (persymmetric property), because of the stationarity of the process. The same persymmetric property remains true for the inverse of covariance matrix $R_{N+1}^{-1} = \{r_{ij}^{-1}\}$. Note that the persymmetric property does not hold in the multidimensional case.

The linear transformation

(2.3.4)
$$\begin{aligned} \xi(0) &= \xi(0), \\ &\cdot \\ &\cdot \\ &\cdot \\ \xi(p-1) &= \xi(p-1), \end{aligned}$$

$$\xi(p) + a_1\xi(p-1) + \dots + a_p\xi(0) = \varepsilon(p),$$

$$\vdots$$

$$\xi(N) + a_1\,\xi(N-1) + \dots + a_p\xi(N-p) = \varepsilon(N),$$

has a Jacobian equal to 1. Further, using the fact that the two sets of random variables $(\xi(0),\dots,\xi(p-1))$, $(\varepsilon(p),\dots,\varepsilon(N))$ are independent Gaussian variables we obtain

$$f_{\xi(0),\dots,\xi(p-1),\varepsilon(p),\dots\varepsilon(N)}(x_o,\dots,x_{p-1}, z_p,\dots,z_N) =$$

$$= (2\pi)^{-\frac{N+1}{2}} |R_p|^{-1/2}\cdot \sigma_\varepsilon^{-(N-p+1)} \exp\Big\{$$

$$-\frac{1}{2}\, \underline{x}_p^* R_p^{-1} \underline{x}_p - \frac{1}{2\sigma_\varepsilon^2} \sum_{i=p}^{N} z_i^2 \Big\},$$

which together with (2.3.4) gives

$$(2.3.5) \qquad f_{\xi(0),\dots,\xi(p-1),\xi(p),\dots,\xi(N)}(x_o,\dots,x_{p-1},x_p,\dots,x_N) =$$

$$=(2\pi)^{-\frac{(N+1)}{2}} |R_p|^{-1/2}\, \sigma_\varepsilon^{-(N-p+1)} \exp\Big\{$$

$$-\frac{1}{2}\, \underline{x}_p R_p^{-1} \underline{x}_p - \frac{1}{2\sigma_\varepsilon^2} \sum_{i=p}^{N} (x_i + a_1x_{i-1}+\dots+a_px_{i-p})^2 \Big\}.$$

Now taking into account the above mentioned persymmetry we obtain

$$r_{ij}^{-1} = r_{ji}^{-1}, \quad r_{ij}^{-1} = r_{N-i,N-j}^{-1},\ a_o = 1,\ a_i = 0 \quad \text{if } i < 0 \text{ or } i > p,$$

and

$$(2.3.6) \qquad r_{ij}^{-1} = \begin{cases} 0 \\ \dfrac{1}{\sigma_\varepsilon^2} \displaystyle\sum_{\ell=0}^{\min(i,p-j+i)} a_\ell a_{\ell+j-i}, & \text{if } i < j. \end{cases}$$

According to the well-known factorization theorem of density functions from (2.3.6) we get the following statement.

Lemma 1. If the Gaussian process $\xi(n)$ is AR type, i.e., satisfies (2.1.27), then the system of sufficient statistics is the following (assuming N > 2p)

$$\Big\{ \sum_{i=p}^{N} \xi(i)\ \xi(i-j),\ j = 0,2,\dots,p;\ \xi(\ell)\ \xi(\ell+k)+\xi(N-\ell)\ \xi(N-\ell-k);\quad k,\ell = 0,\dots,p-1 \Big\}.$$

One may ask whether in the general case is it possible to find a finite system of statistics, when $\xi(n)$ is a component of a multidimensional elementary Gaussian process. The following example shows that such a statistic (containing less than N elements) does not exist in the general case. Indeed, let us consider the moving average process of order two

$$\xi(n) = b_0 \varepsilon(n) + b_1\ \varepsilon(n-1), \tag{2.3.7}$$

where $\varepsilon(n)$ is a Gaussian white noise with $E\varepsilon(n) = 0$, $E\ \varepsilon^2\ (n) = \sigma_\varepsilon^2$. Then we have

$$E\xi(n) = 0,\quad \sigma_\xi^2 = (b_o^2 + b_1^2)\ \sigma_\varepsilon^2,\quad \rho = \frac{E\xi(n)\xi(n-1)}{\sigma_\xi^2} = \frac{b_o b_1}{(b_o^2+b_1^2)},$$

$$E\xi(n)\xi(n+k) = 0, \quad \text{if } |k| \geq 2.$$

The density function of random variables $\xi(1),\dots,\xi(N)$ has the following form

$$f(x_1,\dots,x_N) = (2\pi)^{-\frac{N}{2}}\ \sigma_\xi^{-N}\ |R_N|^{-1/2}\ \exp\Big\{ - \frac{1}{2\sigma_\xi^2} \sum_{i,j=1}^{N} b_{ij}^{-1}\ x_i\ x_j \Big\}, \tag{2.3.8}$$

where

$$R_N^{-1} = \frac{1}{\sigma_\xi^2}\ (b_{ij}^{-1})_{i,j=1}^{N},$$

is the inverse of the covariance matrix R_N. R_N can easily be found on the basis of Example 1 (where the role of the covariance matrix and its inverse is changed), or in the following way. Note that $|R_n|$, $(n \geq 2)$ satisfies the difference equation

$$|R_n| = |R_{n-1}| - \rho^2\ |R_{n-2}|, \tag{2.3.9}$$

and if u_1 and u_2 are the roots of $u^2 - u + \rho^2 = 0$, i.e.,

$$u_1 = \frac{1}{2}(1+\sqrt{1-4\rho^2}),\quad u_2 = \frac{1}{2}(1-\sqrt{1-4\rho^2}),\ \text{then}$$

$$|R_n| = \frac{u_1^{n+1} - u_2^{n+1}}{u_1 - u_2},\ \text{and for}\quad i < j\ .$$

$$\text{(2.3.10)}\quad b_{ij}^{-1} = (-1)^{j-i}\,\rho^{(j-i)}\,|R_{i-1}|\,|R_{N-j}|\,|R_N|^{-1}.$$

(The inverse matrix is, of course, again persymmetric.) Since, for example, the functions b_{iN}^{-1} $(i=1,\ldots,N)$, as functions of ρ, are linearly independent, a system of functions of $\xi(1),\ldots,\xi(N)$ as independent variables, which would form a sufficient statistic, cannot exist.

<u>Example 3.</u> If Q consists of only one Jordan elementary matrix and B_ε is a diagonal matrix

$$Q = \begin{pmatrix} \rho & 1 & 0 & \cdots & 0 \\ 0 & \rho & 1 & \cdots & 0 \\ \vdots & & & & \\ 0 & 0 & 0 & \cdots & 1 \\ 0 & 0 & 0 & \cdots & \rho \end{pmatrix},\quad B_\varepsilon = \begin{pmatrix} \sigma^2_{\varepsilon 1} & 0 & \cdots & 0 \\ 0 & \sigma^2_{\varepsilon 2} & \cdots & 0 \\ \vdots & & & \\ 0 & 0 & \cdots & \sigma^2_{\varepsilon k} \end{pmatrix},$$

$$\text{(2.3.11)}\quad \begin{aligned} \xi^k(n) &= \xi^k(n-1) + \varepsilon^k(n), \\ \xi^{k-1}(n) &= \rho\xi^{k-1}(n-1) + \xi^k(n-1) + \varepsilon^{k-1}(n), \\ &\ \ \vdots \\ \xi^1(n) &= \rho\xi^1(n-1) + \xi^2(n-1) + \varepsilon^1(n). \end{aligned}$$

Then B(0) can be determined using the following method (instead of solving equation (2.1.1.)).
Multiplying the first equation of (2.3.11) by $\varepsilon^k(n)$, by $\xi^k(n-1)$ and by $\xi^k(n)$ and taking the expected values in all cases, we obtain

$$E\xi^k(n)\ \varepsilon^k(n) = \sigma^2_{\varepsilon^k} ,$$

$$E\xi^k(n)\ \xi^k(n-1) = \rho\sigma^2_{\xi^k} ,$$

$$\sigma^2_{\xi^k} = \rho^2\ \sigma^2_{\xi^k} + \sigma^2_{\varepsilon^k} .$$

Again multiplying the first equation in (2.3.11) by $\xi^{k-1}(n)$ and $\xi^{k-1}(n-1)$, respectively, we obtain

$$E\ \xi^k(n)\ \xi^{k-1}(n) = \rho E\ \xi^{k-1}(n)\ \xi^k(n-1),$$

$$E\ \xi^k(n)\ \xi^{k-1}(n-1) = \rho E\ \xi^{k-1}(n-1)\ \xi^k(n-1),$$

respectively (using, that $\varepsilon^k(n)$, $\xi^{k-1}(n)$ are independent). Applying these results and multiplying the second equation in (2.3.11) by $\varepsilon^{k-1}(n)$, $\xi^k(n-1)$, $\xi^k\ (n)$, $\xi^{k-1}(n-1)$, $\xi^k(n)$ we obtain

$$E\ \xi^{k-1}(n)\ \varepsilon^{k-1}(n) = \sigma^2_{\varepsilon^{k-1}},$$

$$E\ \xi^{k-1}(n)\ \xi^k(n-1) = \rho\sigma_{\xi^k\xi^{k-1}} + \sigma^2_{\xi^k}$$

$$E\ \xi^{k-1}(n)\ \xi^k(n) = \sigma_{\xi^k\xi^{k-1}} = \rho^2 E\ \xi^k(n-1)\xi^{k-1}(n-1) + \rho\sigma^2_{\xi^k}$$

$$\sigma_{\xi^k\xi^{k-1}} = \frac{\rho\sigma^2_{\xi^k}}{1-\rho^2} ,$$

$$E\ \xi^{k-1}(n)\ \xi^{k-1}(n-1) = \rho\sigma^2_{\xi^{k-1}} + \sigma_{\xi^k\xi^{k-1}} ,$$

$$\sigma^2_{\xi^{k-1}} = \rho E\ \xi^{k-1}(n)\ \xi^{k-1}(n-1) + E\ \xi^k(n-1)\ \xi^{k-1}(n) + \sigma^2_{\varepsilon^{k-1}} =$$

$$= \rho\left[\rho\sigma^2_{\xi^{k-1}} + \frac{\rho\sigma^2_{\xi^k}}{1-\rho^2}\right] + \rho\frac{\rho\sigma^2_{\xi^k}}{1-\rho^2} + \sigma^2_{\xi^k} + \sigma^2_{\varepsilon^{k-1}} .$$

By this method all the elements of $B(0)$ can be calculated. The conditional density function of variables $\underline{\xi}(1), \underline{\xi}(2),\ldots,\underline{\xi}(N)$ of (2.3.11) under the condition $\underline{\xi}(0) = \underline{x}_0$ has the form

$$c_N \exp\left\{-\frac{1}{2}\sum_{j=1}^{N}\left[\frac{1}{\sigma_\varepsilon^2 k}(x_{kj}-\rho x_{kj-1})^2+\frac{1}{\sigma_\varepsilon^2 k-1}(x_{k-1j}-\rho x_{k-1,j-1}-\right.\right.$$

$$\left.\left.-x_{kj-1})^2+\ldots+\frac{1}{\sigma_\varepsilon^2 1}(x_{1j}-\rho x_{1j-1}-x_{2j-1})^2\right]\right\},$$

from which the system of sufficient statistics may be determined.

2.3.2. Some Auxiliary Theorems. Let (Ω,F,P) be a probability space and let ξ be a random variable on it with $E|\xi| < \infty$. Let $A\varepsilon F$ be a σ-algebra and $\xi_A = E(\xi|A)$. If $\{A_r\}$ is a system of σ-algebras the random elements ξ_{A_r} are uniformly integrable. This means that for any $\varepsilon > 0$ there exists such a number k that

$$(2.3.12) \qquad \int_{\{\omega:E(\xi|A_r) > k\}} E(\xi|A_r)\, P(d\omega) < \varepsilon$$

holds.

The proof of (2.3.12). As $|E(\xi|A_r)| \leq E(|\xi||A_r)$ we get

$$\int_{\{\omega:|\xi A_r|>k\}} |\xi A_r(\omega)|\, P\,(d\omega) \leq \int_{\{\omega:|\xi A_r|>k\}} E(|\xi||A_r)\, P(d\omega) \leq$$

$$\leq \int_{\{\omega:E(|\xi||A_r)>k\}} E(|\xi||A_r)\, P(d\omega) = \int_{\{\omega:E(|\xi||A_r) > k\}} |\xi| P(d\omega)\ .$$

By the assumption that $|\xi(\omega)|$ is integrable we obtain that for any $\varepsilon > 0$ there exists such a $\delta > 0$ that for $P(A) < \delta$

$$\int_A |\xi(\omega)| P(d\omega) < \varepsilon.$$

Now it is enough to find such a number k for which $P\{E(|\xi||A_r) > k\} < \delta$ independently of A_r. From the Chebyshev inequality we obtain

$$P\{E\,(|\xi||A_r) > k\} \leq \frac{1}{k}\,E(E(|\xi||A_r)) = \frac{E|\xi|}{k}\ ,$$

and taking $k = \frac{E|\xi|}{\delta}$ the proof is finished.

Let $d_n = (t_1,t_2,\ldots,t_n) = (t_1^{d_n}, t_2^{d_n},\ldots,t_n^{d_n})$, $0 \leq t_1 < t_2 < \ldots < t_n \leq T$, denote a sequence of divisions, assuming that if $n > m, d_n$

refines d_m, i.e., $d_n \supseteq d_m$ if $n > m$, and if $n \to \infty$

$\sup_i |t_i^{d_n} - t_{t-1}^{d_n}| \to 0$, which we denote by $d_n \to 0$. The density function of random variables $\xi(t_1),\dots,\xi(t_n)$ will be denoted by $f_\xi(x_1,\dots,x_n)$ and let

$$\pi_{d_n}(\xi) = \frac{f_\eta(\xi(t_1),\dots,\xi(t_n))}{f_\xi(\xi(t_1),\dots,\xi(t_n))}$$

denote the ratio of two density functions. The Radon-Nikodym derivative of the measures, generated by the processes $\zeta(t)$ and $\xi(t)$ respectively on $0 \le t \le T$ we denote by

$$\frac{dP_\eta}{dP_\xi}(\xi(t),\ 0 \le t \le T)\ .$$

If the sequence of random variables $\pi_{d_n}(\xi)$ has a limit, denoted by π, in probability, i.e.,

$\lim_{n\to\infty} P\{|\pi_{d_n}(\xi)-\pi| < \varepsilon\} \to 0$, as $n \to \infty$ and $d_n \to 0$, it may happen that the Radon-Nikodym derivative coincides with

$$\pi = \frac{dP_\eta}{dP_\xi}(\xi(t),\ 0 \le t \le T). \tag{2.3.13}$$

Under the condition that π exists we prove the following statements (Lemmas 2-5).

__Lemma 2.__ $E\pi \le 1$.

__Proof.__ From $E\pi_{d_n}(\xi) = 1$ and from Fatou's lemma the proof follows immediately.

__Lemma 3.__ __If__ $P_\eta << P_\xi$ __then__ $E\pi = 1$.

__Proof.__ Let $\tilde{\pi} = \dfrac{dP_\eta}{dP_\xi}$,

then $\pi_{d_n}(\xi) = E(\tilde{\pi}|F^\xi_{d_n})$. As the sequence $\pi_{d_n}(\xi)$ is uniformly integrable $\lim_{n\to\infty} E\pi_{d_n} = E\pi$.

__Lemma 4.__ __If__ $E\pi = 1$, __then__ $P_\eta << P_\xi$ __and__ $\tilde{\pi} = \pi = \dfrac{dP_\eta}{dP_\xi}(\xi)$.

__Proof.__ From the definition $P_\eta << P_\xi$ if and only if when for any $\varepsilon > 0$ there exists such a $\delta > 0$ that for every $d_N=(t_1,\dots,t_N)$ and $A \in F^\xi_{d_N}$ if the inequality $P_\xi(A) < \delta$ holds it follows that $P_\xi(A) < \varepsilon$.

So to prove the lemma it requires that $P_\eta(\eta\in C) = E\chi_{\{\xi\in C\}}$, where $C\in F^\xi_{d_n}$. Let d_n be a sequence of monotone increasing divisions, everywhere dense in $[0,T]$ and $d_n \supseteq d_N$, such that $\pi_{d_n}\to\pi$, as $n\to\infty$. As d_n is a subdivision of d_N

(2.3.14) $$E\,\chi_{\{\xi\in C\}}\cdot\pi \leq \lim_{n\to\infty} E\,\chi_{\{\xi\in C\}}\cdot\pi_{d_n},$$

where $C = A_n$, $A_n\in F^\xi_{d_n}$ and

$$E\chi_{\{\xi\in C\}}\,\pi_{d_n} = E\chi_{\{\xi\in A_n\}}\pi_{d_n} = P_\eta(\eta\in C)$$

From (2.3.14) and the last relation one can obtain

(2.3.15) $$E\chi_{\{\xi\in C\}}\pi \leq P_\eta(\eta\in C).$$

In the same way, instead of C taking the set $\Omega - C$, we get

$$E[1-\chi_{\{\xi\in C\}}]\,\pi\leq 1 - P_\eta(\eta\in C),$$

and now using the equality $E\pi = 1$ it follows that

(2.3.16) $$E\chi_{\{\xi\in C\}}\,\pi \geq P_\eta(\eta\in C).$$

Comparing (2.3.15) and (2.3.16) we get the desired equality.

<u>Lemma 5. If for some</u> $0 < \alpha < 1$ <u>and for any</u> $\varepsilon < 0$

$$E(\pi_{d_n}(\xi))^\alpha < \varepsilon$$

<u>holds, then</u> $P_\eta \perp P_\xi$, <u>i.e., the measures are singular.</u>

<u>Proof.</u> Let d_n such a sequence of divisions for which $E(\pi_{d_n})^\alpha < \varepsilon^2$ and $A = \{\pi_{d_n} < 1/\varepsilon\}$. From the Chebyshev inequality we get

$$1-P_\xi(A) = P\{\pi_{d_n}(\xi) \geq 1/\varepsilon\} \leq \varepsilon.\ E\,\pi_{d_n} = \varepsilon.$$

On the other side

$$P_\eta(A) = \int_A \pi_{d_n}(\xi)P(d\omega) \leq \left(\frac{1}{\varepsilon}\right)^{1-\alpha}\int_A [\pi_{d_n}(\xi)]^\alpha P(d\omega) \leq$$

$$\leq\left(\frac{1}{\varepsilon}\right)^{1-\alpha}\cdot\varepsilon^2 < \varepsilon,$$

which proves the statement.

<u>Theorem 1. Let $w(t)$ be the standard Wiener process with measure P_w and</u>

$$\xi(t) = w(t) + m(t),$$

<u>where $m(0) = 0$ and $m'(t)$ exists with</u> $\int_0^T (m'(t))^2 dt < \infty$. <u>Then π exists and</u>

$$(2.3.17) \qquad \pi = \frac{dP_\zeta}{dP_w}(\zeta) = \exp\{-\frac{1}{2}\int_0^T [m'(t)]^2 dt + \int_0^T m'(t)\, d\zeta(t)\}.$$

<u>Remark 1.</u> As $\pi > 0$ we have $P_w << P_\eta$. If $m(t)$ does not have a derivative (it is not absolutely continuous with respect to the Lebesgue measure) or $m(0) \neq 0$, then $P_\eta \perp P_w$.

<u>Proof of the Theorem.</u> Let d_n: $t_0 = 0 < t_1 < t_1 < \ldots < t_n = T$, then

$$(2.3.18) \qquad \pi_{d_n}(\zeta) = \frac{f_\zeta(\zeta_1,\ldots,\zeta_n)}{f_w(\zeta_1,\ldots,\zeta_n)} = \exp\{-\frac{1}{2}\sum_{i=1}^n \frac{(m(t_i)-m(t_{i-1}))^2}{t_i - t_{i-1}} + \sum_{i=1}^n \frac{m(t_i)-m(t_{i-1})}{t_i - t_{i-1}}(\zeta(t_i) - \zeta(t_{i-1}))\}.$$

If $m(t)$ were not absolutely continuous the first sum on the right should tend to infinity (see Riesz F., Sz-Nagy [1], §36) and in this case

$$E(\pi_{d_n}(\zeta))^{1/2} = \exp\{-\frac{1}{8}\sum_{i=1}^n \frac{(m(t_i)-m(t_{i-1}))^2}{t_i - t_{i-1}}\}$$

should tend to 0. From Lemma 5 it follows that $P_\zeta \perp P_w$.

If $\int_0^s m'(t)\, dt = m(s)$, and $m'(t) \in L^2(0,T)$ then

$$\lim_{d_n \to 0} \sum_{i=1}^n \frac{(m(t_i)-m(t_{i-1}))^2}{t_i - t_{i-1}} = \int_0^T [m'(s)]^2 ds,$$

(see again Riesz F., Sz-Nagy [1] §36). The sum

$$\sum_{i=1}^n \frac{m(t_i)-m(t_{i-1})}{t_i - t_{i-1}}(\zeta(t_i) - \zeta(t_{i-1})) \to \int_0^T m'(t)\, d\bar{\zeta}(t),$$

in probability and in mean square too. We may choose such a subsequence for which the second sum in (2.3.18) is convergent with probability 1.

As $\lim_{n\to\infty} \pi_{d_n}(\eta) = \pi$ exists (with probability 1) using Lemma 4 we can prove that $E\pi = 1$. But this is the consequence that

$\xi = \int_0^T m'(t)\, d\xi(t)$ is a Gaussian random variable with $E\xi = 0$ and

$D^2\xi = \int_0^T [m'(t)]^2 dt$ and

$$E\, e^{\xi - \frac{1}{2} D^2\xi} = 1, \tag{2.3.19}$$

which proves the theorem.

<u>Remark 2.</u> The fact that $\lim_{n\to\infty} \pi_{d_n}(\eta)$ exists follows also from the general martingale convergence theorem, as $\pi_{d_n}(\eta)$ forms a martingale, but we need not use it here, the martingale convergence theorem.

Let $\underline{\xi}(t)$ be an elementary Gaussian process satisfying equation (2.2.1'). We introduce the stochastic process $\underline{\xi}^{d_n}(t)$, where d_n is a monoton increasing sequence of divisions, in the following way

$$\underline{\xi}^{d_n}(0) = \underline{w}(0), \tag{2.3.20}$$

$$\underline{\xi}^{d_n}(t) = \underline{\xi}^{d_n}(t_{i-1}^{n}) + A\underline{\xi}^{d_n} \cdot (t-t_{i-1}) + \underline{w}(t) - \underline{w}(t_{i-1}),$$

$$t_{i-1} < t \le t_i.$$

The process $\underline{\xi}^{d_n}(t)$ is the so called Euler approximation of the process $\underline{\xi}(t)$. The sequence of processes $\{\underline{\xi}^{d_n}(t)\}$, $(n=1,2,\ldots,)$, has two basic properties.

<u>Lemma 6. For any finite set of time points $\theta_1,\ldots,\theta_\ell$ the joint conditional distribution of random vedtors</u> $\underline{\xi}^{d_n}(\theta_1),\ldots,\underline{\xi}^{d_n}(\theta_\ell)$ <u>under the condition $\underline{\xi}^{d_n}(0) = \underline{x}$ tends to the corresponding distribution of $\underline{\xi}(\theta_1),\ldots,\underline{\xi}(\theta_\ell)$.</u>

<u>Lemma 7. Formula (2.3.20) can be understood as a transformation of the space $C_k^x[0,T]$ into itself: to any Wiener sample path corresponds a sample path of the process $\underline{\xi}(t)$. If K is a compactum of $C_k^x[0,T]$, then $\Phi(K)$ is a compactum too, where $C_k^x[0,T]$ is the space</u>

<u>of k-dimensional continuous functions $\underline{x}(t)$, $0 \le t \le T$, $\underline{x}(0) = \underline{x}$, with the uniform metric, denoted by $||\underline{x}||$.</u>

<u>Proof of Lemma 6.</u> As the processes are Gaussian, it is sufficient to show, that the conditional mean value vector and covariance matrix functions of the processes $\underline{\xi}^{d_n}(t)$ under the conditions $\underline{\xi}^{d_n}(0) = \underline{x}$ tend to the corresponding functions of $\underline{\xi}(t)$. This can be obtained by direct calculations on the basis of relation between discrete and continuous time elementary processes.

<u>Proof of Lemma 7.</u> We have to show, on the basis of the theorem of Arzela, the uniform boundedness and the equicontinuity of functions defined by (2.3.20) under the condition that the functions on the right hand side of (2.3.20) have these properties. The uniform boundedness follows from the inequality

$$(2.3.21) \qquad ||\Phi(\underline{x}(t))|| \le ||\ \underline{x}\ (t)|| \ + e^{||A||(T+||x(t)||\ )},$$

taking into account (2.3.21) the equicontinuity follows from the equicontinuity of functions $\underline{x}(t) \in K$, which proves the lemma.

From Levy's theorem on the modulus of continuity of the Wiener process and from Lemma 7 we can derive the following fundamental property.

<u>Lemma 8.</u> <u>For every $\varepsilon > 0$ there exists a compactum K_ε of the space $C_k^x[0,T]$ such that $P_{\xi_n}(K_\varepsilon) > 1 - \varepsilon$, where P_{ξ_n} is the conditional measure generated by the process $\underline{\xi}^{d_n}(t)$ under the condition $\underline{\xi}^{d_n}(0)=\underline{x}$.</u>

<u>Proof of Lemma 8.</u> From Levy's theorem follows the existence of such a K'_ε for the Wiener process. Set $K = \Phi(K'_\varepsilon)$.

<u>Remark 3.</u> Levy's iterated logarithm law states that for the standard Brownian motion w(t)

$$P\{\overline{\lim_{\substack{0 \le t_2 - t_2 \le 1 \\ t_1 - t_2 \downarrow 0}}} \frac{w(t_1) - w(t_2)}{\sqrt{2t \ln 1/t}} = 1\} = 1.$$

The compact subsets of continuous functions are the subsets of uniformly bounded equicontinuous functions.

<u>Remark 4.</u> The reader may find the proof of Lemma 7 in the textbooks for ordinary differential equations (see e.g., P. Henrici [1] Chapter 3) as the main tool of proving existence theorem via

Euler's approximation.

Remark 5. The properties given by Lemmas 6 and 8 and a variant of Prohorov's theorem (see Gikhman-Skorokhod [1] Chapter IX) allows to get necessary and sufficient condition for the weak convergence of the measures P_{ξ_n} to P_ξ. In other terms these properties provide that for every bounded continuous functional $g(\underline{x}(t))$ on $C_k^x[0,T]$

(2.2.22) $$\int_{C_k^x[0,T]} g(\underline{x}(t))\, dP_{\xi_n} \to \int_{C_k^x[0,T]} g(\underline{x}(t))\, dP_\xi .$$

We have collected all the necessary preliminaries to the calculation of the Radon-Nikodym derivative of measure P_{ξ_n} with respect to P_w.

Lemma 9. The measure P_{ξ_n} is absolutely continuous with respect to P_w and

(2.3.23) $$\frac{dP_{\xi_n}}{dP_w}(x(t)) = \pi_{d_n}(\underline{x}(t)) = \exp\Big\{ \sum_{j=1}^{n} (B_w^+ A\underline{x}_{j-1}, \Delta\underline{x}j) - \frac{1}{2}\sum_{j=1}^{n} (A\underline{x}_{j-1}, B^+ A\underline{x}_{j-1})\Delta t_j \Big\},$$

where
$$\Delta\underline{x}_j = \underline{x}(t_j^{d_n}) - \underline{x}(t_{j-1}^{d_n}), \quad \Delta t_j = t_j^{d_n} - t_{j-1}^{d_n}.$$

Proof. Let $d_{n'}$ be a refinement of d_n, $n' > n$, then by direct calculations for the ratio of density functions $f_\xi(\underline{x}(t_1^{d_{n'}}),\dots,\underline{x}(t_{n'}^{d_{n'}}))$ and

$f_w(\underline{x}(t_1^{d_{n'}}),\dots,\underline{x}(t_n^{d\, n'}))$ at the division $d_{n'}$ we get

(2.3.24) $$\frac{f_\xi(\underline{x}(d_{n'}))}{f_w(\underline{x}(d_{n'}))} = \pi_{d_{n'},d_n}(\underline{x}(t)) = \exp\Big\{-\frac{1}{2}\sum_{j=1}^{m}\sum_{i>i_{j-1}}^{i_j}$$

$$\frac{(B_w^+\,\Delta\underline{x}_i^{d_{n'}} - B_w^+\,A\underline{x}_{j-1}^{d_n}\,\Delta t_i^{d_{n'}},\ \Delta\underline{x}_i^{d_{n'}} - A\,\Delta\underline{x}_{j-1}^{d_n}\,\Delta t i^{d_{n'}}}{\Delta t_i^{d_{n'}}} +$$

$$+ \frac{1}{2} \sum_{i=1}^{i_n} \frac{(B_w^+ \Delta x_i^{d_{n'}}, \Delta \underline{x}_i^{d_{n'}})}{\Delta t_i^{d_{n'}}} \Big\} ,$$

where $\Delta t_i^{d_{n'}} = t_i^{d_{n'}} - t_{i-1}^{d_{n'}}$, $\Delta x_i^{d_{n'}} = \underline{x}(t_i^{d_{n'}}) - x(t_{i-1}^{d_{n'}})$, $\underline{x}_j^{d_n} = \underline{x}(t_j^{d_n})$.

As $n' \to \infty$ $\pi_{d_{n'},d_{n'}}$ tends to π_{d_n} in (2.3.23), but $E\pi_{d_n} = 1$ and by Lemma 4 the proof is completed.

2.3.3. The Radon-Nikodym Derivatives With Respect to the Wiener Measure. In the statistical investigations of elementary Gaussian processes with continuous time parameter, similarly to the statistics of independent observations, the maximum-likelihood principle plays an important role. For this purpose it is important to determine the Radon-Nikodym derivative of the measure generated by this process with respect to some standard measure. Theorem 3 in §2.2 suggests that the elementary Gaussian processes with common diffusion matrix generate equivalent measures, and these measures are equivalent to the Wiener-measure with the same local matrix of variance.

Let $C_k[0,T]$ be the metric space of k-dimensional vector-valued continuous functions on the interval $[0,T]$ with the uniform metric $\|\ \|$. It will be convenient to assume $C_k[0,T]$ as direct product of the space $C_k^x[0,T]$ of k-dimensional continuous functions $\underline{x} = \underline{x} = \{\underline{x}(t),\ 0 \le t \le T\}$ with the initial condition $\underline{x}(0) = \underline{x}$ and the k-dimensional Euclidean space R^k.

For the sake of generality we shall consider a Guassian Markov process $\underline{\xi}(t)$ satisfying the stochastic differential equation (2.2.1) and having $f(\underline{x}(0))$ as initial probability density function. Let P_ξ be the probability measure on $C_k[0,T]$ generated by the above process $\underline{\xi}(t)$ and P_w be the "conditional" product of the k-dimensional Lebesque-measure and the measure generated by the Wiener process on the right hand side of (2.2.1). Our main theorem is the following.

Theorem 2. The measures P_ξ and P_w are equivalent and their Radon-Nikodym derivative has the form

(2.3.25)
$$\frac{dP_\xi}{dP_w}(\underline{x}(t)) = f(\underline{x}(0)) \exp\Big\{ \int_0^T (C\,\underline{x}(t), d\underline{x}(t)) - \frac{1}{2} \int_0^T (A\,\underline{x}(t),\ C\,\underline{x}(t))\, dt \Big\} ,$$

<u>where</u> $C = B_W^+A$.

The value of stochastic integral $\int_0^T (\underline{x}(t), d\underline{x}(t))$ can be determined for almost all realizations $\underline{x}(t)$ with respect to P_W.

<u>Remark 1.</u> (2.3.25) may also written in the following form.

$$\frac{dP_\xi}{dP_W}(\underline{x}(t)) = f(\underline{x}(0)) \exp \{Sp[C \int_0^T \underline{x}(t)\, d\underline{x}^*(t)] - \frac{1}{2} Sp[A^*C \int_0^T \underline{x}(t)\, \underline{x}^*(t)\, dt]\}, \tag{2.3.25'}$$

where SpA means the trace of A.

<u>Proof of Theorem 2.3.1.</u> The proof is based on a variant of the invariance principle due to Prokhorov (1956) which will be cited in the course of the proof.

First let us notice that the conditional measures $P_{\xi x}$ and P_{wx} the processes $\underline{\xi}(t)$ and $\underline{w}(t)$ on the space $C_k^{(x)}[0,T]$ under the conditions $\underline{\xi}(0) = \underline{x}$, $w(0) = x$, may be treated in a simpler way. Our theorem will follow from the statement about these conditional measures. We shall prove that the measures $P_{\xi x}$ and P_{wx} are equivalent, and

$$\frac{dP_{\xi x}}{dP_{wx}}(\underline{x}(t)) = \exp \{\int_0^T (C\underline{x}(t), d\underline{x}(t)) - \frac{1}{2}\int_0^T (A\underline{x}(t), C\underline{x}(t))dt\}. \tag{2.3.26}$$

As the terms in the exponent of formula (2.3.23) tend to the integrals

$\int_0^T (C\underline{x}(t), d\underline{x}(t))$ and $-\frac{1}{2}\int_0^T (A\underline{x}(t), C\underline{x}(t))\, dt$ respectively in mean square norm, we can choose a subsequence d_{n_ℓ} of sequence d_n in such a way that the limit $\pi(\underline{x}(t)) = \lim_{\ell\to\infty} \pi_{n_\ell}(\underline{x}(t))$ exists for $\underline{x}(t)$ with probability 1.

Let us consider the compactum K_ε such that

$$P_{\xi_n x}(K_\varepsilon) = \int_{K_\varepsilon} \pi_n(\underline{x}(t))\, dP_{wx} \geq 1 - \varepsilon \tag{2.3.27}$$

for every n. As the elements of K_ε are uniformly bounded functions

$$\frac{1}{2}\left|\sum_{j=1}^{m} (A\, \underline{x}_{j-1}, C\, \underline{x}_{j-1})\, \Delta t_j^{d_n}\right| \leq \frac{1}{2} N_\varepsilon ||A|| \cdot ||C||$$

(where N_ε is the common upper bound for the norms of $\underline{x}(t) \in K_\varepsilon$). So we have

$$(\pi_{d_n}(\underline{x}(t)))^2 \leq e^{2N_\varepsilon ||A|| \cdot ||C||}\, \pi_{d_n}^{(2)}(\underline{x}(t)) \quad \text{for every } \underline{x}(t) \in K_\varepsilon .$$

($\pi_{d_n}^{(2)}$ means the probability density function obtained in the same way for the process $\underline{\xi}^{(2)}(t)$ with parameters 2A and $B_{\underline{w}}$). From this inequality and Lemma 2 we can deduce the uniform integrability of the sequence $\pi_{d_n}(\underline{x}(t))$ on the compactum K_ε with respect to the measure P_{wx}. We notice that the uniform integrability is valid on the whole space $C_k^x[0,T]$ but its verification is not so simple as on the compact subsets of $C_k^x[0,T]$; this is the advantage of the application of Prokhorov's theorem.

From (2.3.27) and on the basis of Fatou's lemma we get

$$\int_{K_\varepsilon} \pi(\underline{x}(t))\, dP_{wx} \geq 1 - \varepsilon .$$

Let g be a non-negative bounded continuous functional on $C_k^x[0,T]$. Using again Fatou's lemma we get

$$\int_{C_k^x[0,T]} g(\underline{x}(t))\, \pi(\underline{x}(t))\, dP_{wx} \leq \lim_{\ell\to\infty} \int_{C_k^x[0,T]} g(\underline{x}(t))\, \pi_{d_{n_\ell}}(\underline{x}(t))\, dP_{wx} . \tag{2.3.28}$$

Using (2.3.27) and (2.3.28) and the uniform integrability of the sequence $\pi_{n_\ell}(\underline{x}(t))$ on K_ε we obtain

$$\overline{\lim_{n\to\infty}} \int_{C_k^x[0,T]} g(\underline{x}(t))\, \pi_{d_n}(\underline{x}(t))\, dP_{wx} \leq \tag{2.3.29}$$

$$\leq \lim_{\ell\to\infty} \int_{K_\varepsilon} (g(\underline{x}(t))\, \pi_{d_n}(\underline{x}(t))\, dP_{wx} + \varepsilon' \leq$$

$$\leq \int_{K_\varepsilon} g(\underline{x}(t))\ \pi(\underline{x}(t))\ dP_{wx} + \ \varepsilon' \leq$$

$$\leq \int_{C_k^x[0,T]} g(\underline{x}(t))\ \pi(\underline{x}(t))\ dP_{wx} + \varepsilon',$$

where ε' is $\max\limits_{\underline{x}(t)\in K_\varepsilon} |g(\underline{x}(t))|$.

Analogous considerations are valid for negative functionals too. So relations (2.3.28) and (2.3.29) involve

$$\text{(2.3.30)} \qquad \lim_{\ell\to\infty} \int_{C_k^x[0,T]} g(\underline{x}(t))\ \pi_{d_{n_\ell}}(\underline{x}(t))\ dP_{wx} = \int_{C_k^x[0,T]} g(\underline{x}(t))\ \pi(\underline{x}(t))dP_{wx},$$

for arbitrary continuous, bounded functional $g(\underline{x}(t))$, i.e., the measure $P_{\xi x}(.) = \int\pi(\underline{x}(t))\ dP_{wx}$ is the weak limit of measures $P_{\xi_n x}$. On the other hand, as we have mentioned, from Lemmas 6 and 8 follows that the sequence $P_{\xi_n x}$ has the weak limit $P_{\xi x}$. But a sequence of measures cannot have two different weak limits, so the measure $\tilde{P}$ generated by the density function $\pi(\underline{x}(t))$ coincides with $P_{\xi x}$. The equivalence of measures $P_{\xi x}$ and P_{wx} follows from the fact that the stochastic integral $\int_0^T \underline{x}(t)\ d\underline{x}(t)$

is finite with probability 1.

<u>Remark 2.</u> In real applications we observe a trajectory of the process $\underline{\xi}(t), 0\leq t\leq T$. By the just proved equivalence of measures P_ξ and P_w the value of the integral $\int_0^T (C\underline{x}(t),\ d\underline{x}(t))$

does not depend on the regarded measure on $C_k[0,T]$, as it is defined as the limit of a sequence of measurable functions on $C_k[0,T]$) with probability 1.

The following formula is often used.

$$(2.3.31)\qquad \frac{dP_{wx}}{dP_{\xi x}}(\xi(t)) = \left(\frac{dP_{\xi x}}{dP_{wx}}(\xi(t))\right)^{-1} =$$

$$= \exp\left\{-\int_0^T (C\underline{\xi}(t), d\underline{\xi}(t)) + \frac{1}{2}\int_0^T (A\underline{\xi}(t), C\underline{\xi}(t))\,dt\right\},$$

$$(2.3.31')\qquad \frac{dP_{\xi A}}{dP_{\xi A_o}}(\xi(t)) = \frac{f_A(\underline{\xi}(0))}{f_{A_o}(\underline{\xi}(0))} \exp\left\{\int_0^T (B_w^+(A-A_o)\underline{\xi}(t), d\underline{\xi}(t)) -\right.$$

$$\left. - \frac{1}{2}\int_0^T (A\underline{\xi}(t), B_w^+ A\underline{\xi}(t))\,dt + \frac{1}{2}\int_0^T (A_o\underline{\xi}(t), B_w^+ A_o\underline{\xi}(t))\,dt\right\},$$

the correctness of which is guaranteed by the above remark.

Remark 3. Using Ito's formula (see Appendix, Theorem 5) we get

$$(2.3.32)\qquad d(B_w^+ A\underline{\xi}(t), \underline{\xi}(t)) = [(B_w^+ A\underline{\xi}(t), A\underline{\xi}(t)) + ((B_w^+ A)^*\underline{\xi}(t), A\underline{\xi}(t)) + \mathrm{Sp}A]dt +$$

$$+ (B_w^+ A\underline{\xi}(t), d\underline{w}(t)) + ((B_w^+ A)^*\underline{\xi}(t), d\underline{w}(t)),$$

or using Ito's formula in the form

$$(2.3.33)\qquad d(\underline{\xi}(t)\, \underline{\xi}^*(t)) = [\underline{\xi}(t)\underline{\xi}^*(t)\, A^* + A\underline{\xi}(t)\, \underline{\xi}^*(t) + B_w]dt +$$

$$+ d\underline{w}(t)\, \underline{\xi}^*(t) + \underline{\xi}(t)\, d\underline{w}^*(t)\ ,$$

we obtain

$$(2.3.34)\qquad \int_0^T (B_w^+ A\underline{\xi}(t), d\underline{\xi}(t)) = \int_0^T (B_w^+ A\underline{\xi}(t), A\underline{\xi}(t))dt + \int_0^T (B_w^+ A\underline{\xi}(t), d\underline{w}(t)) =$$

$$= \frac{1}{2}[(B_w^+ A(\underline{\xi}(T)\underline{\xi}\ (T) - \underline{\xi}(0)\ \underline{\xi}\ (0))] - \frac{1}{2}\, T\ \mathrm{Sp}\ A.$$

We can rewrite (2.3.25) and (2.3.25') by the help of (2.3.34) and we get

(2.3.35) $$\frac{dP_\xi}{dP_w}(\xi(t)) = f(\underline{\xi}(0)) \exp\{-\frac{1}{2}\int_0^T (A\underline{\xi}(t), B_w^+ A\underline{\xi}(t))dt +$$

$$+ \frac{1}{2}[B_w^+ A\xi(T), \xi(T)) - (B_w^+ A\underline{\xi}(0), \underline{\xi}(0))] - \frac{1}{2} T \, Sp\, A\} =$$

$$= f(\underline{\xi}(0)) \exp\{-\frac{1}{2} Sp[A^* B_w A \int_0^T \underline{\xi}(t)\underline{\xi}^*(t)dt] +$$

$$+ \frac{1}{2} Sp[B_w^+ A(\underline{\xi}(T)\underline{\xi}^*(T) - \underline{\xi}(0)\underline{\xi}^*(0)) - TA]\}.$$

Remark 4. For the autoregressive process $\xi(t)$ with the differential equation (2.2.31)

$$d\,\xi^{(p-1)}(t) + [a_1\,\xi^{(p-1)}(t) + \ldots + a_p\,\xi(t)]dt = dw(t)$$

we get

(2.3.36) $$\frac{dP_\xi}{dP_w}(\xi(t)) = f(\underline{\xi}(0)) \exp\{-\frac{1}{2\sigma_w^2}\int_0^T [a_1\,\xi^{(p-1)}(t) + \ldots +$$

$$+ a_p\xi(t)]^2 dt - \frac{1}{2\sigma_w^2} a_1(\xi^{(p-1)}(T))^2 + \ldots +$$

$$+ a_p(\xi(T))^2 - a_1(\xi^{(p-1)}(0))^2 - \ldots - a_p(\xi(0))^2] - \frac{1}{2}a_1 T\}.$$

Remark 5. In the special case when the process $\xi(t) = \xi^1(t)$ and

(2.3.37)
$$d\,\xi^1(t) = \xi^2(t)\,dt,$$
$$\vdots$$
$$d\,\xi^{k-\ell-1}(t) = \xi^{k-\ell}(t)dt,$$
$$d\,\xi^{k-\ell}(t) = (a_{k-\ell,1}\,\xi^k(t) + \ldots + a_{k-\ell,k}\,\xi^1(t))dt + dw_{k-\ell}(t),$$
$$\vdots$$
$$d\,\xi^k(t) = (a_{k,1}\,\xi^k(t) + \ldots + a_{k,k}\,\xi^1(t))\,dt + dw_k(t),$$

and B_w is an $(\ell+1) \times (\ell+1)$ nonsingular matrix $E(\underline{w}(t)\underline{w}^*(t)) = t \cdot B_w$, $\tilde{A} = \{a_{ij}\}$, $(i=k-\ell,\ldots,k; j=1,\ldots,k)$, $\tilde{C} = B_w^{-1} \cdot \tilde{A}$, then in the space $C_\ell[0,T]$ Theorem 2 remains valid, i.e., if $\underline{x}(t) \in C_\ell[0,T]$

(2.3.38) $$\frac{dP_\xi}{dP_w}(\xi(t)) = f(\underline{\xi}(0)) \exp\left\{-\frac{1}{2}\int_0^T (\tilde{A}\underline{\xi}(t), B_w^{-1}\tilde{A}\underline{\xi}(t))dt - \right.$$

$$\left. - \frac{1}{2}\left[(B_w^{-1}\tilde{A}\underline{\xi}(T), \underline{\xi}(T)) - (B_w^{-1}\tilde{A}\underline{\xi}(0), \underline{\xi}(0))\right] - \frac{1}{2} T \operatorname{Sp} A\right\}.$$

<u>Remark 6.</u> The proof of Theorem 2 may be carried out in the case when $\underline{\xi} = \{\underline{\xi}(t), F_t\}$ is a diffusion type process, i.e.,

(2.3.39) $$\underline{\xi}(t) = \int_0^t \underline{\alpha}(s, \underline{\xi}(s))ds + \underline{w}(t),$$

where $\underline{\alpha}(t, \underline{\xi})$ is a measurable vector functional not depending on the future, $\underline{\alpha}(t, \underline{\xi})$ is F_t measurable (non-anticipating) and

$$P\left\{\int_0^T (\underline{\alpha}(s, \underline{\xi}(s), \underline{\alpha}(s, \underline{\xi}(0))) ds < \infty\right\} = 1,$$

(see Liptser, Shiryaev's book [1]).

The concrete formulas, as consequences of Theorem 2, suitable for computational purposes are given in the following examples.

<u>Example 1.</u> In the one dimensional stationary case, when

(2.3.40) $$d\xi(t) = -\lambda\xi(t)\, dt + dw(t),$$

and

$$E\, w(t) = 0, \quad E(dw(t))^2 = \sigma_w^2\, dt,$$

$$E\, \xi(t) = 0, \quad E\, \xi^2(t) = \sigma_w^2/2\lambda,$$

the Radon-Nikodym derivative has the following form

(2.3.41) $$\frac{dP_\xi}{dP_w}(x(t)) = \sqrt{\frac{\lambda}{\pi}}\frac{1}{\sigma_w} \exp\left\{-\frac{\lambda^2}{2\sigma_w^2}\int_0^T x^2(t)dt + \frac{\lambda T}{2} - \frac{\lambda}{2\sigma_w^2}\left[\right.\right.$$

$$\left.\left. x^2(T) + x^2(0)\right]\right\}.$$

To get this formula from the ratio $\pi_{d_n}(\xi)$ of probability density functions of Gaussian variables

$\{\xi(\frac{k}{n}),\ w(\frac{k}{n})\}$, $k = 1,\dots,n$, when n tends to infinity is also possible (see Lemmas 2-4). This ratio can be calculated using the relation

$$\xi\left(\frac{k}{n}\right) = e^{-\frac{\lambda}{n}}\,\xi\left(\frac{k-1}{n}\right) + \frac{\varepsilon(k)}{\sqrt{n}}\,, \quad D^2\varepsilon(k) = \sigma_w^2\left(1-e^{\frac{-2\lambda}{n}}\right)\cdot\frac{n}{2\lambda}\,.$$

Example 2. Let $\xi(t)$ be as above, and let $\eta(t) = \xi(t) + m$. Let P_0 and P_m be the probability measures generated by processes $\xi(t)$ and $\eta(t)$ on C_1 respectively. Then using the "chain-rule" we get

$$\text{(2.3.42)} \qquad \frac{dP_m}{dP_o}(x(t)) = \exp\left\{-\frac{\lambda m}{\sigma_w^2}\left[X(0)+x(T)+\lambda\int_o^T x(t)dt+m\left(1+\frac{\lambda T}{2}\right)\right]\right\}.$$

Example 3. Let $\underline{\xi}(t) = (\xi^1(t), \xi^2(t))$ be a two dimensional elementary Gaussian process with the parameters

$$A = \begin{pmatrix} -\lambda & -\omega \\ \omega & -\lambda \end{pmatrix}, \qquad B_w = \sigma_w^2 \cdot \begin{pmatrix} 1 & 0 \\ 0 & 1 \end{pmatrix},$$

where $w^1(t)$ and $w^2(t)$ are independent. Then

$$\text{(2.3.43)} \qquad f_A(\underline{x}(0)) = \frac{\lambda}{\pi\sigma_w^2}\exp\left\{-\frac{\lambda}{\sigma_w^2}\left[(x^1(0))^2 + (x^2(0))^2\right]\right\},$$

as $AB(0) + B(0)A = -B_w$ (see (2.2.3)) has the above solution, and

$$\text{(2.3.44)} \qquad \frac{dP_\xi}{dP_w}(x(t)) = \frac{\lambda}{\pi\sigma_w^2}\exp\left\{-\frac{\lambda^2+\omega^2}{2\sigma_w^2}\int_o^T\left[(x^1(t))^2+(x^2(t))^2\right]dt -\right.$$

$$-\frac{\lambda}{2\sigma_w^2}\left[(x^1(T))^2 + (x^2(T))^2 + (x^1(0))^2 + (x^2(0))^2\right] +$$

$$\left. + \lambda T + \frac{\omega}{\sigma_w^2}\int_o^T\left[x^1(t)\,dx^2(t) - x^2(t)\,dx^1(t)\right]\right\}.$$

If we consider the complex valued process $x(t) = |x(t)|\,e^{i\theta(t)}$, where $x(t) = x^1(t) + ix^2(t)$, $|x(t)|^2 = ((x^1(t))^2 + (x^2(t))^2$ then using the relations

$$\sum_j\left[x(t_j)\overline{x}(t_{j-1}) - x(t_{j-1})\,\overline{x}(t_j)\right] = -2\sum_j\left[x^2(t_j)\,(x^1(t_j)-x^1(t_{j-1})) -\right.$$

$$\left. - x^1(t_j)\;((x^2(t_j) \;\; -x^2(t_{j-1}))\right]$$

and

$$\sum_j |x(t_j)||x(t_{j-1})| [e^{i(\theta(t_j)-\theta(t_{j-1}))} - e^{i(\theta(t_{j-1})-\theta(t_j))}] =$$

$$= \sum_j |x(t_j)||x(t_{j-1})| 2i \sin(\theta(t_j)-\theta(t_{j-1})) \sim$$

$$\sim 2i \sum_j |x(t_j)|^2 (\theta(t_j) - \theta(t_{j-1})),$$

we get the following relations

$$(2.3.45) \qquad \int_0^T [x^1(t)\, dx^2(t) - x^2(t)\, dx^1(t)] = \int_0^T |x(t)|^2 d\theta,$$

and

$$(2.3.46) \qquad \frac{dP_\xi}{dP_w}(x(t)) = \frac{\lambda}{\pi\sigma_w^2} \exp\{- \frac{\lambda^2+\omega^2}{\sigma_w^2} \int_0^T |x(t)|^2 dt +$$

$$+ \frac{\omega}{\sigma_w^2} \int_0^T |x(t)|^2 d\theta + \lambda T - \frac{\lambda}{2\sigma_w^2}[|x(T)|^2 + |x(0)|^2]\}.$$

2.3.4. Unobservable Components. For discrete time processes representation B (see (2.1.44)) shows that stationary Gaussian processes with rational spectral density generally can be described as multidimensional processes with unobservable components. The autoregressive case is an exception, where the components are observable and a finite set of sufficient statistics exists (see Lemma 1 in this section). The example described by the process (2.3.7) shows that generally we have not a set of sufficient statistics and so the statistical investigations, e.g., parameter estimation and hypothesis testing, will be complicated. Nevertheless, the density functions and, in the continuous case, the Radon-Nikodym derivatives can be determined. The method which we use here is based on the theorem of Radon-Nikodym derivatives of the processes and the filtering theory of partially observable random processes. The fundamental theorems on Radon-Nikodym derivatives and the filtering equations we shall not prove here, we use them in that form as they are stated in the book of Liptser-Shiryaev [1] (see also Remark 6 in § 2.3.3). Here we shall not give the formulas in the discrete case.

We shall use the following theorem, which can be deduced from Girsanov's Theorem (see Liptser-Shiryaev [1], Th. 7.15). We consider the Ito process $(\xi(t),F_t)$, $0 \le t \le T$, with the differential

(2.3.47) $$d\xi(t) = \beta(t,\omega)\, dt + dw(t),$$

and let $P_\xi(.)$, $P_w(.)$ be the measures corresponding to the processes $\xi(t)$ and $w(t)$ respectively.

Theorem 3. Let $\beta(t,\omega)$ be a continuous (in the mean square) Gaussian process. Then $P_\xi \sim P_w$ and

(2.3.48) $$P\{\int_0^T \alpha^2(t,\xi)\, dt < \infty\} = P\{\int_0^T \alpha^2(t,w)\, dt < \infty\} = 1,$$

(2.3.49) $$\frac{dP_\xi}{dP_w}(t,\xi) = \exp\{\int_0^t \alpha(s,\xi)\, d\xi(s) - \frac{1}{2}\int_0^t \alpha^2(s,\xi)\, ds\},$$

(2.3.50) $$\frac{dP_w}{dP_\xi}(t,w) = \exp\{-\int_0^t \alpha(s,w)\, dw(s) + \frac{1}{2}\int_0^t \alpha^2(s,w)\, ds\},$$

where the functional $\alpha(t,\xi) = E(\beta(t,\omega)|F_t^\xi)$ for almost all $t\in[0,T]$.

Before using Theorem 3 in the general case let us consider the system (2.2.60), Example 2 in §2.2.2, where for simplicity let $\sigma_w^2 = 1$. For (2.3.49) we have to get

$$\alpha(t,\xi) = E(\xi^2(t)|F_t^{\xi^1}),$$

where $\xi^2(t)$ is unobservable, $\xi^1(t)$ is observable. Let $\gamma(t) = E(\alpha(t,\xi) - \xi^2(t))^2$, then on the basis of Theorem 10.3 in Liptser-Shiryaev [1] we get

(2.3.51) $$d\alpha(t) = -[2\theta^2\xi^1(t) + 2\theta\alpha(t)]dt + [\gamma(t)+\theta(\sqrt{2}-2)][d\xi^1(t) - \alpha(t)\, dt],$$

(2.3.52) $$\dot{\gamma}(t) = -2\sqrt{2}\theta\gamma(t) - \gamma^2(t),$$

or

(2.3.52′) $$d\alpha(t) = -[2\theta^2\xi^1(t)+(\theta\sqrt{2}+\gamma(t))\alpha(t)]dt + [\gamma(t)+\theta(\sqrt{2}-2)]\, d\xi^1(t),$$

with the initial conditions (see Appendix B, Theorem 1)

$$\alpha(0) = E(\xi^2(0)|\ \xi^1(0)) = \xi^1(0)\frac{\xi^1(0)\ \xi^2(0)}{E(\xi^1(0))^2},$$

$$\alpha(0) = E\ (\xi^2(0) - \alpha(0))^2 = E\ (\xi^2(0))^2 - \frac{(E\ \xi^1(0)\ \xi^2(0))^2}{E(\ \xi^1(0))^2} .$$

In order to find the values

$$\begin{pmatrix} E\ (\xi^1(0))^2 & E\xi^1(0)\ \xi^2(0) \\ E\ \ \xi^1(0)\xi^2(0) & E(\xi^2(0))^2 \end{pmatrix} = B(0) = \begin{pmatrix} b_{11} & b_{12} \\ b_{21} & b_{22} \end{pmatrix},$$

we shall take advantage of elementary Gaussian process $(\xi^1(t), \xi^2(t))$ that $B(0)$ is the unique solution of the system of equations (2.2.3), i.e.,

$$AB(0) + B(0)\ A^* = -\ B_w$$

with

$$A = \begin{pmatrix} 0 & 1 \\ -2\theta^2 & -2\theta \end{pmatrix}, \qquad B_w = \begin{pmatrix} 1 & \theta(\sqrt{2}-2) \\ \theta(\sqrt{2}-2) & \theta^2(\sqrt{2}-2)^2 \end{pmatrix} .$$

We find from this that

$$b_{12} = b_{21} = -\ 1/2\ , \qquad b_{11} = \frac{1}{2\theta}\ , \ b_{22} = \theta(2-\sqrt{2}),$$

and

$$(2.3.53) \qquad \alpha(0) = -\ \theta\xi^1(0), \qquad \gamma(0) = \theta\frac{3-2\sqrt{2}}{2} .$$

The Riccati equation (2.3.52) can be solved (it is an Euler equation) and we get

$$(2.3.54) \qquad \ln\ \frac{\gamma(t)}{\gamma(0)} = -\ 2\sqrt{2}\theta t - \int_0^t \gamma(s)ds,\ \gamma(t) = \gamma(0)e^{-2\sqrt{2}\theta t - \int_0^t \gamma(s)ds}$$

$$(2.3.55) \qquad \gamma(t) = \ [\gamma^{-1}(0)\ e^{2\sqrt{2}\theta t} + \frac{1}{2\sqrt{2}\theta}\ e^{2\sqrt{2}\theta t} - \frac{1}{2\sqrt{2}\theta}\ ]^{-1} =$$

$$= \frac{\theta(6\sqrt{2} - 8\ )}{e^{2\sqrt{2}\theta t}\ (3+2\sqrt{2}) - (3 - 2\sqrt{2})}$$

Further it is easy to verify that the solution of (2.3.52′) has the following form

$$(2.3.56) \qquad \alpha(t) = e^{-\theta\sqrt{2}t - \int_0^t\gamma(s)ds}\{\alpha(0) + \int_0^t e^{\theta\sqrt{2}s+\int_0^s\gamma(n)dn}\ [-2\theta^2\xi^1(s)ds+$$

$$+ \ (\gamma(s) + \theta(\sqrt{2}\ -2))\ d\xi^1(s)\ ]\}.$$

Now using the fact (see (2.3.54))

$$\frac{\gamma(t)}{\gamma(0)}\ e^{\theta\sqrt{2}t} = e^{-\theta\sqrt{2}t - \int_0^t \gamma(s)\,ds}$$

we get

$$\alpha(t) = \frac{\gamma(t)}{\gamma(0)}\ e^{\theta\sqrt{2}t}\ \{\alpha(0) - 2\ \theta^2\int_0^t e^{-\theta\sqrt{2}s}\ \frac{\gamma(s)}{\gamma(0)}\ \xi^1(s)\ ds +$$

$$+\ e^{-\theta\sqrt{2}t}\ \frac{\gamma(0)}{\gamma(t)}\Big[\ \theta(\sqrt{2}-2)\ +\ \gamma(t)\Big]\xi^1(t) - \Big[\theta(\sqrt{2}-2)+\gamma(0)\Big]\xi^1(0)\ +$$

$$-\int_0^t e^{-\theta\sqrt{2}s}\ \frac{\gamma(0)}{\gamma(s)}\ [(\theta\sqrt{2}+\gamma(s))(\theta(\sqrt{2}-2)+\gamma(s)+\dot{\gamma}(s)]\xi^1(s)ds\},$$

using equations (2.3.52), (2.3.53) the last term can be rewritten

$$(2.3.56')\quad \alpha(t) = \frac{\gamma(t)}{\gamma(0)}\ \ e^{\theta\sqrt{2}t}\ \{-\theta\xi^1(0)-2\theta^2\int_0^t e^{-\theta\sqrt{2}s}\ \ \frac{\gamma(0)}{\gamma(s)}\ \xi^1(s)ds\ +$$

$$+\ e^{-\theta\sqrt{2}t}\ \frac{\gamma(0)}{\gamma\ (t)}\ \Big[\theta(\sqrt{2}-2)+\gamma(t)\ \Big]\ \xi^1(t) - \Big[\theta(\sqrt{2}-2)+\gamma(0)\ \Big]\ \xi^1(0)+$$

$$-\int_0^t e^{-\theta\sqrt{2}s}\ \ \frac{\gamma(0)}{\gamma(s)}\Big[\theta^2(\sqrt{2}-2\sqrt{2})-\ 2\theta\ \gamma(s)\Big]\xi^1(s)\ ds\} =$$

$$=\gamma(t)\ 2\theta\int_0^t e^{\theta\sqrt{2}(t-s)}\ \ [1-\frac{\theta(2-\sqrt{2})}{(s)}]\ \xi^1(s)\ ds\ +$$

$$+\ [\gamma(t)\ -\ \theta(2-\sqrt{2})\ ]\ \xi^1(t)\ -\ \frac{1}{2}\ \ \theta\frac{\gamma(t)}{\gamma(0)}\ e^{\theta\sqrt{2}t}\ \ \xi^1(0).$$

Finally to have the Radon-Nikodym derivative we derive

$$\int_0^t \alpha(s,\ \xi^1(s))\ d\xi^1(s),\qquad \frac{1}{2}\int_0^t \alpha^2(s,\ \xi^1(s))\ ds,$$

which can be done by simple calculations. We obtain

$$(2.3.57)\quad \frac{dP_\xi}{dP_W}\ (t,\xi)\ =\ \exp\ \{\int_0^t \alpha(s,\xi(s))d\xi(s)\ -\ \frac{1}{2}\int_0^t \alpha^2(s,\xi(s))\ ds\} =$$

$$=\ \exp\Big\{2\theta\gamma(t)\xi(t)\ \ \int_0^t \frac{e^{\theta\sqrt{2}(t-s)}}{\gamma(s)}[\gamma(s)-\theta(2-\sqrt{2})\ ]\xi(s)\ ds\ +$$

$$-\ \frac{1}{2}\int_0^t (\xi(s))^2[\gamma^2(s)+2\sqrt{2}\ \theta\gamma(s)+\ \theta^2 2\sqrt{2}]ds\ -\ \int_0^t \xi(n)\ \Big\{2\theta(1\ +$$

$$-\theta(2-\sqrt{2}))\gamma(n)+2\gamma^2(n)] \int_0^n \frac{e^{\theta\sqrt{2}(n-s)}}{\gamma(s)} (\gamma(s)-\theta(2-\sqrt{2}))\xi(s)ds +$$

$$-\frac{1}{2}\theta\frac{\gamma(n)}{\gamma(0)} e^{\theta\sqrt{2}n}\xi(0) [\gamma(n)-\theta(2-\sqrt{2})]\} dn +$$

$$-2\theta^2 \int_0^t \gamma^2(n) \quad [\int_0^n \frac{e^{2\theta(n-s)}}{\gamma(s)} [\gamma(s)-\theta(2-\sqrt{2})]\xi(s) ds]^2 dn +$$

$$+\theta^2\xi(0) \int_0^t \frac{\gamma^2(n)}{\gamma(0)} e^{\theta\sqrt{2}n} \int_0^n \frac{e^{\theta\sqrt{2}(n-s)}}{\gamma(s)} (\gamma(s)-\theta(2-\sqrt{2}))\xi(s)dsdn +$$

$$+\int_0^t (\gamma(s)-\theta(2-\sqrt{2}))\xi(s)d\xi(s) - \frac{1}{2}\theta\frac{\gamma(t)}{\gamma(0)} e^{\sqrt{2}\theta t}\xi(0) [\xi(t) +$$

$$-\xi(0) + \frac{1}{4}\theta\gamma(t)t\xi(0)]\} .$$

The last form shows that if θ is unknown no sufficient statistic exists. Asymptotically good parameter estimates can be achieved by taking only some terms in the exponent.

Now let us return to the general case. On the basis of Lemma 2 in §2.2.1 and representation (2.2.40) we assume that there are some unobservable components. For simplicity we assume that the unobservable components are $\xi^{p-q+1}(t), \ldots, \xi^p(t)$ and the representation

$$\text{(2.3.58)} \qquad d\xi^{p-q}(t) = \sum_{i=1}^{p} di \cdot \xi^i(t)\, dt + dw(t),$$

$$d\,\xi^i(t) = \lambda_i\, \xi^i(t)\, dt + dw(t), \qquad i = p-q+1,\ldots,p,$$

holds, where not all d_i are $0(i=p-q,\ldots,p)$.

To calculate

$$\text{(2.3.59)} \qquad \alpha_i(t) = E(\xi^i(t)\,|F_t^{\xi^{p-q}}), \quad i = p-q+1,\ldots,p,$$

$$\text{(2.3.60)} \qquad \gamma_{ij}(t) = E(\alpha_i(t) - \xi^i(t))(\alpha_j(t)- \xi^j(t)), \quad i,j=p-q+1,\ldots,p,$$

we use Theorem 10.3 in Liptser-Shiryaev [1], in this special case. We have the following equations for $\underline{\alpha}(t)^* = (\alpha_{p-q+1}(t),\ldots, \alpha_p(t))$ and $\gamma(t) = (\gamma_{ij}(t))$

$$(2.3.61)\quad d\,\underline{\alpha}(t) = a_1\underline{\alpha}(t)\,dt + [\underline{b}_1 + \gamma(t)\,\underline{A}_1]\,[d\xi^{p-q}(t) -$$

$$-(\underline{A}_1^*\,\underline{\alpha}(t) + d_{p-q}\cdot\,\xi^{p-q}(t))\,dt] =$$

$$=\{[a_1-\underline{b}_1\underline{A}_1^* - \gamma(t)\underline{A}_1\underline{A}_1^*]\underline{\alpha}(t) - [\underline{b}_1+\gamma(t)\underline{A}_1]d_{p-q}\,\xi^{p-q}(t)\,dt +$$

$$+ [\underline{b}_1 + \gamma(t)\,\underline{A}_1]\,d\,\xi^{p-q}(t)\},$$

$$(2.3.62\quad \dot{\gamma}(t) = \frac{d\gamma(t)}{dt} = 2[a_1-\underline{b}_1\underline{A}_1^*\,]\gamma(t) - (t)\underline{A}_1\underline{A}_1^*\gamma(t),$$

where

$$a_1 = \begin{pmatrix} \lambda_{p-q+1} & 0 & \cdots & 0 \\ 0 & \lambda_{p-q+2} & \cdots & 0 \\ \cdot & & & \\ \cdot & & & \\ \cdot & & & \\ 0 & & & \lambda_p \end{pmatrix}, \qquad \underline{b}_1 = \begin{pmatrix} 1 \\ 1 \\ \cdot \\ \cdot \\ \cdot \\ 1 \end{pmatrix}_{qx1},$$

$$\underline{A}_1 = \begin{pmatrix} d_{p-q+1} \\ \cdot \\ \cdot \\ \cdot \\ d_p \end{pmatrix}_{qx1},$$

Equation (2.3.62) is an Euler type with solution

$$(2.3.63)\quad \gamma(t) = e^{-2[\underline{b}_1\underline{A}_1^*-a_1]t}\,[\gamma^{-1}(0)+\underline{A}_1\underline{A}_1^*\int_0^t e^{-2[\underline{b}_1\underline{A}_1^*-a_1]n}\,dn]^{-1},$$

which can be checked by immediate calculation (see Appendix (A1.21)). Further for $\underline{\alpha}(t)$ we get, as it can easily be calculated,

$$(2.3.64)\quad \underline{\alpha}(t) = e^{\int_0^t[a_1-\underline{b}_1\underline{A}_1^*-\gamma(s)\underline{A}_1\underline{A}_1^*]ds}\,\{\underline{\alpha}(0) +$$

$$+\int_0^t e^{\int_0^s[\underline{b}_1\underline{A}_1^*+\gamma(n)\underline{A}_1\underline{A}_1^*-a_1]dn}\,[-(\underline{b}_1+\gamma(s)\underline{A}_1)d_{p-q}\cdot\,\xi^{p-q}(s)ds+$$

$$+(\underline{b}_1 + \gamma(s)\,\underline{A}_1)\,d\,\xi^{p-q}(s)]\},$$

with the initial values

$$(2.3.65)\qquad \underline{\alpha}(0) = E\left(\begin{matrix}\xi^{p-1+1}(0)\\ \cdot\\ \cdot\\ \cdot\\ \xi^{p}(0)\end{matrix}\;\middle|\; \xi^{p-q}(0)\right)$$

$$(2.3.66)\qquad \gamma\,(0) = (E(\,\xi^{p-q+i}(0) - \alpha_i(0)(\xi^{p-q+j}(0) - \alpha_j(0))).$$

In order to find these values in (2.3.65) and (2.3.66) we shall take again advantage of elementary Gaussian process $(\xi^{p-q}(t),\ldots\ \xi^{p}(t))$ that $B(0) = (E\xi^{p-q+i}(0)\xi^{p-q+j}(0))$ is the unique solution of the system of equations (2.2.3)

$$AB(0) + B(0)A^{*} = -B_w$$

where

$$A = \begin{pmatrix} d_{p-q} & d_{p-q+1} & \cdots & d_p\\ 0 & \lambda_{p-q+1} & \cdots & 0\\ \cdot & & & \\ \cdot & & & \\ \cdot & & & \\ 0 & & & \lambda_p \end{pmatrix}, \qquad B_w = \begin{pmatrix} 1 & \cdots & 1\\ \cdot & & \\ \cdot & & \\ \cdot & & \\ 1 & \cdots & 1 \end{pmatrix}$$

Using formula (2.3.49), (2.3.58), (2.3.63), and (2.3.64) we get that in the general case the Radon-Nikodym derivative exists and we can state the following theorem.

Theorem 4. Let the stationary Gaussian process $\xi(t)$ permit the spectral representation given by (2.2.34), then $\frac{dP_\xi}{dP_w}$ exists, has the form (2.3.49) where $\underline{\alpha}(t)$ is given by (2.3.63) - (2.3.66), if $Q(z) \neq$ const. no sufficient statistic exists for the set of unknown parameters $(b_0,\ldots,b_q)$, $(a_1,\ldots,a_p)$.

Remark 1. To get asymptotically effective estimators we have to investigate each term in (2.3.63) when $t \to \infty$.

Remark 2. To complete our discussion we give the sufficient statistics for processes with observable components. We recall that for an elementary Gaussian process $\underline{\xi}^{*}(t) = (\xi^{1}(t),\ldots,\ \xi^{p}(t))$, where all components are observable and with the representation

$$d\underline{\xi}(t) = A\underline{\xi}(t)dt + d\underline{w}(t)$$

where $\underline{w}^*(t) = (w^1(t),\dots,w^p(t))$ is the Wiener process, $E\underline{\xi}(0) = \underline{0}$, $E\underline{\xi}(0)\underline{\xi}^*(0) = B(0)$, $E\underline{w}(t) = \underline{0}$, $E\underline{w}(t)\,\underline{w}^*(t) = B_w t$, (with invertible B_w), (2.3.25) holds. Using formula (2.3.34), or (2.3.35), the set of sufficient statistics is the following

$$(2.3.67)\qquad \{\int_0^T \underline{\xi}(s)\,\underline{\xi}^*(s)\,ds,\ \underline{\xi}(T)\,\underline{\xi}^*(T),\ \underline{\xi}(0)\,\underline{\xi}^*(0)\}.$$

For the autoregressive process (2.2.31) one can obtain the set of sufficient statistics from (2.3.36) as

$$(2.3.68)\qquad \{\int_0^T[\xi^{(i)}(s)]^2\,ds,\ [\xi^{(i)}(T)]^2,\ [\xi^{(i)}(0)]^2,\ i = 0,1,\dots,p\}.$$

Remark 3. Note that if

$$\zeta(t) = \xi(t) + e^{i\omega_0 t}$$

with unknown parameter ω_0, $E\xi(t) = 0$, $\sigma_w^2 = 1$, and $\xi(t)$ is an autoregressive process, then even if it is of order one we have no sufficient statistics for ω_0. Indeed, from (2.3.41)

$$\frac{dP_\zeta}{dP_w}(\zeta(t),T) = \frac{2\lambda}{\sqrt{2\pi}}\,e^{-\lambda(\zeta(0)-1)^2}\exp\{-\frac{\lambda^2}{2}\int_0^T(\zeta(t)-e^{i\omega_0 t})^2\,dt +$$

$$-\lambda\int_0^T\xi(t)d\xi(t))\} = \sqrt{\frac{2}{\pi}}\,\lambda\,\exp\{-\frac{\lambda^2}{2}[\int_0^T\zeta^2(t)\,dt - 2\int_0^T\zeta(t)\,e^{i\omega_0 t}\,dt +$$

$$\frac{1}{2i\,\omega_0}[e^{2i\omega_0 T}-1]] - \frac{\lambda}{2}[(\zeta(T)-e^{i\omega_0 T})^2 + (\zeta(0)-1)^2 + \frac{\lambda}{2}T\} =$$

$$= \sqrt{\frac{2}{\pi}}\,\lambda\,\exp\{-\frac{\lambda^2}{2}\int_0^T\zeta^2(t)\,dt + \lambda^2\int_0^T e^{i\omega_0 t}\,\zeta(t)\,dt - \frac{\lambda^2}{4i\omega_0}[e^{2i\omega_0 T}-1]+$$

$$+\frac{\lambda}{2}T - \frac{\lambda}{2}[(\zeta(T)-e^{i\omega_0 T})^2 + (\zeta(0)-1)^2]\}.$$

We prove the following statement.

Theorem 5. Let $\xi(t)$ be the one dimensional elementary Gaussian process, i.e.,

$$d\xi(t) = -\lambda\xi(t)\,dt + dw(t)$$

where $\lambda > 0$ and $Edw = 0$, $E(dw(t))^2 = dt$. We have (see (2.3.41))

$$\frac{dP_{\xi^o}}{dP_{w_o}}(\xi) = \exp\{-\lambda \int_o^t \xi(s)d\xi(s) - \frac{\lambda^2}{2}\int_o^t \xi^2(s)\,ds\} =$$

$$= \exp\{-\lambda\int_o^t \xi(s)\,dw(s) + \frac{1}{2}\lambda^2 \int_o^t \xi^2(s)\,ds\} = \zeta^{-1}(t).$$

Then the process $\zeta(t)$ is a martingale with respect to $(P_{\xi 0}, F_t)$ and the process $\xi(t)$ is a Wiener process with respect to the measure P_{w0}.

Proof. As $\zeta(t)$ is a Radon-Nikodym derivative we obtain $(s < t)$

(2.3.69) $$E_{P_\xi}(\zeta(t)|F_s) = E_{P_\xi}[\exp(\lambda \int_o^s \xi(n)\,dw(n) - \frac{\lambda^2}{2}\int_o^s \xi^2(n)dn +$$

$$\lambda\int_s^t \xi(n)dw(n) - \frac{\lambda^2}{2}\int_s^t \xi^2(n)dn)|F_s] = \zeta(s)\; E_{P_\xi}(e^{\lambda \int_s^t \xi(n)dw(n) -}$$

$$- \frac{\lambda^2}{2}\int_s^t \xi(n)\;dn|F_s) = \zeta(s)\; E_{P_w}(1|F_s) = \zeta(s),$$

and by the Ito formula

$$\zeta(t) = 1 + \lambda \int_o^t \xi(s)\;\zeta(s)\;dw(s).$$

To prove that with respect to P_{w0} the process $\xi(t)$ is a Wiener process it suffices to establish that

(2.3.70) $$E_{P_w}[\exp(i\alpha(\xi(t) - \xi(s)))|F_s] = e^{-\frac{\alpha^2}{2}(t-s)}$$

Using the property of conditional expectation

(2.3.71) $$E_{P_w}[e^{i\alpha(\xi(t)-\xi(s))}|F_s] = \zeta^{-1}(s)\; E_{P_\xi}[e^{i\alpha(\xi(t)-\xi(s))}\zeta(t)|F_s].$$

By the Ito formula

(2.3.72) $$\zeta(t)\; e^{i\alpha(\xi(t)-\xi(s))} = \zeta(s) + \lambda \int_s^t e^{i\alpha(\xi(n)-\xi(s))}\xi(s)\;\zeta(s)dw(n) +$$

$$+ i\alpha \int_s^t e^{i\alpha(\xi(n)-\xi(s))}\zeta(n)\;dw(n) - \frac{\alpha^2}{2}\int_s^t e^{i\alpha(\xi(n)-\xi(s))}\zeta(n)dn.$$

But by the properties of stochastic integrals (see (2.1.15) and (2.1.16) in the deterministic case)

$$E_{P_\xi}[\int_s^t e^{i\alpha(\xi(n)-\xi(s))}\xi(n)\zeta(n)\, dw(n)|F_s] = 0,$$

$$E_{P_\xi}[\int_s^t e^{i\alpha(\xi(n)-\xi(s))}\zeta(n)dw(n)|F_s] = 0,$$

and

(2.3.73) $$\zeta^{-1}(s)\, E_{P_\xi}[e^{i\alpha(\xi(t)-\xi(s))}\zeta(t)|F_s] =$$

$$= -\frac{\alpha^2}{2}\int_s^t \zeta^{-1}(s)\, E_{P_\xi}[e^{i\alpha(\xi(n)-\xi(s))}\xi(n)|F_s]\, dn.$$

The last equation has the only solution

(2.3.74) $$\zeta^{-1}(s)\, E_{P_\xi}[e^{i\alpha(\xi(t)-\xi(s))}\zeta(t)|F_s] = e^{-\frac{\alpha^2}{2}(t-s)},$$

from (2.3.71) and (2.3.74) there follows the theorem.

<u>Remark 1.</u> The theorem remains valid in the multidimensional case too, i.e., the process $\underline{\xi}(t)$ with the differential

(2.3.75) $$d\underline{\xi}(t) = A\,\underline{\xi}(t)\,dt + d\underline{w}(t)$$

is a Wiener process with respect to the measure P_w.

<u>Remark 2.</u> If $\underline{\xi}(t)$ has the differential (2.3.75) then the process

(2.3.76) $$\underline{\xi}(t) = e^{-At}\,\underline{\xi}(t), \qquad t \geq 0,$$

is Gaussian with independent increments and with covariance matrix

(2.3.77) $$D(t) = e^{-tA}B(0)e^{-tA^*} - B(0),$$

$$D'(t) = -e^{-tA}[B(0)A^* + AB(0)]e^{-tA^*}.$$

E.g., in the one dimensional case

$$\zeta(t) = \frac{\sqrt{2\lambda t}}{\sigma_w}\,\xi(\frac{1}{2\lambda}\log t)$$

is a Brownian motion.

CHAPTER 3

THE MAXIMUM LIKELIHOOD ESTIMATORS AND THEIR DISTRIBUTIONS IN THE ONE DIMENSIONAL CASE

3.1 Basic Principles of Statistical Estimation Theory

In the preceding chapters, we discussed the structure of the multivariate elementary Gaussian systems. In this and the following chapters we consider the estimation of the coefficients in the differential and difference equations of the given processes. Before the detailed discussion of the estimations of the parameters of the elementary Gaussian processes we recall some general results on estimation theory.

Let $\underline{x}^* = \{x_1,\dots,x_n\}$ denote a sequence of not necessarily independent random variables, i.e., n consecutive observations from a discrete time stochastic process $\xi(n)$ with the joint probability density function $f(x_1,\dots,x_n;\ \alpha)$, depending on a parameter α. We call $\underline{x}$ a sample or realization of the process $\xi(n)$. The elements $(i = 1,2,\dots,n)$ may be vector valued. The sample $\underline{x}$ is an element (point) of the saple space X.

In mathematical statistics the concept of the sample space (X,F) consisting of the set of all sample points X and σ-algebra F of its subsets is primary. In our case, in the problems of elementary Gaussian processes, X is the space of sequences $x = (\underline{x}_1,\underline{x}_2,\dots\underline{x}_n)$, where $\underline{x}_i \in R^k$, or the space of continuous functions $x = (\underline{x}(t)), 0 \le t \le T$.

Let (Y,B) be another measure space. Any F/B measurable mapping $y = g(x)$ of the space X into Y is called a statistic. If $\underline{x}$ is a realization then $y = g(x)$ is a function, or functional, of this realization. Examples of statistics are

$$\xi(0) + \xi(T),\quad \frac{1}{N}\sum_{i=1}^{N} \xi(i),\quad \int_o^T \xi(t)dt,\quad \int_o^T \xi^2(t)\ dt.$$

The theory of estimation is one of the most important parts of mathematical statistics. We assume that the family $\{P_\alpha : \alpha \in \Theta\}$ of probability measure on (X,F), depending on the parameter α is dominated by a certain σ-finite measure μ (i.e., $P_\alpha \ll \mu, \alpha \in \Theta$), the Radon-Nikodym derivative $f_\alpha = \frac{dP_\alpha}{d\mu}$ will be called probability density function. The statistic $\hat{\alpha}(x) = \hat{\alpha}(x_1,\dots,x_n)$, or

$\hat{\alpha}(x) = \hat{\alpha}(x(t), 0 \leq t \leq T)$, is called an unbiased estimator of α if $E_\alpha \hat{\alpha} = \alpha$ for all $\alpha \in \Theta$. The statistic $\tilde{\alpha}(x)$ is called sufficient for α (or for the family $P_\alpha, \alpha \in \Theta$) if for each $A \in F$ a version of conditional probability $P_\alpha(A|\tilde{\alpha}(x))$ not depending on can be chosen.

The sequence of the statistics $\alpha_n(x)$, $n = 1,2,\ldots$, is called a consistent estimate of α if $\alpha_n(x) \to \alpha$ in P_α probability for all $\alpha \in \Theta$. $\alpha_n(x)$ is a strongly consistent estimate of α if it tends with probability 1 to α for all $\alpha \in \Theta$.

For fixed $\underline{x}$ we are led to compare $f(\underline{x},\alpha)$ for various values of the parameters α. $f(\underline{x},\alpha)$ viewed as a function of α is called the likelihood function of $\underline{x}$, and often is denoted by $L(\alpha, \underline{x})$. In the continuous time case the likelihood function is defined in an analogous way

$$L(\alpha, \underline{x}) = \frac{dP_\alpha}{d\mu} (\underline{x}(s), 0 \leq s \leq t),$$

or

$$L(\alpha, \underline{x}) = f_\alpha (\underline{x}(s), 0 \leq s \leq t)$$

as a function of α.

Let $\hat{\alpha}(x_1,\ldots,x_n)$ be an estimator of the unknown parameter α, $\alpha \in \Theta$.

Let us introduce the "local coordinates" $(\zeta_1, \zeta_2,\ldots,\zeta_{n-1})$ on the hypersurface $\hat{\alpha}(x_1,\ldots,x_n) = \text{const}$. If the mapping $x_1,x_2,\ldots,x_n \longleftrightarrow \alpha, \zeta_1,\ldots,\zeta_{n-1}$ is one to one and continuously differentiable then the joint density function can be written in the following way

$$f(x_1(\zeta_1,\ldots, \zeta_{n-1}, \hat{\alpha}),\ldots, x_n(\zeta_1,\ldots,\zeta_{n-1}, \hat{\alpha}))|J| = f(\zeta_1,\ldots,\zeta_{n-1},\hat{\alpha})$$

where $|J|$ is the Jacobian of the mapping. If in addition $|J|$ does not depend on α and $g(\hat{\alpha},\alpha)$ denotes the probability density function of the estimator $\hat{\alpha}$ for a given α and $h(\zeta_1,\ldots,\zeta_{n-1}|\hat{\alpha},\alpha)$ denotes the conditional density function then we get

$$f(x_1,\ldots, x_n; \alpha) = g(\hat{\alpha},\alpha)\, h(\zeta_1,\ldots, \zeta_{n-1}|\hat{\alpha};\alpha). \tag{3.1.1}$$

Further we assume that the following <u>regularity conditions</u> hold, the partial derivatives

$$\frac{\partial f}{\partial \alpha}, \quad \frac{\partial g}{\partial \alpha}, \quad \frac{\partial h}{\partial \alpha}$$

exist and

$$\left|\frac{\partial f}{\partial \alpha}\right| < F_o(x_1,\ldots,x_n), \quad \left|\frac{\partial g}{\partial \alpha}\right| < G_o(x_1,\ldots,x_n), \left|\frac{\partial h}{\partial \alpha}\right| < H_o(\zeta_1,\ldots,\zeta_{n-1},\hat{\alpha})$$

hold uniformly with respect to α, where

$F_o(x_1,\ldots,x_n)$, $G_o(x_1,\ldots x_n)$, $H_o(\zeta_1,\ldots, \zeta_{n-1},\hat{\alpha})$, $\hat{\alpha}(x_1,\ldots,x_n)$.

$\cdot G_o(x_1,\ldots,x_n)$

have a final first moment. Under these regularity conditions the so called Cramer-Rao inequality

$$(3.1.2) \qquad \frac{\left(1 + \dfrac{d_\alpha E\ (\hat{\alpha}-\alpha)}{d\alpha}\right)^2}{\int_{-\infty}^{\infty}\cdots\int_{-\infty}^{\infty}\left(\dfrac{d \log L\ (\alpha,\ x_1,\ldots,x_n)}{d\alpha}\right)^2 f(x_1,\ldots,x_n,\alpha)\,dx_1\ldots dx_n} =$$

$$= \frac{\left(1 + \dfrac{dE_\alpha(\hat{\alpha}-\alpha)}{d\alpha}\right)^2}{E_\alpha\left(\dfrac{d}{d\alpha}\log L(\alpha,\underline{x})\right)^2} \leq E_\alpha\ (\hat{\alpha} - \alpha)^2,$$

holds. In (3.1.2) we can put the Radon-Nikodym derivatives in the continuous time case.

The equality sign in (3.1.2) will be attained if and only if the following two conditions hold:

A) $h(\eta_1,\ldots,\eta_{n-1}|\hat{\alpha},\alpha)$ does not depend on α, i.e., $\hat{\alpha}$ is a sufficient statistic of α. In this case it is said that the factorization criterion

$$(3.1.1') \qquad f(\underline{x};\ \alpha) = g(\hat{\alpha},\alpha)\ h(\underline{\zeta}|\hat{\alpha})$$

holds.

B) $$\frac{d}{d\alpha}\log g(\hat{\alpha},\alpha) = k\ (\alpha)\ (\hat{\alpha}-\alpha)\ .$$

If α satisfies A and B, $\hat{\alpha}$ is said to be an <u>efficient</u> statistic (or estimator). Only one unbiased, i.e., $E_\alpha\ \hat{\alpha} = \alpha$ and efficient estimator exists. The efficient estimator $\hat{\alpha}$ is the unique nontrivial solution of the likelihood equation

$$\frac{d}{d\alpha} \log L\ (\alpha,\underline{x}) = 0.$$

This follows from condition B). A solution $\bar{\alpha}$ is called trivial if $k(\alpha) = 0$. The likelihood function in most of our problems will have an exponential form. In many problems we take the logarithm of the likelihood function and the maximum likelihood estimator $\hat{\alpha}(\underline{x})$ of α, i.e., the value of α which maximizes $L(\alpha, \underline{x})$, is obtained as a root of the likelihood equation.

Let $S(\alpha) = \frac{d}{d\alpha} \log L(\alpha,\underline{x})$ then $S(\alpha)$ as a function of n and α has the following form (if the regularity condition holds)

$$(3.1.3) \qquad S_n(\alpha) = \frac{\frac{d}{d\alpha} L\ (\alpha,\ x_1,\ldots,x_n)}{L(\alpha,x_1,\ldots,x_n)}\ ,\ \int \frac{d}{d\alpha} L_n(\alpha)d\mu = \int S_n(\alpha)dP = 0,$$

and

$$(3.1.4) \qquad E\ (\log L_n(\alpha)\,|F^x_{n-1}) \leqq \log L_{n-1}(\alpha),$$

as

$$\int_{\Lambda_{n-1}} \frac{dP_n}{d\mu_n}\, d\mu = \int_{\Lambda_{n-1}} \frac{dP_{n-1}}{d\,\mu_{n-1}}\, d\mu$$

for every $\Lambda_{n-1} \in F^x_{n-1}$, with $S_0 = 0$ and $ES_n(\alpha) = 0$. The L_n process forms a martingale.

$I_n(\alpha) = E(S_n(\alpha))^2$ is called the Fisher information in $\underline{x}^* = (x_1,\ldots,x_n)$.

If

$$I_n^{-1}(\alpha)\ [\frac{d}{d\alpha} S_n(\alpha)] = I_n^{-1}(\alpha)[\frac{d^2}{d\alpha^2} \log L_n\ (\alpha,\underline{x})]$$

tends to - 1 in probability we investigate the asymptotic efficiency of the maximum likelihood estimator $\hat{\alpha}$.

Expanding by Taylor's series $\log L_n(\alpha,\underline{x})$ in α at $\hat{\alpha}$ we obtain

$$(3.1.5) \qquad \log L_n(\alpha,\underline{x}) = \log L_n(\hat{\alpha},\underline{x}) + \frac{(\hat{\alpha}_n-\alpha)}{2}\ [\frac{d^2}{d\alpha^2} \log L_n(\alpha,\underline{x})]_{\alpha=\hat{\alpha}} \pm o(1)$$

or

$$(3.1.5') \qquad L_n(\alpha,\underline{x}) = L_n(\hat{\alpha},\underline{x}) \exp\{\frac{1}{2}(\hat{\alpha}_n-\alpha)[\frac{d^2}{d\alpha^2}\log L_n(\alpha,\underline{x})]_{\alpha=\hat{\alpha}} +$$

$$+\sigma(1)\} = L_n(\hat{\alpha},\underline{x}) \exp\{-\frac{I_n(\alpha)}{2}(\hat{\alpha}_n-\alpha)^2 + \sigma(1)\}.$$

This shows that the maximum likelihood estimator $\hat{\alpha}$ satisfies the factorization criterion of sufficiency only asymptotically.

We say that an estimator α^* of α is asymptotically efficient if

$$(3.1.6) \qquad (I_n(\alpha))^{1/2}[\alpha^*-\alpha-\frac{S_n(\alpha)}{I_n(\alpha)}] \to 0, \qquad n\to\infty,$$

in probability. To compare various estimates $\hat{\alpha} = \hat{\alpha}(x)$ of the unknown parameter $\alpha\in\Theta$ we introduce non-negative loss functions $w(\alpha,\hat{\alpha})$ and average loss

$$R(\alpha,\hat{\alpha}) = E_\alpha\, w(\alpha, \hat{\alpha}(x)).$$

The most commonly used function is $w(\alpha,\beta) = |\alpha-\beta|^2$ or in the multi-dimensional case $w(\underline{\alpha},\underline{\beta}) = ||\,\underline{\alpha}-\underline{\beta}\,||^2 = (\underline{\alpha}-\underline{\beta})^*(\underline{\alpha}-\underline{\beta})$. The Fisher information matrix $I(\underline{\alpha}) = (I_{ij}(\alpha))$ is defined in the following way

$$I_{ij}(\underline{\alpha}) = E_\alpha[\frac{\partial}{\partial\alpha_i}\log L(\underline{\alpha},\underline{x})][\frac{\partial}{\partial\alpha_j}\log L(\underline{\alpha},\underline{x})].$$

For an unbiased estimator $\hat{\alpha}(x)$ in the one-dimensional case the Cramer-Rao inequality states (see (3.1.2))

$$(3.1.2') \qquad E_\alpha\,(\hat{\alpha}-\alpha)^2 \geq \frac{1}{I_n(\alpha)}, \quad \alpha\in\Theta.$$

In the multivariate case (under similar conditions of regularity) the following Cramer-Rao matrix inequality holds for an unbiased estimator $\hat{\alpha}$

$$(3.1.7) \qquad E_\alpha(\underline{\alpha}-\hat{\underline{\alpha}})(\underline{\alpha}-\hat{\underline{\alpha}})^* \geq I^{-1}(\underline{\alpha}).$$

<u>Confidence Regions.</u> In the preceding we defined the measure of accuracy for an estimate $\underline{\alpha}$ of the true value $\underline{\alpha}$ by the risk function $E_\alpha(\underline{\alpha}-\hat{\underline{\alpha}})^*(\underline{\alpha}-\hat{\underline{\alpha}})$ and we get the lower bounds in (3.1.2) and (3.1.7) for this value. As in the statistics of stochastic processes the asymptotic behavior of estimates depends not only on the sample size $0\leq t\leq T$, but on the parameters too we shall give the accuracy with

respect to the confidence regions (interval estimates of unknown parameters) or in other words we shall give the distributions of point estimates. If we have a single unknown parameter then the construction of confidence intervals is the same as in case of independent random variables. For the mean m of the one dimensional elementary Gaussian process $\xi(t)$ we get the maximum likelihood estimator (see(3.2.2))

$$\hat{m} = \frac{\xi(0) + \xi(T) + \lambda \int_0^T \xi(t)dt}{2 + \lambda T}$$

which is normally distributed with $E\hat{m} = m$ and variance $\sigma_w^2 / (\lambda T+1)\lambda$. The confidence limits for m on level β are

$$m - z_\beta \frac{\sigma_w}{\sqrt{\lambda(\lambda T+1)}} < m < m + z_\beta \frac{\sigma_w}{\sqrt{\lambda(\lambda T+1)}}$$

where

$$\frac{1}{\sqrt{2\pi}} \int_{-z_\beta}^{z_\beta} e^{-t2/2} dt = \beta. \text{ If } \lambda \to 0,\ T \to \infty,\ \lambda T = \kappa < K$$

we get an infinite confidence interval. If the time is discrete we shall get for the mean the following confidence interval (by the maximum likelihood estimator $\tilde{m}$, see §3.5)

$$\tilde{m} - z_\beta \frac{\sigma_\varepsilon}{\sqrt{1-\rho^2}} < \frac{1+\rho}{\sqrt{2+(n-2)(1-\rho)}} \quad m < \tilde{m} + z_\beta \frac{\sigma_\varepsilon}{\sqrt{1-\rho^2}} \frac{\sqrt{1+\rho}}{\sqrt{2+(n-2)(1-\rho)}}$$

where z_β is as above and

$$\tilde{m} = \frac{\xi(0) + \xi(n) + (1-\rho) \sum_1^{n-1} \xi(i)}{2+(n-2)(1-\rho)}$$

We see that when $\rho \to 1$ ($n(1-\rho)$ is bounded), we have again infinite confidence interval. Even in engineering practice it is accepted for the same accuracy (the same length of confidence interval) instead of n independent observations (in the discrete time case) we must have $\frac{n}{(1-\rho)^2}$ observations, which may be very great if $\rho \sim 1$.

In the case when we have two unknown parameters and one realization of the stochastic process $\xi(t)$, on $0 \le t \le T$, there may be that to reduce the error in one parameter component must be accompanied by increased uncertainties in the magnitude of the other component, as it is the case in the one dimensional elementary process with unknown m and λ (see §3.4). This <u>uncertainty principle</u>

for stochastic processes was stated by A.N. Kolmogorov in the late forties (in Jerevan, 1947). Later we shall deal with the problem on infinite confidence intervals. Their construction is similar to the one used in finite case.

We shall now turn our attention to the selection of a region in the parameter space and the specification of the probability that the estimated values of a set of parameters will lie within the selected region. In this fashion, it is possible to label the estimates with a measure of confidence.

The problem of the construction of confidence intervals for a single parameter α is the following. Let P_α be a set of probability measures $\alpha \in \Theta$, and β a real number $0 \leq \beta \leq 1$. We seek functions $\underline{h}$, $\overline{h}$ defined on the sample space $(x_1,\ldots,x_n) = \underline{x}$ with $\underline{h} < \overline{h}$ such that for each $\alpha \in \Theta$, i.e., uniformly in α

$$P_\alpha(\{\underline{x}: \alpha \in [\underline{h}(\underline{x}), \overline{h}(\underline{x})]\}) \geq \beta.$$

Then the interval $(\underline{h},\overline{h})$ is called a 100β percent confidence interval. If there exists a function $T(\underline{x},\alpha)$ which is strictly monotone in α for each $\underline{x}$ and $a = T(\underline{x},\alpha)$ is solvable for each a and $\underline{x}$ and the random variables T_α defined for each α by the way $\underline{x} \to T(\underline{x},\alpha)$ are assumed to have the same distribution function P_α independent of α. Then a confidence interval for α can always be constructed.

Generally, the notion of confidence sets may be also introduced. Let β be a real number with $0 \leq \beta \leq 1$ and

$$\beta \leq \inf_{\alpha \in \Theta} P_\alpha(K_\alpha)$$

where

$$K_\alpha = \{\underline{x}: \underline{\alpha} \in K(\underline{x})\},$$

$$K(\underline{x}) = \{\underline{\alpha}: \underline{x} \in K_\alpha\},$$

then $K(\underline{x})$ is called a confidence set for $\underline{x}$ with confidence level β, because

$$P_\alpha \{\underline{x}: \underline{\alpha} \in K(\underline{x})\} \geq \beta.$$

If Θ is an interval on R_1 and $K(x)$ are also intervals we get the confidence intervals.

The confidence set $K(x)$ covers the true parameter α with probability at least as large as β.

The function

$$k(\alpha, \alpha') = P_{\alpha}\ (\{\underline{x}\colon \alpha' \in K(\underline{x})\})$$

is the accuracy function.
A confidence set $K(\underline{x})$ with confidence level β for $\alpha \in \Theta$ is called unbiased when the function $k(\underline{\alpha},\underline{\alpha}')$ has the following properties

$$k(\alpha,\alpha') \geq \beta, \qquad \alpha \in \Theta$$

$$k(\alpha,\alpha) \geq k\ (\alpha,\alpha'), \qquad \alpha,\ \alpha' \in \Theta, \quad \alpha \neq \alpha'.$$

3.2 The Unknown Mean

Let us first assume that $\eta(t) = \xi(t) + m$ where $\xi(t)$ is a continuous time parameter one-dimensional elementary Gaussian process with known parameters σ_w and λ i.e., $d\xi(t) = - \lambda\xi(t)dt + \sigma_w dw(t)$. It remains to estimate the unknown expectation m by obeserving the process $\eta(t)$ on the time interval $[0,T]$. We have already computed (See §2.3.3, (2.3.42)) the Randon-Nikodym derivative of the measure $P_{\lambda,m}$ generated by the process $\eta(t)$ with respect to the measure $P_{\lambda,0}$ generated by $\xi(t)$ (on the space of the continuous functions defined on $[0,T]$):

$$(3.2.1) \qquad \frac{dP_{\lambda,m}}{dP_{\lambda,0}} = \exp\{ - \frac{\lambda}{\sigma_w^2}\ (m[\eta(0)+\eta(T)+ \lambda\int_0^T \eta(t)dt]+ m^2(1+ \frac{\lambda T}{2}))\}.$$

Hence we get the maximum likelihood estimator for m

$$(3.2.2) \qquad \hat{m} = \frac{\eta(0)+\eta(T)+\lambda \int_0^T \eta(t)\ dt}{2 + \lambda T}\ .$$

Which is unbiased and

$$(3.2.3) \qquad E(\hat{m} - m)^2 = \frac{\sigma_w^2}{\lambda(\lambda T+2)}\ .$$

The random variable $\hat{m}$ is normally distributed (see Appendix B, Lemma 4). The last formula can be derived from the relations $(E\xi(t)=0)$

$$(3.2.4) \qquad E(\frac{\xi(0)+\xi(T)}{2})^2 = \sigma_w^2\ \frac{1+e^{-\lambda T}}{4\lambda}\ ,$$

$$(3.2.5)\qquad E\left(\int_0^T \xi(t)dt\right)^2 = \int_0^T\int_0^T E\xi(t)\xi(s)dt\,ds = \int_0^T\int_0^T \frac{\sigma_w^2}{2\lambda} e^{-\lambda|t-s|}dt\,ds =$$

$$= \frac{\sigma_w^2}{\lambda^2}\left(T+\frac{e^{-\lambda T}-1}{\lambda}\right) = \frac{\sigma_w^2}{\lambda^3}\left(e^{-\lambda T}+\lambda T-1\right),$$

$$(3.2.6)\qquad E\left(\frac{\xi(0)+\xi(T)}{2}\right)\left(\frac{1}{T}\int_0^T \xi(t)\,dt\right) = \sigma_w^2\,\frac{1-e^{-\lambda T}}{2T\lambda^2}\quad .$$

On the basis of the introduction of this chapter we can state that $\hat{m}$ is a sufficient statistic and the unique efficient estimator of m. From the above results (3.2.4) - (3.2.6) it follows that under the conditions $T = 1$, $\sigma_w^2 = 1$ the estimator $\hat{m}_1 = \frac{\eta(0)+\eta(T)}{2}$ is better then the estimator $\hat{m}_2 = \frac{1}{T}\int_0^T \eta(s)\,ds$, when $0 < \lambda < 2$. The last statement means

$$(3.2.7)\qquad D^2\hat{m}_1 = \frac{1+e^{-\lambda}}{4\lambda} < D^2\hat{m}_2 = \frac{\lambda+e^{-\lambda}-1}{\lambda^3}, \qquad 0 < \lambda < 2,$$

which can be easily checked by direct calculation.

A common used estimator for the unknown expectation is the arithmetic mean value

$$(3.2.8)\qquad \frac{1}{T}\int_0^T \eta(t)\,dt \approx \frac{1}{N+1}\sum_0^N \eta\left(\frac{i\cdot T}{N+1}\right) = \bar{\xi}_N.$$

As the variance of $\bar{\xi}_N$ tends to the variance of $\frac{1}{T}\int_0^T \eta(t)\,dt$ the above (3.2.7) inequality shows that for a fixed λ the variance of $\bar{\xi}_N$ cannot monotonically decrease when $N \to \infty$, i.e., ξ_N has a minimum variance at some fixed $N_0(=N_0(\lambda))$.

3.3. The Unknown λ

Let us now suppose that $m = 0$. Since σ_w can exactly be estimated with probability 1 by observing $\xi(t)$ on an arbitrary time interval we assume that $\sigma_w = 1$. The transformation

$$(3.3.1)\qquad t' = t/T\ ,\ \xi'(\cdot) = \frac{\xi(\cdot)}{\sigma_w\sqrt{T}}\ ,$$

enables us to treat the special case $T = 1$ and $\sigma_w = 1$ only, instead of the general case; here $\lambda' = \lambda T = \kappa$ and therefore in the case of a known m the realizations of the process are characterized by only one parameter; this is independent of the choice of a time unit. In what follows we shall often assume that the transformation (3.3.1) has been made we shall simply write κ. In the theory of diffusion processes λ is called as the drift parameter, but in the theory of ordinary differential equations it is called damping (or decay) parameter. Let us recall the Randon-Nikodym derivative of the measure P_λ generated by $\xi(t)$ on the product of the real line and the space of all continuous functions defined on $[0,T]$ with respect to the product of the Lebesgue measure and the conditional Wiener measure P_w ($\sigma_w = 1$) (see Example 1 in §2.3.3 and (2.3.41)):

$$(3.3.2)\qquad \frac{dP_\lambda}{dP_w}(\cdot) = \sqrt{\frac{\lambda}{\pi}} \quad \exp\{-\lambda s_1^2 - \frac{1}{2}\lambda^2 T\, s_2^2 + \frac{1}{2}\lambda T\},$$

where

$$(3.3.3)\qquad s_1^2 = \frac{1}{2}[\xi^2(0) + \xi^2(T)],$$

$$s_2^2 = \frac{1}{T}\int_0^T \xi^2(t)\,dt.$$

Hence the maximum likelihood estimator $\hat{\lambda}$ of the parameter λ (the unique positive solution of the likelihood equation) can be written as

$$(3.3.4)\qquad \lambda = \frac{-[s_1^2 - 1/2T] + \sqrt{(s_1^2 - \frac{1}{2}T)^2 + Ts_2^2}}{2\,Ts_2^2} .$$

In order to determine the probability distribution function of the estimator λ we give an explicit formula for the joint characteristic function $\psi(\alpha_1,\alpha_2)$ of the sufficient statistics s_1^2, Ts_2^2:

Theorem 1. Let us introduce the notations $\kappa = \lambda T$ and $\Lambda = \sqrt{\kappa^2 - 2T^2 i\alpha_2}$, then

$$(3.3.5)\qquad \psi(\alpha_1, \alpha_2) = E\exp(i\alpha_1, s_1^2 + i\alpha_2 s_2^2) =$$

$$= \frac{2\sqrt{\kappa}\; e^{\kappa/2}\sqrt{\Lambda}}{[(\kappa - i\alpha_1 T + \Lambda)^2 e^{\Lambda} - (\kappa - i\alpha_1 T - \Lambda)^2 e^{-\Lambda}]^{1/2}}$$

Proof. Let $u(T,x)$ be the following conditional characteristic function:

$$u(T,x) = E\{\exp(i\alpha_1 s_1^2 + i\alpha_2 Ts_2^2) \mid \xi(0) = x\}.$$

Then the function

$$u_1(T,x) = u(T,x)\, e^{-i\alpha_1\frac{x^2}{2}}$$

is a functional of the diffusion process $\xi(t)$ satisfying the conditions of Kolmogorov equations (see Gihman, Skorokhod [1]), consequently the following equation holds:

$$(3.3.6)\qquad \frac{\partial u_1}{\partial T} = \frac{1}{2}\,\frac{\partial^2 u_1}{\partial x^2} - \lambda x\,\frac{\partial u_1}{\partial x} + x^2\, i\alpha_2 u_1,$$

with the boundary condition

$$u_1(0,x) = e^{i\,\alpha_1\frac{x^2}{2}}$$

The function $v(T,\gamma) = \int_{-\infty}^{\infty} e^{-\gamma x^2} u_1(T,x)\,dx$ satifies the equation

$$(3.3.7)\qquad \frac{\partial v}{\partial T} = -\frac{\partial v}{\partial \gamma}\,[2\,\gamma^2 - 2\lambda\gamma + i\alpha_2] + (\lambda-\gamma)\,v,$$

with the boundary condition

$$v(0,\gamma) = [\gamma - \frac{i\alpha_1}{2}]^{-1}\,.$$

Let us observe that

$$(3.3.8)\qquad \psi(\alpha_1,\alpha_2) = v(T,\lambda - \frac{i\alpha_1}{2}).$$

By setting

$$z(s,\gamma) = \int_0^{\infty} e^{-sT}\, v(T,\gamma)\, dT,$$

equation (3.3.7) can be rewritten

$$(3.3.9)\qquad s\cdot z(s,\gamma) - \frac{1}{\gamma - \frac{i\alpha_1}{2}} = -\frac{\partial z}{\partial \gamma}[2\gamma^2 - 2\lambda\gamma + i\alpha_2] + (\lambda-\gamma)\,z(s,\gamma),$$

i.e.,

$$(3.3.10)\qquad \frac{\partial z}{\partial \gamma} = z\,\frac{\lambda-\gamma-s}{2\gamma^2 - 2\lambda\gamma + i\,\alpha_2} + \frac{1}{(\gamma - \frac{i\alpha_1}{2})(2\gamma^2 - 2\lambda\gamma + i\alpha_2)}\,.$$

The solution of equation (3.3.10) is

$$z(s,\gamma) = \exp\{\int_0^{\gamma} \frac{\lambda-y-s}{2y^2 - 2\lambda y + i\alpha_2}\,dy\}[c - \int_0^{\gamma} \frac{1}{u - \frac{i\alpha_1}{2}}\,\exp\{\int_0^{u} \frac{\lambda-y-s}{2y^2 - 2\lambda y + i\alpha_2}\,dy\}du].$$

Let γ_1,γ_2 be the roots of equation

$2\gamma^2 - 2\lambda\gamma + i\alpha_2 = 0$, then

(3.3.11) $$z(s,\gamma) = \exp\{ a(s,\gamma_1,\gamma_2)\ln(\gamma-\gamma_1)+b(s,\gamma_1,\gamma_2)\ln(\gamma-\gamma_2)\}\cdot$$
$$\cdot\Big[c- \int_0^{\gamma} \frac{1}{u-\frac{i\alpha_1}{2}} \exp\{a\cdot\ln(u-\gamma_1)-b\cdot\ln(u-\gamma_2)\}du\}\Big],$$

where

$$a(s,\gamma_1,\gamma_2) = -\frac{1}{2}\Big(1+\frac{\lambda+\gamma_2-s}{\gamma_1+\gamma_2}\Big), \quad b(s,\gamma_1,\gamma_2) = -\frac{1}{2}\,\frac{\lambda+\gamma_2-s}{\gamma_1+\gamma_2} .$$

From (3.3.11) we get an explicit formula for $v(T,\gamma)$

(3.3.12) $$v(T,\gamma) = \frac{(\gamma_1-\gamma_2)^{1/2} e^{1/2(\lambda T-T(\gamma_1-\gamma_2))}}{\Big[(\gamma-\gamma_2)(\gamma_1-\frac{i\alpha_1}{2})+(\gamma-\gamma_1)(\frac{i\alpha_1}{2}-\gamma_2)\, e^{-2T(\gamma_1-\gamma_2)}\Big]^{1/2}}.$$

This formula together with relation (3.3.8) proves the Theorem.

If we know the probability distribution of the family of random variables $\zeta_\lambda = \lambda x^2 T s_2^2+\lambda x s_1^2$, then the probability distribution of the estimator $\hat{\lambda}$ can be easily determined since

$$P_\lambda\ \{\hat{\lambda}>\lambda x\} = P_\lambda\{\zeta_\lambda < 1/2 + 1/2\ T\lambda x\} .$$

<u>Corollary.</u> For T= 1 the distribution of the random variable has the following characteristic function

$$\psi_{\zeta_\lambda(x)}(\alpha) =$$

(3.3.13) $$= \frac{2\, e^{\lambda/2}[1-2i\alpha x^2]^{1/4}}{\Big[(1-i\alpha x+\sqrt{1-2i\alpha x^2})^2 e^{\lambda\sqrt{1-2i\alpha x^2}}-(1-i\alpha x-\sqrt{1-2i\alpha x^2})^2 e^{-\lambda\sqrt{1-2i\alpha x^2}}\Big]^{1/2}}$$

<u>Theorem 2. The random variable</u> $2\zeta_\lambda|x$ <u>has an asymptotic distribution</u> $\chi^2(1)$ <u>when</u> $\lambda \to 0$.

<u>Proof.</u> From formula (3.3.13) by direct calculation we get

$$\psi_{\zeta_\lambda}(\alpha)\to \frac{1}{(1-i\alpha x)^{1/2}}, \text{ when } \lambda \to 0.$$

This proves the theorem, which means that

$$\lim_{\lambda\to 0} P\{\frac{2\zeta_\lambda}{x} < z^2\} = \sqrt{\frac{2}{\pi}} \int_0^z e^{-\frac{n^2}{2}}\, dn.$$

Table 1.

In the table the quantiles $z_p(\lambda)$ of the maximum likelihood estimator $\hat{\lambda}$ are given, i.e. $P_\lambda\{\hat{\lambda} > z_p\} = p$. In brackets the values $x = \frac{z}{\lambda}$ are given.

λ \ p	0,001	0,01	0,025	0,05	0,1	0,9	0,95	0,975	0,99	0,999
0	0 (637000)	0 (6370)	0 (1020)	0 (255,0)	0 (63,60)	0 (0,369)	0 (0,260)	0 (0,199)	0 (0,151)	0 (0,092)
0,01	10,60 (1060)	4,232 (423,2)	2,274 (227,4)	1,170 (117,0)	0,4734 (47,34)					
0,05	11,195 (263,9)	6,330 (126,6)	4,0065 (80,13)	2,5130 (50,26)	1,3375 (26,75)					
0,1	14,38 (143,8)	7,344 (73,44)	4.879 (48,79)	3,268 (32,68)	1,908 (19,08)					
0,2	15,664 (78,32)	8.468 (42,34)	5,902 (29,51)	4,154 (20,77)	2,624 (13,12)					
0,3	16,488 (54,96)	9.207 (30,69)	6,561 (21,87)	4,746 (15,82)	3,120 (10,40)					
0,4	17,080 (42,70)	9,756 (24,39)	7,080 (17,70)	5,208 (13,02)	3,517 (8,793)					
0,5	17,670 (35,34)	10,230 (20,46)	7,515 (15,03)	5,605 (11,21)	3.8610 (7,722)	0,2085 (0,417)	0,1510 (0,302)			
0,6	18,108 (30,18)	10,638 (17,73)	7,896 (13,16)	5,9532 (9,922)	4,1676 (6,946)					
0,7	18,522 (26,46)	11,011 (15,73)	8,239 (11,77)	6,2713 (8,959)	4,4471 (6,353)					
0,8	18,896 (23,62)	11,360 (14,20)	8,560 (10,70)	6,5648 (8,206)	4,7103 (5,887)					
0,9	19,260 (21,40)	11,682 (12,98)	8,8587 (9,843)	6,8409 (7,601)	4,9446 (5,494)					
1	19.60 (19,60)	11,98 (11,98)	9,140 (9,140)	7,103 (7,103)	5,188 (5.188)	0,445 (0,445)	0,332 (0,332)	0,269 (0,269)	0,205 (0,205)	0,130 (0,130)
1,5	21,060 (14,04)	13,3080 (8,872)	10,3845 (6,923)	8,2590 (5,506)	6.2325 (4,155)	0,7005 (0,467)	0,5325 (0,355)	0,4275 (0,285)	0,3360 (0,224)	0,2145 (0,143)
2	22,32 (11,16)	14,462 (7,231)	11,426 (5,713)	9,272 (4,636)	7,156 (3,278)	0,972 (0,486)	0,750 (0,375)	0,606 (0,303)	0,480 (0,240)	0,310 (0,155)
2,5	23,4750 (9,390)	15,515 (6,206)	12,4575 (4,983)	10,2050 (4,082)	8,0100 (3,204)	1,2575 (0.503)	0,9850 (0,394)	0,8000 (0,320)	0,6500 (0,256)	0,4125 (0,165)
3,	24,342 (8,114)	16,506 (5,502)	13,389 (4,463)	11,082 (3,694)	8,811 (2,937)	1,557 (0,519)	1,233 (0,411)	1,111 (0,337)	0,807 (0,269)	0,525 (0,175)
3,5	25,5850 (7,310)	17,4440 (4,984)	14,2800 (4,080)	11,9105 (3,403)	9,5970 (2,742)	1,8655 (0,533)	1,4910 (0,426)	1,2355 (0.353)	0,9975 (0,285)	0,6405 (0,183)
4	26,576 (6,644)	18,352 (4,588)	15,136 (3,784)	12,732 (3,183)	10,352 (2,588)	2,180 (0,545)	1,760 (0,440)	1,468 (0,367)	1,192 (0,298)	0,776 (0,194)
4,5	27,5355 (6,119)	19,2285 (4,273)	15,9705 (3,549)	13,5225 (3,005)	11,0835 (2,463)	2.5020 (0,556)	2,0430 (0.454)	1,7100 (0,380)	1,3905 (0,309)	0,9135 (0,203)

λ \ p	0,001	0,01	0,025	0,05	0,1	0,9	0,95	0,975	0,99	0,999
5	28,470 (5,694)	20,090 (4,018)	16,795 3,359)	14,395 (2,879)	11,755 (2,351)	2,835 (0,567)	2,325 (0,0465)	1,965 (0,393)	1,615 (0,323)	1,070 (0,214)
5,5	29,3865 (5,343)	20,9275 (3,805)	17,5835 (3,197)	15,0535 (2,737)	12,5125 (2,275)	3,1680 (0,576)	2,6235 (0,477)	2,4640 (0,448)	1,8315 (0,333)	1,2265 (0,223)
6	30,318 (5,053)	21,750 (3,625)	18,366 (3,061)	15,792 (2,632)	13,282 (2,212)	3,510 (0,585)	2,922 (0,487)	2,490 (0,415)	2,061 (0,347)	1,392 (0,232)
6,5	31,1610 (4,794)	22,5615 (3,471)	19,1295 (2,943)	16,5295 (2,543)	13,8905 (2,137)	3,8545 (0,593)	3,2305 (0,497)	2,7690 (0,426)	2.3140 (0,356)	1,5730 (0,242)
7	32,025 (4,575)	23,359 (3,337)	19,894 (2,842)	17.248 (2,464)	14,574 (2,082)	4,207 (0,601)	3.542 (0,506)	3,052 (0,436)	2,555 (0,365)	1,764 (0,252)
7,5	32,8882 (4,385)	24,1500 (3,220)	20,6475 (2,753)	17,9550 (2,394)	15,2475 (2,033)	4,5600 (0,608)	3,8550 (0,514)	3,3375 (0,445)	2,8200 (0,376)	1,9575 (0,261)
8	33,728 (4,216)	24,920 (3,115)	21,384 (2,673)	18,672 (2,334)	15,912 (1,989)	4,920 (0,615)	4,176 (0,522)	3,632 (0,454)	3,088 (0,386)	2,168 (0,271)
8,5	34,5610 (4,066)	25,6870 (3,022)	22,1170 (2,602)	19,3715 (2,279)	16,5750 (2,950)	5,2700 (0,620)	4,5050 (0,530)	3,9270 (0,462)	3,3490 (0,394)	2,3630 (0,278)
9	35,388 (3,932)	26,451 (2,939)	22,689 (2,521)	20,070 (2,230)	17,226 (1,914)	5,643 (0,627)	4,842 (0,538)	4,230 (0,470)	3,618 (0,402)	2,592 (0,288)
9,5	36,1855 (3,809)	27,1700 (2,860)	23,5790 (2,482)	20,7480 (2,184)	17,8790 (1,882)	6,0135 (0,633)	5,1870 (0,546)	4.5315 (0,477)	3,8950 (0,410)	2,812 (0,296)
10	37,04 (3,704)	27,55 (2,755)	24,28 (2,428)	21,47 (2,147)	18,53 (1,853)	6,38 (0,638)	5,50 (0,550)	4,84 (0,484)	4,20 (0,420)	3,04 (0,304)
20	52,200 (2,610)	42,040 (2,102)	37,800 (1,890)	34,4360 (1,7218)	30,8960 (1,5448)	14,1780 (0,7089)	12,7140 (0,6357)	11,580 (0,579)	10,380 (0.519)	8,320 (0,416)
30	66,270 (2,209)	55,230 (1,841)	50,520 (1,684)	46,7370 (1,5579)	42,7080 (1,4236)	22,4310 (0,7477)	20,4960 (0,6832)	18,960 (0,632)	17,340 (0,578)	14,400 (0,480)
40	79,800 (1,995)	67,920 (1,698)	62,800 (1,570)	58.6800 (1.4670)	54,2320 (1,3558)	30,9400 (0,7735)	28,5920 (0,7148)	26,720 (0,668)	24,680 (0.617)	21,000 (0,525)
50	92,900 (1,858)	80,3200 (1,6064)	74,8400 (1,4968)	70,3950 (1,4079)	65,5800 (1,3116)	39,6150 (0,7923)	36,9000 (0,7380)	34,7100 (0,6972)	32,350 (0,647)	27,950 (0,559)
60	105,780 (1,763)	92,520 (1,542)	86,6820 (1,4447)	81,9540 (1,3659)	76,7940 (1,2799)	48,4140 (0,8069)	45,3660 (0,7561)	42,8880 (0,7148)	40.200 (0,670)	30,300 (0,585)
70	118,370 (1,691)	104,510 (1,493)	98,3770 (1,4054)	93,3800 (1,3340)	87,9130 (1,2559)	57,3020 (0,8186)	53,9560 (0,7708)	51,2120 (0,7316)	48,2300 (0,689)	42,490 (0,607)
80.	130,800 (1,635)	116,40 (1,455)	109,960 (1,3745)	104,7120 (1,3089)	98,9600 (1,2370)	66,2722 (0,8284)	62,6240 (0,7828)	59,3320 (0,7454)	56,400 (0,0705)	50,080 (0,326)
90	143,100 (1,590)	128,160 (1,424)	121.4550 (1,3495)	115,6950 (1,2885)	109,9350 (1,2215)	75,2940 (0,8366)	71,3880 (0,7932)	68,1570 (0,7573)	64,620 (0,718)	57,870 (0,643)
100	155,32 (1,5532)	139,79 (1,3979)	132.86 (1,3286)	127,15 (1,2715)	120,86 (1,2086)	84,37 (0,8737)	80,21 (0.8021)	76,8 (0,768)	73,0 (0,730)	65,7 (0,657)
500	609,000 (1,218)	580,000 (1,160)	567,0500 (1,1341)	555,800 (1.1116)	543.15 (1,0833)	462.05 (0,9241)	451,4500 (0.9029)	442,600 (0,8852)	432,600 (0,8652)	412,00 (0,824)
1000	1149 (1,149)	1110,6 (1,1106)	1092,6 (1.0926)	1077.3 (1,0773)	1060.00 (1,0600)	945,3 (0,9453)	929.9 (0.9299)	917,00 (0,9170)	902.3 (0,9023)	872,00 (0,872)
10 000	10477 (1,0477)	10336 (1,03360)	10282,1 (1,02821)	10236,3 (1,02333)	10183,9 (1,01839)	9821,4 (0,98214)	9771,1 (0.97711)	9727,6 (0,97276)	9377,5 (0,93775)	9274 (0,9574)

Theorem 3. $\kappa = \lambda T$ has an asymptotic normal distribution when $\kappa \to \infty$, i.e.,

$$\lim_{\chi \to \infty} P\{\hat{\kappa} < \kappa + z\sqrt{\kappa}\} = \frac{1}{\sqrt{2\pi}} \int_{-\infty}^{z} e^{-n^2/2}\, dn, \quad |\psi_\zeta(\alpha) - e^{-\alpha^2/2}| < \frac{c}{\sqrt{\kappa}},$$

with parameters $E\hat{\kappa} - \kappa = 0$, $D^2\hat{\kappa} = \kappa$, $\zeta = \frac{\hat{\kappa} - \kappa}{\sqrt{\kappa}}$.

The statement of this theorem immediately follows from formula (3.3.13). In Table 1 we give the quantiles $z_p(\kappa)$ of the maximum likelihood estimator $\hat{\kappa}$, i.e.,

$$P_\kappa\{\hat{\kappa} > z_p\} = p.$$

The distribution function is tabulated for p = 0.001, 0.01, 0.025, 0.05, 0.1, 0.9, 0.95, 0.975, 0.99, 0.999 and for κ = 0.01, 0.05, 0.1-1, (by stepsize 0.1), 1-10 (by stepsize 0.5), 10-100 (by stepsize 10) and κ = 500, 1000, 10000.

Computations of the values of distribution function where carried out by the inverse Laplace transform formula. The calculation for one integral takes time 1-25 minutes on computer URAL-2. The programs were written in Assembler Language. For given κ and p the quantiles $z_p(\kappa)$ were calculated by iteration. Here we do not give the numerical method by means of which Table 1 was obtained.

For values of $\kappa = \lambda T$ which are netiher too large nor too small we must use the statistics S_1^2, S_2^2 to estimate κ. Confidence intervals for κ can be constructed by using (3.3.4) and determing the distribution of the random variable $\kappa^2 S_2^2/y^2 + \kappa S_1^2/y$ as we did it. For an arbitrarily chosen level p and for κ the equation

$$P_\kappa\{\hat{\kappa} > z\} = p,$$

has a unique solution $z = \psi_p(\kappa)$. Its inverse function

$$\psi^{-1}(z) = \Phi_p(z)$$

can also be uniquely determined and therefore gives the limits of a confidence interval, which means that identically in κ

$$P_\kappa\{\kappa \leq \Phi_p(\hat{\kappa})\} \equiv p.$$

For $\kappa \to \infty$ or $\kappa \to 0$, these limits are determined by the corresponding limit distribution (see Theorems 2 and 3). Note that the normal approximation works only when $\lambda \sim 1000$!

Example 1. Using Table 1 when T = 150 and $\hat{\lambda}=0.02$ ($\hat{\kappa}=3.0$) we get the following symmetric (in p) confidence limits on p = 0.05 level (using linear interpolation in columns p = 0.025 and p =0.975).

$$\kappa_{\text{Lower bound}} = 0.03, \qquad \kappa_{\text{Upper bound}} = 7.00$$

At the same time the normal approximation gives

$$\kappa_{\text{Lower bound}} = 3 - 1.96\sqrt{3} = -0.39, \quad \kappa_{\text{Upper bound}} = 6.39,$$

where the lower bound does not give anything as $\kappa > 0$ must be in the stationary case.

Example 2. Using Table 1 we get for the empirical value $\hat{\rho} = 0.5$ with n = 40 (number of experiments) the following symmetric (in p) confidence limits on p = 0.05 level (assuming $\lambda = -n \log \rho$):

$$\rho_{\text{Lower bound}} = 0.375 , \qquad \rho_{\text{Upper bound}} = 0.684$$

The same limits by normal approximation are: 0.301 and 0.914. In case $\hat{\rho} = 0.9$ and $n = 40$ we get (on the same level)

$$\rho_{\text{Lower bound}} = 0.819 , \qquad \rho_{\text{Upper bound}} = 0.995$$

The normal approximation gives 0.780 and 1.02 and the upper bound is unacceptable in the stationary case.

3.4 Two Unknown Parameters

In this section we shall prove that if λ, m both are unknown we cannot estimate them well, in the sense that λ has lower confidence limits which are 0 and the variance of m is great. When the parameters m and λ are unknown, we can rewrite the Radon-Nikodym derivative in the following form (see (2.3.41))

$$(3.4.1) \quad \frac{dP_\lambda}{dP_w}(\cdot)=\sqrt{\frac{\lambda}{\pi}} \cdot \frac{1}{\sigma_w} \exp\{-\frac{\lambda}{\sigma_w^2}[\, s_1^2+ \frac{1}{2}\kappa s_2^2+(m-m_1)^2+\frac{\kappa}{2}(m-m_2)^2-\frac{1}{2}\sigma_w^2 T]\},$$

where $\kappa = \lambda T$,

$$(3.4.2) \quad m_1 = \frac{\xi(0)+\xi(T)}{2} , \qquad m_2 = \frac{1}{T}\int_0^T \xi(t)\, dt,$$

$$s_1^2 = \frac{1}{2}\{[\xi(0)-m_1]^2+[\xi(T)-m_1]^2\} = \frac{1}{4}[\xi(T)-\xi(0)]^2,$$

$$s_2^2 = \frac{1}{T}\int_0^T [\xi(t)-m_2]^2\, dt.$$

From (3.4.1) we can conclude that the system m_1, m_2, s_1^2, s_2^2 forms a sufficient statistic.

In the same way as in 3.3 we can get the characteristic function of random variables (assuming $m=0$) m_1, m_2, $\chi_1^2= \frac{1}{2}(\xi^2(0)+ \xi^2(T))$, $\chi_2^2 = \int_0^T \xi^2(t)dt$,

$$(3.4.3) \quad \psi(\alpha_1,\alpha_2,\alpha_3,\alpha_4)=E \exp\{i(\alpha_1 m_1+\alpha_2\chi_1^2+\alpha_3 m_2+\alpha_4\chi_2^2)\} =$$

$$= \frac{2\sqrt{\lambda}\, e^{\kappa/2}\, \Lambda^{1/2}}{\psi(T\ (\alpha_2,\alpha_4))^{1/2}} \exp \frac{1}{2}\{- \frac{\alpha_1\alpha_3 +\alpha_3^2}{\Lambda}-\sigma_w^2 T + (\frac{i\alpha_1}{2} -T\sigma_w^2\frac{\alpha_2\alpha_3 + \frac{i\alpha_3\lambda}{\sigma_w^2}}{\Lambda}) \ .$$

$$\cdot \Big[\Big(\frac{i\alpha_1\sigma_w^2}{2} (1+e^{-\Lambda}) - i\alpha_3\sigma_w^2 \frac{1-e^{-\Lambda}}{\Lambda}\Big) \frac{(\kappa - T\sigma_w^2 i\alpha_2 + \Lambda)e^{\Lambda} - (\kappa - T\sigma_w^2 i\alpha_2 - \Lambda)}{T\psi(\alpha_2,\alpha_4)} +$$

$$+ \Big(\frac{i\alpha_1\sigma_w^2}{2}(1+e^{\Lambda}) - i\alpha_3\sigma_w^2 \frac{1-e^{\Lambda}}{\Lambda}\Big) \frac{(\kappa - T\sigma_w^2 i\alpha_2 + \Lambda) - (\kappa - T\sigma_w^2 i\alpha_2 - \Lambda)e^{-\Lambda}}{T\psi(\alpha_2,\alpha_4)} \Big]\Big\},$$

where

$$\Lambda = \sqrt{\kappa^2 - 2T^2\, \sigma_w^2\, i\alpha_4},$$

(3.4.4) $$\psi(\alpha_2,\alpha_4) = \frac{1}{T^2}\left[(\kappa - T\sigma_w^2 i\alpha_2 + \Lambda)^2 e^{\Lambda} - (\kappa - T\sigma_w^2 i\alpha_2 - \Lambda)^2\, e^{-\Lambda}\right].$$

If $\alpha_1 = \alpha_3 = 0$ we get Theorem 1 in §3.3. The characteristic function of m_1 and m_2 is

(3.4.5) $$\psi(\alpha_1,\alpha_3) = \exp\{-\tfrac{1}{2}(\alpha_1^2 \frac{1+e^{-\kappa}}{4\lambda}\sigma_w^2 + \alpha_1\alpha_3 \frac{1-e^{-\kappa}}{T\lambda^2}\sigma_w^2 + \alpha_3^2 \frac{T\lambda + e^{-\kappa} - 1}{\kappa^2\lambda}\sigma_w^2)\}$$

which may be compared by (3.2.4) - (3.2.6). Further when $\kappa = \lambda T \to 0$, the characteristic function of $(\sqrt{\lambda} m_1, \sqrt{\lambda}\, m_2, \lambda\, \chi_1^2, \lambda\, \chi_2^2)$ has the form

(3.4.6) $$\frac{(1+\frac{\kappa}{2})\quad e^{-\frac{(\alpha_1+\alpha_3)^2}{2(1-\sigma_w^2 i\alpha_2)}\sigma_w^2 - \frac{\kappa\sigma_w^2}{12}\alpha_3^2}}{\{1-\sigma_w^2 i\alpha_2 + \frac{\kappa}{2}[(1-\sigma_w^2 i\alpha_2)^2 + 1 - 2\alpha_w^2 \frac{i\alpha_4}{T}]\}^{1/2}} + \sigma(\kappa),$$

and therefore when $\kappa \to 0$ the random variables m_1 and s_1^2 form an asymptotically sufficient statistic.

The maximum likelihood equations of m and λ are as follows:

(3.4.7) $$\frac{\sigma_w^2}{2\lambda}(1+\lambda T) - s_1^2 - \lambda T s_2^2 - (m_1 - m)^2 - \lambda T(m_2 - m)^2 = 0,$$

(3.4.8) $$2(m - m_1) + \lambda T(m - m_2) = 0.$$

From (3.4.8) we see that the maximum likelihood estimators $\hat{m}$, $\hat{\kappa} = \hat{\lambda} T$ are related by

$$\hat{m} = \frac{2m_1 + \hat{\kappa} m_2}{2 + \hat{\kappa}}$$

but the solution for κ is a root of a cubic equation.

Theorem 1. For $\kappa = \lambda T \to \infty$ the estimators m_2 and $\hat{\lambda} = \frac{\sigma_w^2}{2s_2^2}$ (see (3.4.2)) of m and λ respectively are simultaneously efficient, and the distribution function of the random vector

(3.4.9) $$\frac{m_2 - m}{\sqrt{\frac{2\sigma_w^2}{\lambda\kappa^2}(e^{-\kappa}+\kappa-1)}}, \qquad \frac{\hat{\lambda}-\lambda}{\sqrt{\frac{\lambda}{T}}},$$

tends to the normal distribution with parameters $(0,\ 0,\ \begin{pmatrix}1 & 0\\ 0 & 1\end{pmatrix})$.

Proof. Simple calculations yield

$$E\, m_2 = m, \qquad E(m_2-m)^2 = \frac{\sigma_w^2(\kappa+e^{-\kappa}-1)}{\lambda\kappa^2},$$

(3.4.10) $$E\, s_2^2 = \frac{\sigma_w^2}{2\lambda} - \frac{\sigma_w^2}{\lambda\kappa}\left[1+\frac{1}{\kappa}(e^{-\kappa}-1)\right],$$

$$E(s_2^2-E\, s_2^2)^2 = \frac{\sigma_w^4}{4\lambda^2\kappa}\{1+\frac{1}{\kappa}(e^{-2\kappa}-1) + \frac{8}{\kappa^3}(\kappa+e^{-\kappa}-1)^2 -$$

$$- \frac{4}{\kappa^2}(4\kappa+2\kappa e^{-\kappa}-7+8e^{-\kappa}-e^{-2\kappa})\}.$$

The random variables m_2 and $\frac{1}{T}\int_0^T(\xi(t)-m)^2\,dt$ tend to m and $\sigma_\xi^2 = \frac{\sigma_w^2}{2\lambda}$ respectively when $\kappa = \lambda T \to \infty$ and from (3.4.3) we have that m_2-m, $\frac{1}{T}\int_0^T(\xi(t)-m)^2\,dt$ has the characteristic function (if $\kappa \to \infty$)

(3.4.11) $$\exp\{-\frac{1}{2}\frac{\alpha_3^2\sigma_w^2}{\kappa\lambda} + o(\frac{1}{\kappa})\}\exp\{i\alpha_4\frac{\sigma_w^2}{2\lambda} - \frac{1}{2}\alpha_4^2\frac{\sigma_w^4}{4\lambda^2\kappa} + o(\frac{1}{\kappa})\}$$

which shows that they are asymptotically normally distributed and independent. According to a lemma of Cramer (see Cramer [1], §33.3) the asymptotic distributions of the vectors

(3.4.12) $$\frac{m_2 - m}{\sqrt{\frac{\sigma_w^2}{\lambda\kappa^2}(\kappa+e^{-\kappa}-1)}}, \qquad \frac{\frac{1}{T}\int_0^T(\xi(t)-m)^2dt - \frac{\sigma_w^2}{2\lambda}}{\frac{\sigma_w^2}{\sqrt{4\lambda^2\kappa}}}$$

and

(3.4.13) $$\frac{(m_2 - m)\sqrt{\kappa}}{\frac{\sigma_w}{\sqrt{\lambda}}}, \qquad \frac{s_2^2 - \frac{\sigma_w^2}{2\lambda}}{s_2^2\sqrt{\frac{1}{\kappa}}}$$

coincide, which proves the theorem.

In statistical investigations of independent random variables (observations) it is well known that if $\xi_1, \xi_2, \ldots, \xi_n$ are normally distributed with parameters (m, σ^2), where both of them are unknown, then with an arbitrary degree of confidence, a finite confidence interval can be constructed, e.g., by the help of t-statistics (see Cramer [1], p. 653). This means that there exists functions $\overline{h}(x_1, \ldots, x_n)$ and $\underline{h}(x_1, \ldots, x_n)$, also for an arbitrary degree of confidence $\beta > 1/2$, for which

$$P\{\overline{h}(\xi_1, \ldots, \xi_n) \geq m\} \geq \beta, \quad P\{\underline{h}(\xi_1, \ldots, \xi_n) < m\} \geq \beta$$

holds uniformly with respect to m and σ. The functions $\overline{h}(\cdot)$ and $\underline{h}(\cdot)$ are independent of σ.

This is not the situtation in the case of stationary Gaussian Markov process. We assume that T = 1 and $\sigma_w = 1$. Let us take positive functional $\kappa(\xi)$ for the lower confidence limig of κ, and $\overline{\mu}(\xi)$, $\underline{\mu}(\xi)$ real-valued functionals as upper and lower confidence limits for m. We assume that all these functionals are continuous on R_ξ in the C[0,1] metric, but $\overline{\mu}$ and $\underline{\mu}$ may assume values $+\infty$ and $-\infty$. The continuity induced by the topology of the real line, closed by points $-\infty$ and ∞. First we have the following assertion, which says that no nonzero lower limit can be constructed for the parameter κ with any degree of confidence.

Theorem 2. Let $\beta > 0$, and let $\kappa(\xi)$ be a positive functional defined in the space R_ξ and continuous in the C[0,1] metric, with the property that $\kappa(\xi) \to \infty$ if $\sup|\xi(t)| \to \infty$. Let it safisfy for any m and κ the condition $P\{\kappa \geq \kappa(\xi)\} \geq \beta$.
Then

$$P\{\kappa(\xi) = 0\} \geq g(\kappa, \beta), \tag{3.4.14}$$

where the positive function $g(\cdot)$ does not depend on the choice of functional and $g(\kappa, \beta) \to 1$ as $\kappa \to 0$.

For parameter m the following statement says that if m, κ are unknown it is impossible to construct finite confidence intervals using continuous functions. We assume that $\overline{\mu}$ and $\underline{\mu}$ has the property that for a real value c

$$\overline{\mu}(\xi + c) = \overline{\mu}(\xi) + c \text{ and } \underline{\mu}(\xi + c) = \underline{\mu}(\xi) + c.$$

Theorem 3. Let $\beta > 1/2$, and let $\underline{\mu}(\xi)$, $\overline{\mu}(\xi)$ be real valued functionals (which may assume values $-\infty$ or $+\infty$) on the space R_ξ, which are continuous in the C[0,1] metric and which satisfy the conditions

(3.4.15)
$$P\{m \geq \underline{\mu}(\xi)\} \geq \beta, \qquad P\{m < \overline{\mu}(\xi)\} \geq \beta,$$

for any m and κ $(-\infty < m < \infty,\ \kappa > 0)$. Then

(3.4.16)
$$P\{\overline{\mu}(\xi) = \infty\} \geq f(\kappa,\beta), \qquad P\{\underline{\mu}(\xi) = -\infty\} \geq f(\kappa,\beta),$$

where $f(\kappa,\beta)$ does not depend on the choice of these functionals, and $f(\kappa,\beta) \to 1/2$ as $\kappa \to 0$.

Before the proof we give some simulation results to illustrate the situation and to have a picture on function $g(\kappa,\beta)$. Let us take the following estimators ($T = 1$, $\sigma_w^2 = 1$)

$$\tilde{m}_1 = \frac{1}{N}\sum_1^N \xi_i, \qquad \tilde{\lambda}_1 = \frac{1}{\frac{2}{N}\sum_1^N(\xi_i - \tilde{m}_1)^2},$$

(3.4.17) $\tilde{m}_2$, $\tilde{\lambda}_2$ the maximum likelihood estimators,

$$\tilde{m}_3 = \frac{\xi(0) + \xi(1)}{2}, \qquad \tilde{\lambda}_3 = \frac{2}{(\xi(1) - \xi(0))^2},$$

where $\xi_i = \xi(\frac{i}{N})$ $(i=1,2,\dots N)$, $\xi_0 = \xi(0)$. N was taken between 60 and 100 and n (the number of samples) was 1000. The simulation was carried out on CDC 3300 (for details see Benczúr and Arató, 1972). In Tables 2-4 the empirical quantiles are given and as a first guess we find that e.g.,

$g(\kappa,0.05) \approx 1$ if $\kappa < 0.5$, (i.e., $P(\kappa(\xi) = 0) \approx 1$ if $\kappa \leq 0.5$ on level $\beta = 0.05$)
$g(\kappa,0.5) \approx 1$ if $\kappa < 4$,
$g(\kappa,0.9) \approx 1$ if $\kappa < 9$,
$g(\kappa,0.95) \approx 1$ if $\kappa < 12$, (i.e., $P(\kappa(\xi)=0) \approx 1$ if $\kappa \leq 12$ on level $\beta = 0.95$)

Plotting the quantiles z_p of Table 3 in Figure 1 we can easily find the lower confidence limit for a given κ. E.g., for $\kappa = 3.5$ we get $\kappa(\xi) = 0$, with probability 1, if $\beta > 0.5$ It seems that

$$g(\kappa,\beta) \approx e^{-c_\beta \kappa}, \text{ when } \kappa \to 0,$$

but it is not proved.

Table 2.

In the table the values z_p are given for which $P^{(n)}_{\lambda,m}\{\tilde{\lambda}_1>z_p\}=p$. m,λ are unknown and estimated.

λ \ p	0,05	0,1	0,9	0,95	Empirical mean
0,00001	12,58	10,04	1,15	0,80	5,09
0,0001	12,98	10,49	1,39	0,88	5,18
0,01	13,40	10,74	1,47	1,00	5,30
0,1	14,62	10,76	1,50	1,00	5,41
1	13,47	11,35	1,74	1,33	5,72
2	13,87	11,87	2,13	1,78	6,51
5	19,14	16,43	3,80	2,97	9,46
10	23,18	22,42	7,22	5,95	14,05

Table 3.

In the table the values z_p are given which $P^{(n)}_{\lambda,m}\{\tilde{\lambda}_2>z_p\}=p$. m,λ are unknown and estimated.

λ \ p	0,05	0,1	0,9	0,95	Empirical mean
0,00001	12,68	9,77	0,87	0,55	4,75
0,0001	13,25	10,24	0,94	0,59	4,86
0,01	13,85	10,41	0,96	0,61	5,00
0,1	13,95	10,87	0,99	0,68	5,12
1	14,33	11,34	2,40	1,01	5,63
2	14,78	12,24	1,86	1,38	6,54
5	19,94	17,36	3,61	2,89	9,56
10	26,74	23,61	7,40	6,38	14,74

Table 4.

In the table the values z_p are given for which $P^{(n)}_{\lambda,m}\{\tilde{\lambda}_3>z_p\}=p$. m,λ are unknown and estimted.

λ \ p	0,05	0,1	0,9	0,95	Empirical mean
0,00001	398	116	0,69	0,49	815
0,0001	423	128	0,71	0,50	3100
0,01	423	150	0,75	0,51	3200
0,1	710	179	0,85	0,57	5106
1	1168	326	1,11	0,79	10772
2	2039	426	1,58	1,23	-
5	3531	602	3,37	2,26	-
10	3950	1331	5,71	4,92	-

Proof of Theorem 3. Because of the symmetry it is sufficient to prove the statement for $\overline{\mu}(\xi)$. For a bounded functional $\overline{\mu}$ the inequality $P\{m < \overline{\mu}(\xi)\} \geq \beta$ cannot hold true for all m and κ, because when $\overline{\mu}(\xi) \leq K < \infty$ we have

$$P_{K,\kappa}\{K < \overline{\mu}(\xi)\} = 0$$

For sufficiently large values of c there exist $\xi_0(t) \geq -K > -\infty$, independent of $\overline{\mu}$, such that $\overline{\mu}(\xi) \leq c$ when $\xi(t) \leq \xi_0(t)$, for all $0 \leq t \leq 1$.

Let

$$\Gamma = \{\xi : \overline{\mu}(\xi) < c\}, \qquad \Gamma_1 = \{\xi : -\kappa^{-1+\delta} \leq \xi \leq \xi_0\},$$

where $0 < \delta < 1/2$. Evidently $\Gamma \supset \Gamma_1$, $P(\Gamma) \geq P(\Gamma_1)$ and

$$P_{\kappa,c}\{c < \overline{\mu}(\xi)\} = 1 - P_{\kappa,c}(\Gamma) \leq 1 - P_{\kappa,c}(\Gamma_1). \tag{3.4.18}$$

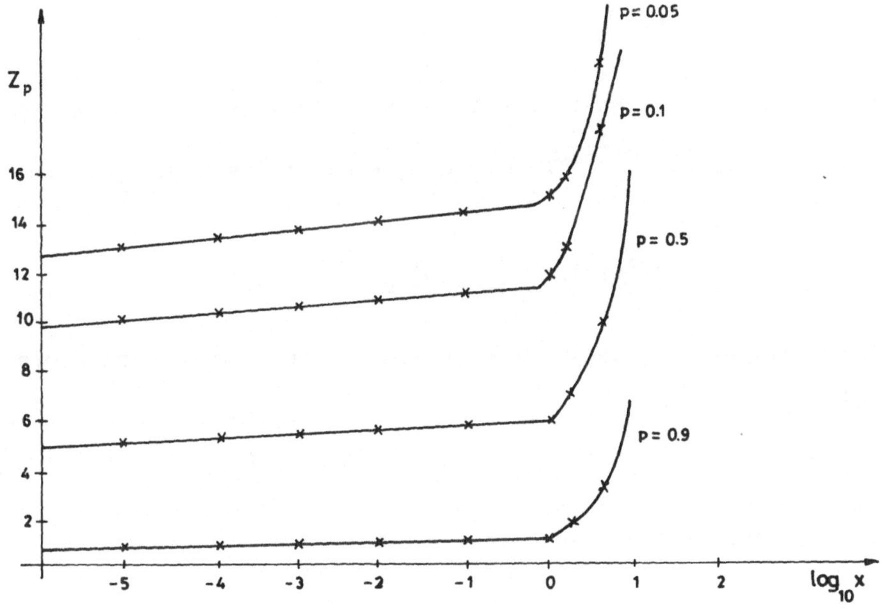

Figure 1. Empirical Quantiles of Maximum Likelihood Estimator

By using

$$\frac{dP}{dP_W} = \sqrt{\frac{\kappa}{\pi}} \exp\{ -\kappa(\xi_o-c)^2 - \frac{1}{2}[\kappa\{(\xi(1)-c)^2 - (\xi(0)-c)^2\} + - \kappa + \kappa^2 \int_o^1 (\xi(t)-c)^2 \, dt\},$$

we get

(3.4.19)

$$P_{\kappa,c}(\Gamma_1) = \int_{\Gamma_1} \frac{dP}{dP_W} dP_W \geq (1-\frac{\kappa^{2\delta}}{2}) \int_{\Gamma_1} \sqrt{\frac{\kappa}{\pi}} e^{-\kappa(x_o-c)^2 - \frac{\kappa}{2}\,(x_1-c)^2-(x_o-c)^2} \, dP_W,$$

Let

$$\Gamma_2 = \{\xi: -\kappa^{-1+\delta} \leq \xi \leq \xi_o, 0<t\leq 1; -\kappa^{-1+\delta}+\kappa^{-\varepsilon} < \xi(0) \leq \xi_o(0)-\kappa^{-\varepsilon}\},$$

where $0 < \varepsilon < \delta/2$, ξ is arbitrary, and

$$\Gamma_3 = \{ \xi: \sup_{0\leq t\leq 1} | \xi(t)-\xi(0)| < \kappa^{-\varepsilon}, -\kappa^{-1+\delta} + \kappa^{-\varepsilon} < \xi(0) \leq \xi_o(0)-\kappa^{-\varepsilon}\}.$$

By using the formula

$$\text{(3.4.20)} \qquad \int_{\Gamma_3} dP_W \geq 1 - 2\kappa^{\varepsilon} \sqrt{\frac{2}{\pi}} \, e^{-\frac{\kappa^{-2\varepsilon}}{2}},$$

which is valid for Wiener processes (see Appendix (B2.25)) we get the inequality

$$\text{(3.4.21)} \quad \int_{\Gamma} e^{-\frac{\kappa}{2}\,(x_1-c)^2-(x_o-c)^2} dP_W \geq e^{-\kappa^{-\varepsilon}(\kappa^{\delta}+|c|\kappa)} \, dP_W \geq$$

$$\geq e^{-\kappa^{-\varepsilon}(\kappa^{\delta}+|c|\kappa)} \, (1-2\kappa^{\varepsilon}\sqrt{\frac{2}{\pi}} \, e^{-\frac{\kappa^{-2\varepsilon}}{2}}).$$

Let $\Phi(x)$ denote the normal distribution function with parameters (0,1); then

$$\text{(3.4.22)} \quad \int_{\Gamma_2} \sqrt{\frac{\kappa}{\pi}} \, e^{-\kappa(x_o-c)^2} dx_o = \Phi\{\sqrt{2\kappa}(\xi_o(0)-c-\kappa^{-\varepsilon})\} - \Phi\{\sqrt{2\kappa}(-\kappa^{-1+\delta}-c+\kappa^{-\varepsilon})\}.$$

From (3.4.19), (3.4.21) and (3.4.22) we obtain

(3.4.23)

$$P_{\kappa,c}(\Gamma_1) \geq (1-\frac{\kappa^{2\delta}}{2})(1-\kappa^{\delta-\varepsilon})(1-2\kappa^{\varepsilon}\sqrt{\frac{2}{\pi}} e^{-\frac{\kappa^{-2\varepsilon}}{2}})[\Phi(\sqrt{2\kappa}(\xi_o(0) - c - \kappa^{-\varepsilon})\}-$$

$$-\Phi\{\sqrt{2\kappa}\ (-\kappa^{-1+\delta}-c+\kappa^{-\varepsilon})\}] .$$

Hence as $\kappa \to 0$ we have

$$P_{\kappa,c}\{ c < \bar{\mu}(\xi)\} = 1 - P_{\kappa,c}\{\Gamma_1\} \leq \frac{1}{2} + \varepsilon_0 \quad , \quad \kappa < \kappa_0(\varepsilon_0),$$

and for an arbitrarily small $\varepsilon_0 > 0$, which proves the theorem.

Theorem 3 can be reworded as follows: When the parameters m and κ of a stationary Gaussian Markov process are unknown, it is impossible to construct finite confidence intervals for m using continuous functionals.

Corollary. From the proof provided we can see that for any $\varepsilon > 0$ there exists $\Lambda(\varepsilon)$ such that for small values of κ

$$\sup_{m,\kappa<\kappa_0} P_{\kappa,m}\{ \bar{\mu}(\xi) > m \} \leq \frac{1}{2} + \Lambda\kappa_0^{1/2-\varepsilon}.$$

Thus we can make an estimate of the behavior of the function $f(\kappa,\alpha)$. The proof of Theorem 2 is similar to that of Theorem 3.

3.5 The Discrete Time Case

3.5.1. Single Parameters. In practice, even if continuous sampling is available, in order to compute stochastic integrals

$$\int_0^T \xi(t)\,dt, \qquad \int_0^T \xi^k(t)\,dt,$$

have to be approximated by appropriate finite sums on a equidistant partition d_n, $0 = t_0 < t_1 < \ldots < t_n = T$ of $[0,T]$. The natural, and essential, question is whether the resulting estimate of

$$\rho = e^{-\frac{\lambda T}{n}} = \frac{E(\xi(t_i)-m)(\xi(t_{i-1})-m)}{\sigma_\xi^2} ,$$

is the best one that could be obtained in view of discrete sampling? If $\hat{\rho}_n(T)$ is the likelihood equation estimate obtained by discrete sampling (and $\hat{\lambda}_n(T) = -\frac{n}{T}\log\hat{\rho}$) then if $n \to \infty$ we shall prove that

$$\hat{\lambda}_n(T) \to \lambda$$

in probability and

$$\sqrt{\frac{n}{T}}\,(\hat{\lambda}_n-\hat{\lambda}), \quad \sqrt{\frac{n}{T}}\,(\hat{\lambda}_n(T)-\lambda)$$

are bounded in probability where $\hat{\lambda}$ is the maximum likelihood estimator in the continuous time case (see Theorem 6 below).
Let

$$(3.5.1) \qquad \zeta(n+1) = \rho\zeta(n) + \varepsilon(n+1),$$

with $E\varepsilon(n) = E\zeta(n) = 0$, $\sigma_\varepsilon^2 = (1-\rho^2)\sigma_\zeta^2$. If $\xi(n) = \zeta(n)+m$ we have three unknown parameters (m, σ_ξ^2, ρ) or $(m, \sigma_\varepsilon^2, \rho)$. Using (2.3.2) we have the density function of a realization $\xi(1),\dots,\xi(n)$ in the form

$$(3.5.2) \quad f_{\xi(1),\dots,\xi(n)}(x_1,\dots,x_n) = (2\pi)^{-n/2}\sigma_\xi^{-n}(1-\rho^2)^{-\frac{n-1}{2}} \exp\{$$

$$-\frac{1}{2\,\sigma_\xi^2(1-\rho^2)}(\underline{x}-m)^* B_n^{-1}(\underline{x}-m)\},$$

$$(3.5.3) \qquad B_n^{-1} = \begin{pmatrix} 1 & -\rho & 0 & 0 & \dots & 0 & 0 \\ -\rho & 1+\rho^2 & -\rho & 0 & \dots & 0 & 0 \\ \cdot & & & & & & \\ \cdot & & & & & & \\ \cdot & & & & & & \\ 0 & 0 & 0 & 0 & \dots & 1+\rho^2 & -\rho \\ 0 & 0 & 0 & 0 & \dots & -\rho & 1 \end{pmatrix}$$

The conditional density function has the form

(3.5.4)
$$f(x_2,\dots,x_n \mid \xi(1)=x_1) = (2\pi)^{-\frac{n-1}{2}}\sigma_\varepsilon^{-(n-1)} \exp\{-\frac{1}{2\sigma_\varepsilon^2}\sum_2^n (x_i-\rho x_{i-1}-m(1-\rho))^2\}.$$

Before investigating maximum likelihood estimators, let us deal with the logarithmic derivative of the density. Let the unknown parameters be m, σ_ξ^2 and ρ, and let us introduce the following notation:

$$(3.5.5) \quad R_n^{(1)} = \frac{\partial \log f}{\partial m} = \frac{1}{\sigma_\xi^2}\{(x_1-m)+\frac{1}{1+\rho}\sum_2^n (x_i-m-\rho(x_{i-1}-m))\},$$

$$R_n^{(2)} = \frac{\partial \log f}{\partial \sigma_\xi^2} = -\frac{n}{2\sigma_\xi^2} +$$

$$+\frac{1}{2\sigma_\xi^4}[(x_1-m)^2 + \frac{1}{1-\rho^2}\sum_2^n (x_i-m-\rho(x_{i-1}-m))^2],$$

$$R_n^{(3)} = \frac{\partial \log f}{\partial \rho} = \frac{\rho(n-1)}{1-\rho^2} - \frac{\rho}{\sigma_\xi^2(1-\rho^2)} \sum_2^n [x_i-m-\rho(x_{i-1}-m)]^2 +$$

$$+ \frac{1}{\sigma_\xi^2(1-\rho^2)} \sum_2^n [x_i-m-\rho(x_{i-1}-m)](x_{i-1}-m).$$

In the case of the unknown parameters m, σ_ε^2 and ρ the corresponding derivatives will have the following form:

$$(3.5.6) \quad H_n^{(1)} = \frac{\partial \log f}{\partial m} = \frac{1-\rho}{\sigma_\varepsilon^2} [(1+\rho)(x_1-m) + \sum_2^n (x_i-m-\rho(x_{i-1}-m))],$$

$$H_n^{(2)} = \frac{\partial \log f}{\partial \sigma_\varepsilon^2} = -\frac{n}{2\sigma_\varepsilon^2} +$$

$$+ \frac{1}{2\sigma_\varepsilon^4} [(1-\rho^2)(x_1-m)^2 + \sum_2^n (x_i-m-\rho(x_{i-1}-m))^2]$$

$$H_n^{(3)} = \frac{\partial \log f}{\partial \rho} = -\frac{\rho}{1-\rho^2} + \frac{1}{\sigma_\varepsilon^2}[\rho(x_1-m)^2 + \sum_2^n (x_i-m-\rho(x_{i-1}-m))(x_{i-1}-m).$$

After simple but lengthy calculations, we obtain

$$(3.5.7) \quad D^2R_n^{(1)} = \text{Var } R_n^{(1)} = \frac{2+(n-2)(1-\rho)}{(1+\rho)\ \sigma_\xi^2},$$

$$D^2R_n^{(2)} = \text{Var } R_n^{(2)} = \frac{n}{2\sigma_\xi^4},$$

$$D^2R_n^{(3)} = \text{Var } R_n^{(3)} = \frac{(1+\rho^2)(n-1)}{(1-\ {}^2)^2},$$

$$E\ R_n^{(1)}\ R_n^{(2)} = E\ R_n^{(1)}\ R_n^{(3)} = 0,$$

$$E\ R_n^{(2)}\ R_n^{(3)} = -\frac{(n-1)\rho}{\sigma_\xi^2(1-\rho^2)},$$

and similarly

$$(3.5.8) \quad D^2\ H_n^{(1)} = \text{Var } H_n^{(1)} = \frac{(1-\rho)[2+(n-2)(1-\rho)]}{\sigma_\varepsilon^2},$$

$$D^2\ H_n^{(2)} = \text{Var } H_n^{(2)} = \frac{n}{2\ \sigma_\varepsilon^4},$$

$$D^2 H_n^{(3)} = \mathrm{Var}\ H_n^{(3)} = \frac{n-1}{1-\rho^2} + \frac{2\rho^2}{(1-\rho^2)^2} ,$$

$$E\ H_n^{(1)}\ H_n^{(2)} = E\ H_n^{(1)}\ H_n^{(3)} = 0,$$

$$E\ H_n^{(2)}\ H_n^{(3)} = \frac{\rho}{\sigma_\varepsilon^2(1-\rho^2)} .$$

As can be seen, when all three parameters m, σ_ξ^2, ρ or m, σ_ε^2, ρ are unknown the determination of the maximum likelihood estimators is very time-consuming, and a successful investigation of their asymptotic behavior can hardly be expected. Instead, we shall use an idea first arrived at by Wald (see (3.1.5)-(3.1.7)), according to which we start with a study of the asymptotic behavior of the quantities $R_n^{(1)}$, $R_n^{(2)}$, $R_n^{(3)}$ or $H_n^{(1)}$, $H_n^{(2)}$, $H_n^{(3)}$ and only subsequently demonstrate that the solution of the system of equations (when normalized by the corresponding variances) has the same distribution, uniformly in unknown parameters, as the quantities $R_n^{(1)}/\sqrt{\mathrm{Var}\ R_n^{(1)}}$, $R_n^{(2)}/\sqrt{\mathrm{Var}\ R_n^{(2)}}$ and $R_n^{(3)}/\sqrt{\mathrm{Var}\ R_n^{(3)}}$. The normalizing factors, which are to be multiplied by, will be precisely $\sqrt{\mathrm{Var}\ R_n^{(1)}}$, $\sqrt{\mathrm{Var}\ R_n^{(2)}}$ and $\sqrt{\mathrm{Var}\ R_n^{(3)}}$ respectively (see (3.1.5), (3.1.5')).

When considering the variances in (3.5.7) and (3.5.8) we immediately obtain the result that for ρ close to one the maximum likelihood estimation of m is not consistent in the case of (3.5.7), while in the case of (3.5.8) the variance does not remain finite.

When only a single parameter is unknown we obtain the following estimates. Let m be unknown. Its maximum likelihood estimate (cf., (3.2.2)) will be

$$(3.5.9) \qquad \hat{m} = \frac{x_1 + x_n + (1-\rho)\sum_{2}^{n-1} x_i}{2+(1-\rho)(n-2)}$$

where m is normally distributed with parameters

$$m, \sigma_\xi \sqrt{\frac{1+\rho}{2+(n-2)(1-\rho)}}$$

Let m = 0; then σ_ξ^2 has the following maximum likelihood estimate:

$$(3.5.10) \qquad \hat{\sigma}_\xi^2 = \frac{1}{n(1-\rho^2)} [(1-\rho^2)x_1^2 + \sum_{2}^{n} (x_i - \rho x_{i-1})^2].$$

The estimate $\hat{\sigma}_\xi^2$ has a χ^2 distribution with expectation σ_ξ^2 and characteristic function

$$f(\alpha) = \left(1 - \frac{2\sigma_\varepsilon^2 \, i\alpha}{n(1-\rho^2)}\right)^{-\frac{n}{2}} = \left(1 - \frac{2\,\sigma_\xi^2 \, i\alpha}{n}\right)^{-\frac{n}{2}}. \tag{3.5.11}$$

Let $m = 0$; let σ_ξ^2 (or σ_ε^2) be known and ρ unknown. To obtain a maximum likelihood estimate we must solve a cubic equation, while on the basis of conditional density function we obtain the following estimate from (3.5.4):

$$\hat{\rho} = \frac{\sum_{2}^{n} x_i x_{i-1}}{\sum_{1}^{n-1} x_i^2} \tag{3.5.12}$$

The ergodic theorem (Appendix B2, Theorem 3) yields

$$\frac{1}{n-1} \sum_{1}^{n-1} \xi^2(i) \to \sigma_\xi^2, \tag{3.5.13}$$

in the mean square sense and with probability 1, while the random variable

(3.5.14)

$$\sum_{2}^{n} \xi(i)\xi(i-1) - \rho\Sigma\xi^2(i) = \sum_{2}^{n}\xi(i-1)[\xi(i) - \rho\xi(i-1)] = \sum_{2}^{n} \xi(i-1)\varepsilon(i)$$

has the variance $(n-1)\cdot(1-\rho^2)\cdot\sigma_\xi^2 = (n-1)\,\sigma_\varepsilon^2$. Hence it follows that the variance of the random variable

$$\sqrt{n-1}\,(\hat{\rho}-\rho) = \sqrt{n-1}\,\frac{\sum_{2}^{n} \xi(i)\xi(i-1) - \rho \sum_{1}^{n-1} \xi^2(i)}{\sum_{1}^{n-1} \xi^2(i)} =$$

$$= \frac{\sum_{2}^{n}\xi(i-1)\,\varepsilon(i)}{\sqrt{n-1}\cdot \frac{1}{n-1}\sum_{1}^{n-1} \xi^2(i)}$$

asymptotically equals $1 - \rho^2$.

The estimate $\hat{\rho}$ has a distribution which is asymptotically normal for any fixed value of ρ; this follows e.g., from Theorem 4, Appendix B2. This uniform asymptotic normality, however, holds true only for the interval $-1 + \varepsilon < \rho < 1 - \varepsilon$ (arbitrary $\varepsilon > 0$). Thus confidence intervals (upper and lower estimates) for ρ can only be constructed in an open interval $(-1,1)$.

In case of two unknown parameters we shall mention only the case dealing with m and ρ. When $\sigma_\xi^2 = 1$ it can be shown (see Theorems 1 and 2 below) that maximum likelihood estimates have an asymptotically normal distribution uniformly in the interval $-\infty < m < \infty, -1 < \rho < 1$ with a covariance matrix

$$\begin{pmatrix} \sigma_\xi \sqrt{\dfrac{1+\rho}{2+(n-2)(1-\rho)}} & 0 \\ 0 & \dfrac{1-\rho^2}{\sqrt{(n-1)(1+\rho^2)}} \end{pmatrix} .$$

The proof is based on the fact that the quantities $R_n^{(1)}$ and $R_n^{(2)}$ have asymptotically a normal distribution uniformly in the corresponding interval. However, when $\sigma_\varepsilon^2 = 1$ (this case is closer to physical reality as well as to the continuous case) the uniform asymptotic normality of the distribution in the relevant interval $-\infty < m < \infty$, $-1 < \rho < 1$ does not hold (see below, Theorem 3).

The distinction between the cases $\sigma_\xi^2 = 1$ and $\sigma_\varepsilon^2 = 1$ becomes clear when we compare the corresponding variances in (3.5.7) and (3.5.8).

3.5.2. Distribution of the Derivatives of the Likelihood Function. To investigate the asymptotic behavior of maximum likelihood estimators, let us first investigate the properties of the distribution of the random vector $(R_n^{(1)}, R_n^{(2)}, R_n^{(3)})$ for $n \to \infty$. We shall use the following notation:

$$\tilde{R}_n^{(1)} = \frac{R_n^{(1)}}{DR_n^{(1)}} , \quad \tilde{R}_n^{(2)} = \frac{R_n^{(2)}}{DR_n^{(2)}} , \quad \tilde{R}_n^{(3)} = \frac{R_n^{(3)}}{DR_n^{(3)}} .$$

Theorem 1. The characteristic function $f_n(t_1,t_2,t_3)$ of the random vector $(\tilde{R}_n^{(1)}, \tilde{R}_n^{(2)}, \tilde{R}_n^{(3)})$ converges as $n \to \infty$ for any t_1, t_2 and t_3 uniformly to the characteristic function of the normal distribution with expectation (0,0,0) and with correlation matrix

$$\begin{pmatrix} 1 & 0 & 0 \\ 0 & 1 & -\rho \dfrac{2}{1+\rho^2} \\ 0 & -\rho \dfrac{2}{1+\rho^2} & \end{pmatrix}$$

convergence is uniform when $-\infty < m < \infty$, $0 < \sigma_\xi^2 \le K < \infty$ and $-1 < \rho < 1$, where K is an arbitrary fixed constant.

Proof. The form ot the matrix (3.5.3) and formulas (3.5.5) and (3.5.7) give the following characteristic function of $(\tilde{R}_n^{(1)}, \tilde{R}_n^{(2)}, \tilde{R}_n^{(3)})$:

(3.5.15) $$f_n(t_1,t_2,t_3) = E\exp\{it_1\tilde{R}_n^{(1)}+it_2\tilde{R}_n^{(2)}+it_3\tilde{R}_n^{(3)}\} =$$

$$= c_n \int\ldots\int \exp\{-\frac{1}{2\sigma_\xi^2(1-\rho^2)}(\underline{x}-m)^*B_n^{-1}(\underline{x}-m) +$$

$$+ it_1\tilde{R}_n^{(1)}+it_2\tilde{R}_n^{(2)}+it_3\tilde{R}_n^{(3)}\}\,dx_1,\ldots,dx_n =$$

$$= c_n \exp\{-it_2\sqrt{\frac{n}{2}} + it_3\,\rho\sqrt{\frac{n-1}{1+\rho^2}}\}\int\ldots\int\exp\{-\frac{1}{2}[\underline{Y}^*A_n\underline{Y} - \underline{Y}\,\underline{\Lambda}]\,dy_1\ldots dy_n$$

where

(3.5.16) $$c_n = (2\pi)^{-\frac{n}{2}}\sigma_\xi^{-n}(1-\rho^2)^{-\frac{n-1}{2}},$$

$$\underline{Y} = \underline{X} - m,$$

(3.5.17)
$$A_n = \begin{pmatrix} a_1 & b & 0 & \ldots & 0 \\ b & a & b & \ldots & 0 \\ 0 & b & a & \ldots & 0 \\ \cdot & & & & \cdot \\ \cdot & & & & \cdot \\ \cdot & & & & \cdot \\ 0 & & \ldots & a & b \\ 0 & & \ldots & b & a_1 \end{pmatrix}$$

and

$$a_1 = \frac{1}{\sigma_\xi^2(1-\rho^2)}\left[1-\frac{2it_2}{\sqrt{2n}}+\frac{2i\rho t_3}{\sqrt{(n-1)(1+\rho^2)}}\right],$$

(3.5.18)
$$a = \frac{1}{\sigma_\xi^2(1-\rho^2)}\left[(1+\rho^2)\left(1-\frac{2it_2}{\sqrt{2n}}\right)+\frac{4i\rho t_3}{\sqrt{(n-1)(1+\rho^2)}}\right],$$

$$b = -\frac{1}{\sigma_\xi^2(1-\rho^2)}\left[\rho\left(1-\frac{2it_2}{2n}\right)+\frac{it_3(1+\rho^2)}{(n-1)(1+\rho^2)}\right],$$

$$\underline{\Lambda} = \begin{pmatrix} c \\ c(1-\rho) \\ \cdot \\ \cdot \\ \cdot \\ c(1-\rho) \\ c \end{pmatrix}, \qquad c = \frac{2it_1}{\sigma_\xi \sqrt{(1+\rho)\,[\,2+(n-2)(1-\rho)]}}.$$

Let numbers $d_i\,(i=1,\ldots,n)$ be chosen so that

(3.5.19)
$$\begin{aligned} a_1 d_1 + bd_2 &= \frac{c}{2}, \\ bd_3 + ad_2 + bd_1 &= (1-\rho)\frac{c}{2}, \\ &\cdot \\ &\cdot \\ &\cdot \\ bd_{n-1} + ad_{n-2} + bd_{n-3} &= (1-\rho)\frac{c}{2}, \\ bd_{n-1} + a_1 d_n &= \frac{c}{2}. \end{aligned}$$

Under the transformation $y_i = z_i - d_i$, the expression $\underline{Y}^* A_n \underline{Y} - \underline{Y}^* \underline{\Lambda}$ becomes

(3.5.20)
$$(z_1-d_1,\ldots,z_n-d_n)\, A_n \begin{pmatrix} z_1-d_1 \\ \cdot \\ \cdot \\ \cdot \\ z_n-d_n \end{pmatrix} - D_n$$

where

(3.5.21)

$$D_n = a_1 d_1^2 + a(d_2^2+\ldots+d_{n-1}^2) + a_1 d_n^2 + 2b(d_1 d_2+\ldots+d_{n-1}d_n) =$$

$$= \frac{c}{2}(d_1+d_n) + \frac{(1-\rho)c}{2} \sum_{2}^{n-1} d_i.$$

(3.5.22)
$$d_i = d + \theta_1 u_1^i + \theta_2 u_2^i, \quad (i=1,2,\ldots,n),$$

where

$$d = \frac{(1-\rho)c}{2(a+2b)} , \tag{3.5.23}$$

and u_1 and u_2 are zeros of the equation $bu^2 + au + b = 0$. The quantities θ_1 and θ_2 can be determined using the first and the last equations of (3.5.19):

(3.5.24)

$$\theta_1 = \frac{a_1u_2^n + bu_2^{n-1} - (a_1u_2+bu_2^2)}{(a_1u_1+bu_1^2)(a_1u_2^n+bu_2^{n-1})-(a_1u_2+bu_2^2)(a_1u_1^n+bu_1^{n-1})} \cdot$$

$$\cdot \left(\frac{c}{2} - d(a_1+b)\right),$$

$$\theta_2 = \frac{a_1u_1^n+bu_1^{n-1} - (a_1u_1 + bu_1^2)}{(a_1u_1+bu_1^2)(a_1u_2^n+bu_2^{n-1})-(a_1u_2+bu_2^2)(a_1u_1^n+bu_1^{n-1})} \cdot$$

$$\cdot \left(\frac{c}{2} - d(a_1+b)\right)$$

On the basis of a new solution, say $z_i - d_i = x_i$, the characteristic function (3.5.15) becomes

$$f_n(t_1,t_2,t_3) = c_n \exp\{ - it_2\sqrt{\tfrac{n}{2}} + it_3\, \rho\sqrt{\tfrac{n-1}{1+\rho^2}} + \frac{D_n}{2} \} \cdot \tag{3.5.25}$$

$$\cdot \int\dots\int \exp\{ - \tfrac{1}{2}\, \underline{X}^*A\, \underline{X} \}dx_1,\dots,dx_n \cdot$$

Since (see, for example, Cramer 1 , p. 136)

$$\int\dots\int \exp\{- \tfrac{1}{2}\, \underline{X}^*A_n\, \underline{X}\}\, dx_1,\dots,dx_n = (2\pi)^{n/2}|A_n|^{-1/2} ,$$

it follows that

$$f_n(t_1,t_2,t_3) = (2\pi)^{n/2}\, c_n|A_n|^{-1/2} \exp\{ -it_2 \sqrt{\tfrac{n}{2}} + it_3\rho\cdot \sqrt{\frac{n-1}{1+\rho^2}} + \frac{D_n}{2} \}. \tag{3.5.26}$$

From (3.5.17) it becomes apparent that

(3.5.27) $$|A_n| = a_1|\tilde{A}_{n-2}| - 2b^2a_1\,|\tilde{A}_{n-3}| + b^4|\tilde{A}_{n-4}|,$$

where $|\tilde{A}_n|$ satisfies the difference equation

(3.5.28) $$|\tilde{A}_n| = a|\tilde{A}_{n-1}| - b^2|\tilde{A}_{n-2}|.$$

It can now easily be shown that

(3.5.29) $$|\tilde{A}_n| = \alpha_1\, v_1^n + \alpha_2\, v_2^n,$$

where v_1 and v_2 are roots of the equation $v^2 - av + b^2 = 0$, while from the conditions $|\tilde{A}_1| = a$ and $|\tilde{A}_2| = a^2 - b^2$ we get

(3.5.30) $$\alpha_1 = \frac{v_1}{v_1-v_2}, \qquad \alpha_2 = \frac{v_2}{v_2-v_1}.$$

By substitution in (3.5.29) and (3.5.27) we get the following two expressions:

(3.5.31) $$|\tilde{A}_n| = \frac{v_1^{n+1} - v_2^{n+1}}{v_1-v_2},$$

(3.5.32)
$$|A_n| = \frac{1}{v_1-v_2}\left[\, v_1^{n-3}\,(a_1v_1-b^2)^2 - v_2^{n-3}(a_1v_2-b^2)^2\,\right] =$$

$$= \frac{v_1^{n-3}(a_1v_1-b^2)^2}{v_1-v_2}\left[\, 1 - \left(\frac{v_2}{v_1}\right)^{n-3} \frac{(a_1v_2-b^2)^2}{(a_1v_1-b^2)^2}\,\right].$$

To simplify calculations, let us write

$$f_n^{(1)}\,(t_1,t_2,t_3) = \exp\left\{\frac{D_n}{2}\right\},$$

$$f_n^{(2)}(t_1,t_2,t_3) = |A_n|^{-1/2}\,\sigma_\xi^{-n}(1-\rho^2)^{-\frac{n-1}{2}}\exp\left\{-\,it_2\sqrt{\frac{n}{2}} + it_3\rho\sqrt{\frac{n-1}{1+\rho^2}}\right\},$$

and let us deal with the asymptotic behavior of the two functions separately. In what follows, M_i and $\overline{M}_i$ are constants, while the variables t_1, t_2 and t_3 belong to an arbitrary finite interval $T_1 \times T_2 \times T_3$ which is uniformly bounded for $\rho \in (-1,1)$ and $\sigma_\xi^2 \in (0,K)$.

From (3.5.18) we obtain

(3.5.33)

$$v_1 = \frac{1}{\sigma_\xi^2(1-\rho^2)} \left[(1+\rho^2)\left(1-\frac{2it_2}{\sqrt{2n}} + \frac{4it_3\rho}{\sqrt{(n-1)(1+\rho^2)}}\right) + \right.$$

$$\left. + (1-\rho^2)\sqrt{\left(1-\frac{2it_2}{\sqrt{2n}}\right)^2 + \frac{4t_3}{(n-1)(1+\rho^2)}}\ \right],$$

$$v_2 = \frac{1}{\sigma_\xi^2(1-\rho^2)} \left[(1+\rho^2)\left(1-\frac{2it_2}{\sqrt{2n}}\right) + \frac{4it_3\rho}{\sqrt{(n-1)(1+\rho^2)}} \right] +$$

$$- (1-\rho^2)\sqrt{\left(1-\frac{2it_2}{\sqrt{2n}}\right)^2 + \frac{4t_3}{(n-1)(1+\rho^2)}}\ ]$$

For a sufficiently large n we obtain

$$\left(1-\frac{2it_2}{\sqrt{2n}}\right)^{-2} = 1 + \frac{4it_2}{\sqrt{2n}} + \frac{M_1}{n},$$

and

$$\left[1 + \frac{4t_3^2}{(n-1)(1+\rho^2)}\left(1+\frac{4it_2}{\sqrt{2n}} + \frac{M_1}{n}\right)\right]^{1/2} = 1+\frac{2t_3^2}{(n-1)(1+\rho^2)} + \frac{M_2}{n^{3/2}},$$

and therefore (3.5.33) can be written as

(3.5.33′)

$$v_1 = \frac{1}{\sigma_\xi^2(1-\rho^2)} \left[1-\frac{2it_2}{\sqrt{2n}} + \frac{2it_3\rho}{\sqrt{(n-1)(1+\rho^2)}} + \frac{t_3^2(1-\rho^2)}{(n-1)(1+\rho^2)} + \frac{M_2}{n^{3/2}}\right],$$

$$v_2 = \frac{1}{\sigma_\xi^2(1-\rho^2)} \left[1-\frac{2it_2}{\sqrt{2n}} + \frac{2it_3\rho}{\sqrt{(n-1)(1+\rho^2)}} - \frac{t_3^2(1-\rho^2)}{(n-1)(1+\rho^2)} + \frac{M_2}{n^{3/2}}\right].$$

Using the series $\log(1+x) = x - x^2/2 + \theta x^3/3$, where $|\theta| < 1$ for $|x| < 1/2$, on the basis of (3.5.33′) we obtain

$$v_1^{\frac{n-1}{2}} = \sigma_\xi^{-(n-1)}(1-\rho^2)^{-\frac{n-1}{2}} \exp\left\{-\frac{n-1}{2}\left[-\frac{2it_2}{\sqrt{2n}} + \frac{2\rho it_3}{\sqrt{(n-1)(1+\rho^2)}} + \right.\right.$$

$$\left.\left. + \frac{t_3^2(1-\rho^2)}{(n-1)(1+\rho^2)} - \frac{4\rho t_2t_3}{\sqrt{2n(n-1)(1+\rho^2)}} + \frac{1}{2}\left(\frac{2t_2^2}{n} + \frac{4\rho^2t_3^2}{(n-1)(1+\rho^2)}\right) + \frac{M_3}{n^{3/2}}\right]\right\},$$

and therefore

$$(3.5.34)\quad v_1^{\frac{n-1}{2}} = \sigma_\xi^{-(n-1)}(1-\rho^2)^{\frac{-n-1}{2}} \exp\left\{it_2\frac{n-1}{2\sqrt{2n}} - it_3\rho\sqrt{\frac{n-1}{1+\rho^2}} - \right.$$

$$-\frac{t_3^2(1-\rho^2)}{2(1+\rho^2)} + \frac{2\rho t_2 t_3}{\sqrt{\frac{2n}{n-1}}\,(1+\rho^2)} - \frac{t_2^2}{2}\frac{n-1}{n} - \frac{\rho^2 t_3^2}{1+\rho^2}\}(1+\frac{M_3}{\sqrt{n}}).$$

It can easily be calculated that

(3.5.35)

$$a_1v_1^2-b^2 = \frac{1}{\sigma_\xi^4(1-\rho^2)}\Big[(1-\frac{2it_2}{\sqrt{2n}})^2 + 2it_3(1-\frac{2it_2}{\sqrt{2n}})\frac{\rho}{\sqrt{(n-1)(1+\rho^2)}} + \frac{M_4}{n}\Big],$$

$$v_1-v_2 = \frac{1}{\sigma_\xi^2}\Big[1-\frac{2it_2}{\sqrt{2n}}+\frac{M_5}{n}\Big],$$

$$v_1(v_1-v_2)^{1/2} = \frac{1}{\sigma_\xi^3(1-\rho^2)}\Big[1+\frac{M_6}{\sqrt{n}}\Big],$$

$$a_1v_2^2-b^2 = \frac{1}{\sigma_\xi^4(1-\rho^2)}\cdot\frac{M_7}{n}.$$

From (3.5.35), (3.5.34) and (3.5.32) we get

(3.5.36)

$$f_n^{(2)}(t_1,t_2,t_3) = \exp\{-it_2\sqrt{\frac{1}{2n}} - \frac{t_2^2}{2}\frac{n-1}{n} - \frac{t_3^2}{2} + \frac{2\rho t_2t_3}{\sqrt{2\frac{n}{n-1}}\,(1+\rho^2)}\}(1+\frac{M_8}{\sqrt{n}}).$$

The asymptotic behavior of $f_n^{(1)}(t_1,t_2,t_3)$ can be discussed as follows. From (3.5.21) - (3.5.24) we obtain

$$D_n = \frac{c}{2}(2d+\theta_1u_1+\theta_2u_2+\theta_1u_1^n+\theta_2u_2^n) + \frac{c(1-\rho)}{2}\Big[(n-2)d +$$

$$+\theta_1u_1^2\frac{1-u_1^{n-2}}{1-u_1} + \theta_2u_2^2\frac{1-u_2^{n-2}}{1-u_2}\Big] = \frac{c^2(1-\rho)}{4(a+2b)}\Big[2+(n-2)(1-\rho)\Big] +$$

$$+\frac{c^2}{4(a+2b)}\Big[a+2b-(1-\rho)(a_1+b)\Big]g_n(t_1,t_2,t_3),$$

where

$$g_n(t_1,t_2,t_3) = \frac{\theta_1}{\frac{c}{2}-d(a_1+b)}\Big[1+u_1^{-(n-1)}+(1-\rho)\frac{1-u_1^{n-2}}{1-u_1}\Big]+$$

$$+ \frac{\theta_2}{\frac{c}{2} - d(a_1+b)} \left[1+u_2^{-(n-1)} + (1-\rho)u_2^{-(n-2)} \frac{1-u_2^{n-2}}{1-u_2} \right] .$$

Thus

(3.5.37)

$$D_n = - t_1^2(1+ \frac{\overline{M}_1}{\sqrt{n}}) - \frac{t_1^2}{2+(1-\rho)(n-2)} (1+ \frac{\overline{M}_1}{\sqrt{n}}) \frac{2it_3}{\sqrt{(n-1)(1+\rho^2)}} \cdot$$

$$\cdot \frac{1}{\sigma_\xi^2(1+\rho)} g_n(t_1,t_2,t_3).$$

Because u_1 and u_2 satisfy the relations

(3.5.38) $$u_1 = - \frac{v_2}{b}, \quad u_2 = - \frac{v_1}{b}, \quad u_1u_2 = 1, \ |u_1| \leq 1, \ |u_2| \geq 1,$$

it is easy to show that

(3.5.39)

$$a_1u_2+b = \frac{1}{b} \frac{1}{\sigma_\xi^2(1-\rho^2)} \left[(1- \frac{2it_2}{\sqrt{2n}})^2 + 2it_3(1- \frac{2it_2}{\sqrt{2n}}) \frac{\rho}{\sqrt{(n-1)(1+\rho^2)}} + \frac{\overline{M}_2}{n} \right],$$

$$a_1u_1+b = \frac{1}{b} \cdot \frac{1}{\sigma_\xi^4(1-\rho^2)} \cdot \frac{\overline{M}_3}{n},$$

$$a_1 + bu_2 = a_1 - v_1 = \frac{\overline{M}_4}{\sigma_\xi^2 . n},$$

$$a_1+bu_1 = a_1 - v_2 = \frac{1}{\sigma_\xi^2} \left[1 - \frac{2it_2}{\sqrt{2n}} + \frac{\overline{M}_5}{n} \right],$$

$$\frac{v_2}{v_1} = \left[\rho - \frac{\overline{M}_6}{\sqrt{n}} \right],$$

$$b = \frac{1}{\sigma_\xi^2(1-\rho^2)} (-\rho + \frac{\overline{M}_7}{\sqrt{n}}), \quad \overline{M}_7 \neq 0 \text{ if } \rho = 0.$$

Let the uniformly bounded quantity $1/\sigma_\xi^4 (1-\rho^2)(b^2-v_1a_1)$ be denoted by $\overline{M}_8$ and the quantity $n\, \sigma_\xi^4(1-\rho^2)(b^2-v_2a_1)$ by $\overline{M}_3$; then

(3.5.40)

$$\frac{(a_1u_2+b) - (a_1+bu_2)\, u_2^{-(n-1)}}{(a_1+bu_1)(a_1u_2+b) - u_1^{n-2}\, u_2^{-n+2}\, (a_1+bu_2)(a_1u_1+b)} =$$

$$= \sigma_\xi^2 \frac{1 - \bar{M}_{10} \, (-\rho + \frac{\bar{M}_7}{\sqrt{n}}) \, (\frac{\bar{M}_4}{n}) \, \bar{M}_8}{1 - \frac{\bar{M}_5}{\sqrt{n}} - \bar{M}_9 \, \frac{\bar{M}_4}{n} \, \frac{\bar{M}_3}{n} \cdot \bar{M}_8} = \sigma_\xi^2 \cdot \bar{M}_{11} \quad ,$$

where

$$\left| \frac{u_1}{u_2} \right|^{n-1} = \bar{M}_9 \leq 1, \quad u_2^{-(n-1)} = \bar{M}_{10} \leq 1.$$

Similarly

$$u_2 \cdot \frac{a_1 + bu_1 - (a_1 u_1 + b) \, u_1^{n-2}}{(a_1 + bu_1)(a_1 u_2 + b) - u_1^{n-2} \cdot u_2^{-n+2} (a_1 + bu_2)(a_1 u_1 + b)} = \sigma_\xi^2 \cdot \bar{M}_{14} \, ,$$

which follows from the relations

$$|u_1| = M_{13} \leq 1, \quad v_1 = \frac{1}{\sigma_\xi^2 (1-\rho^2)} \left[1 + \frac{M_{12}}{\sqrt{n}} \right] ,$$

(3.5.41)

$$\frac{1}{1+\rho} \; \frac{1 + u_1^{n-1} + (1-\rho) \, u_1 \frac{1-u_1^{n-2}}{1-u_1}}{2 + (n-2)(1-\rho)} = \bar{M}_{15} \quad ,$$

$$\frac{1}{1+\rho} \; \frac{1 + u_2^{-(n-1)} + (1-\rho) \, u_2^{-n+2} \frac{1-u_2^{n-2}}{1-u_2}}{2 + (n-2)(1-\rho)} = \bar{M}_{16} \; ,$$

which are uniformly bounded. Thus from (3.5.41), (3.5.40) and (3.5.37) we get

(3.5.42)

$$D_n = - t_1^2 \left(1 + \frac{M_1}{\sqrt{n}}\right) -$$

$$- t_1^2 \left(1 + \frac{\bar{M}_1}{\sqrt{n}}\right) \frac{2it_3}{\sqrt{(n-1)(1+ \;^2)}} \left[\bar{M}_{15} \cdot \bar{M}_{11} + \bar{M}_{16} \cdot \bar{M}_{14}\right],$$

and finally

(3.5.43) $$f_n^{(1)}(t_1,t_2,t_3) = \exp\left(\frac{D_n}{2}\right) = \exp\{ -\frac{t_1^2}{2} \} \left(1 + \frac{\bar{M}_{17}}{\sqrt{n}}\right).$$

From (3.5.26), (3.5.36) and (3.5.43) we obtain the following result for the characteristic function of $(\tilde{R}_n^{(1)}, \tilde{R}_n^{(2)}, \tilde{R}_n^{(3)})$:

(3.5.44) $$f_n(t_1,t_2,t_3) = f_n^{(1)}(t_1,t_2,t_3) \cdot f_n^{(2)}(t_1,t_2,t_3) =$$

$$= \exp -\{\frac{1}{2}(t_1^2+t_2^2+t_3^2+\frac{4\rho t_2 t_3}{\sqrt{2(1+\rho^2)}})\}\cdot(1+\frac{\bar{M}_{17}}{\sqrt{n}}).$$

Hence as $n \to \infty$ the functions $f_n(t_1,t_2,t_3)$ converge in any finite interval of the Cartesian product $T_1 \times T_2 \times T_3$ to the characteristic function of the normal distribution, uniformly in $-\infty < m < \infty$, $-1 < \rho < 1, 0 < \sigma_\xi^2 \leq K < \infty$. The proof of the theorem is complete.

By rearranging the corresponding formulas the rate of convergence can also be determined.

3.5.3. Asymptotic Distribution of Maximum Likelihood Estimates. Using Theorem 1, we can now investigate the asymptotic behavior of the maximum likelihood estimates as given by the solutions of the equations

$$R_n^{(1)} = 0,\ R_n^{(2)} = 0,\ R_n^{(3)} = 0. \tag{3.5.45}$$

Convergence of distributions is understood in the following as weak convergence. We now prove

Theorem 2. The system of equations (3.5.45) has almost surely as $n\to\infty$ a solution $\hat{m}(\xi_1,\dots,\xi_n)$, $\hat{\sigma}_\xi^2(\xi_1,\dots,\xi_n)$, $\hat{\rho}(\xi_1,\dots,\xi_n)$ such that the distribution of the random vector

$$(\hat{m}_n - m)\sqrt{\frac{2+(n-2)(1-\rho)}{\sigma_\xi^2(1+\rho)}},\ (\hat{\sigma}_{\xi,n}^2-\sigma_\xi^2)\frac{1}{\sigma_\xi^2}\sqrt{\frac{n}{2}},\ (\hat{\rho}_n-\rho)\frac{\sqrt{(n-1)(1+\rho^2)}}{1-\rho^2},$$

converges to the distribution of the random vector $(\tilde{R}_n^{(1)}, \tilde{R}_n^{(2)}, \tilde{R}_n^{(3)})$ as $n \to \infty$, and this convergence is uniform in the region $-\infty < m < \infty$, $0 < \sigma_\xi^2 \leq K < \infty$, $-1 < \rho < 1$.

Proof. Let us take

$$L_n^{(1)} = (1+\rho)\,\sigma_\xi^2 R_n^{(1)} = 0, \tag{3.5.46}$$

$$L_n^{(2)} = 2(1-\rho)\,\sigma_\xi^4 R_n^{(2)} = 0,$$

$$L_n^{(3)} = (1-\rho^2)\,\sigma_\xi^2 R_n^{(3)} = 0,$$

which is evidently equivalent to the system (3.5.45). The left-hand sides in (3.5.46) are polynomials in the variables m, σ_ξ^2 and σ; their Taylor series about the true values m_0, $\sigma_{\xi,0}^2$ and ρ_0 of the parameters are as follows:

(3.5.47)

$$L_n^{(1)}\Big|_{m_o,\sigma_o^2,\rho_o} + \frac{\partial L_n^{(1)}}{\partial m}\Big|_{m_o,\sigma_o^2,\rho_o}(m-m_o) + \frac{\partial L_n^{(1)}}{\partial \rho}\Big|_{m_o,\sigma_o^2,\rho_o}\cdot$$

$$\cdot(\rho-\rho_o) + \ldots = 0,$$

$$L_n^{(2)}\Big|_{m_o,\sigma_o^2,\rho_o} + \frac{\partial L_n^{(2)}}{\partial m}\Big|_{m_o,\sigma_o^2,\rho_o}(m-m_o) + \ldots = 0,$$

$$L_n^{(3)}\Big|_{m_o,\sigma_o^2,\rho_o} + \frac{\partial L_n^{(3)}}{\partial m}\Big|_{m_o,\sigma_o^2,\rho_o}(m-m_o) + \ldots = 0.$$

We could calculate with no difficulty the derivatives in (3.5.47), but we shall not write down the corresponding formulas. With the substitution

$$m-m_o = x\sqrt{\frac{\sigma_o^2(1+\rho_o)}{2+(n-2)(1-\rho_o)}},$$

$$\sigma_\xi^2 - \sigma_o^2 = y\,\sigma_o^2\sqrt{\frac{2}{n}},$$

$$\rho-\rho_o = z\,\frac{1-\rho_o^2}{\sqrt{(n-1)(1+\rho_o^2)}},$$

and after dividing the first equation (3.5.47) by $\sigma_o(1+\rho_o)^{1/2}\cdot$ $\cdot(2+(n-2)(1-\rho_o))^{1/2}$ the second by $(1-\rho_o^2)\,\sigma_o^2\sqrt{2n}$ and the third by $(1-\rho_o^2)\cdot\sigma_o^2\,[(n-1)(1+\rho_o^2)]^{1/2}$ we obtain the following system:

(3.5.48)

$$\tilde{R}_n^{(1)} + x\,\frac{\partial L_n^{(1)}}{\partial m}\Big|_o\cdot\frac{1}{2+(n-2)(1-\rho_o)} + \ldots = 0,$$

$$\tilde{R}_n^{(2)} + y\,\frac{\partial L_n^{(2)}}{\partial \sigma_\xi^2}\Big|_o\cdot\frac{1}{n(1-\rho_o^2)} + \ldots = 0,$$

$$\tilde{R}_n^{(3)} + z\,\frac{\partial L_n^{(3)}}{\partial \rho}\Big|_o\cdot\frac{1}{\sigma_o^2(n-1)(1+\rho_o^2)} + \ldots = 0.$$

It can easily be shown that

$$\frac{\partial L_n^{(1)}}{\partial m}\Big|_o\cdot\frac{1}{2+(n-2)(1-\rho_o)} = -1,$$

$$\frac{\partial L_n^{(2)}}{\partial \tilde{\sigma}_\xi^2}\Big|_o\,\frac{1}{n(1-\rho_o^2)} = -1,$$

$$L_n^{(3)}\Big|_0 \frac{1}{\sigma_0^2(n-1)(1+\rho_0^2)} \to -1 ,$$

where convergence should be understood as convergence almost surely as well as uniform convergence on the set $-\infty < m_0 < \infty$, $0 < \sigma_0^2 \leq K < \infty$, $1 < \rho_0 < 1$. The remaining terms in (3.5.48) tend to zero almost surely and uniformly on the above mentioned set in the true-parameter region.

For large n the quantities $|R_n^{(i)}|$ and $|R_n^{(i)}| / |R_n^j|$, are bounded in probability and uniformly in the same set, as follows from Theorem 1. Therefore the system

$$(3.5.49) \qquad \begin{aligned} 1 - \varepsilon_1 + \dots &= 0, \\ 1 - \varepsilon_2 + \dots &= 0, \\ 1 - \varepsilon_3 + \dots &= 0, \end{aligned}$$

where

$$\varepsilon_1 = \frac{x}{\tilde{R}_n^{(1)}}, \qquad \varepsilon_2 = \frac{y}{\tilde{R}_n^{(2)}}, \qquad \varepsilon_3 = \frac{z}{\tilde{R}_n^{(3)}},$$

for large n with probability arbitrarily close to 1 has a solution $\epsilon_1^{(n)}, \epsilon_2^{(n)}, \epsilon_3^{(n)}$ which uniformly on the set $-\infty < m_0 < \infty$, $0 < \sigma_0^2 \leq K < \infty$, $-1 < \rho < 1$ belongs to the interval $(1-\delta, 1+\delta)$ (with arbitrary δ). Hence the limit distribution of the variables

$$x_n = \varepsilon_1^{(n)} \cdot \tilde{R}_n^{(1)}, \qquad y_n = \varepsilon_3^{(n)} \cdot \tilde{R}_n^{(3)}, \qquad z_n = \varepsilon_3^{(n)} \cdot \tilde{R}_n^{(3)},$$

for $n \to \infty$ coincides with the distributions of $\tilde{R}_n^{(1)}$, $\tilde{R}_n^{(2)}$ and $\tilde{R}_n^{(3)}$, because $\epsilon_i^{(n)} = 1$ $(i=1,2,3)$ as $n \to \infty$ uniformly in the region $-\infty < m_0 < \infty$, $0 < \sigma_0^2 < K < \infty$, $-1 < \rho_0 < 1$. The proof of Theorem 2 is complete.

The relations

$$x_n = (\hat{m}_n - m_0) \sqrt{\frac{2+(n-2)(1-\rho_0)}{\sigma_0^2\,(1+\rho_0)}},$$

$$y_n = (\hat{\sigma}_n^2 - \sigma_0^2)\, \frac{1}{\sigma_0^2} \sqrt{\frac{n}{2}},$$

$$z_n = (\hat{\rho}_n - \rho_0)\, \frac{\sqrt{(n-1)(1+\rho_0^2)}}{1-\rho_0^2},$$

show that the estimators $\hat{\rho}_n$ and $\hat{\sigma}_n^2$ are uniformly consistent, which is not true for $\hat{m}_n$.

3.5.4. <u>Results Obtained for Discrete Analogues of the Continuous-Time Case.</u> As we mentioned above, we have a case corresponding to the continuous-time one when the parameters m, σ_ε^2 and ρ are unknown. Here the assertion concerning uniform asymptotic normality of quantities $H_n^{(i)}$ (see (3.5.6)) is not true. Nevertheless, two theorems concerning $H_n^{(1)}$ can be proved; these theorems correspond to Theorems 1 and 2 when $\kappa = (1-\rho^2)\, n \to \infty$ (see also Theorems 2 and 3 in §3.3).

<u>Theorem 3. The distribution of the random variables</u> $\tilde{H}_n^{(i)} = H_n^{(i)} / \operatorname{Var} H_n^{(i)}$ (i=1,2,3) <u>tends to the normal distribution with parameters</u>

$$\left(0,0,0; \quad \begin{vmatrix} 1 & 0 & 0 \\ 0 & 1 & 0 \\ 0 & 0 & 1 \end{vmatrix} \right)$$

<u>as $\kappa \to \infty$.</u>

The proof proceeds in the same manner as in §3.5.3. The quantities v_1 and v_2 in (3.5.33) will, however, have the following form:

$$(3.5.50) \quad v_1 = \frac{1}{2\sigma_\varepsilon^2}\Big[(1+\rho^2)\Big(1-\frac{2it_2}{\sqrt{2(n-1)}}\Big)+\frac{it_3 2\rho\sqrt{1-\rho^2}}{\sqrt{n-1}} + (1-\rho^2)\Big\{\Big(1-\frac{2it_2}{\sqrt{2(n-1)}}\Big)^2-\frac{4i\rho t_3}{\sqrt{(n-1)(1-\rho^2)}}\Big(1-\frac{2it_2}{\sqrt{2(n-1)}}\Big) + \frac{4(1-\rho^2)t_3^2}{n-1}\Big\}^{1/2}\Big],$$

$$v_2 = \frac{1}{2\sigma_\varepsilon^2}\Big[(1+\rho^2)\Big(1-\frac{2it_2}{\sqrt{2(n-1)}}\Big)+\frac{it_3 2\rho\sqrt{1-\rho^2}}{\sqrt{n-1}} - (1-\rho^2)\Big\{\Big(1-\frac{2it_2}{\sqrt{2(n-1)}}\Big)^2-\frac{4i\rho t_3}{\sqrt{(n-1)(1-\rho^2)}}\Big(1-\frac{2it_2}{\sqrt{2(n-1)}}\Big)+\frac{4(1-\rho^2)t_3^2}{n-1}\Big\}^{1/2}\Big]$$

When considering (3.5.34) it can be seen from (3.5.50) that the normality of the asymptotic distribution holds only for $\kappa \to \infty$.

For $\kappa \to \infty$ we obtain estimators equivalent to the maximal likelihood estimators when considering the following:

$$(3.5.51) \quad \hat{m} = \frac{1}{n}\sum_1^n \xi(i), \qquad \hat{\sigma}_\varepsilon^2 = (1-\rho^2)\,\hat{s}_\xi^2$$

$$\hat{\rho} = \frac{1}{(n-1)\,\hat{s}_{\xi}^{2}} \sum_{2}^{n} \zeta(i)\ \zeta(i-1),$$

where

(3.5.52) $$\zeta(i) = \xi(i) - \hat{m}, \qquad \hat{s}^2 = \frac{1}{n} \sum_{1}^{n} \zeta^2(i)$$

Simple calculations give us

(3.5.53) $$E\,\hat{m} = m,\ D^2\,\hat{m} = \frac{\sigma_\xi^2}{n} + \frac{2\rho}{1-\rho}\,\frac{\sigma_\xi^2}{n} + o(\frac{1}{n}),$$

$$E\,(\hat{\rho}-\rho) = 0\ (\frac{1}{n}), \qquad D^2\hat{\rho} = \frac{1-\rho^2}{n} + o(\frac{1}{n}),$$

$$E(\hat{\sigma}_\varepsilon^2-\sigma_\varepsilon^2) = 0\ (\frac{1}{n}),\ D^2\hat{\sigma}_\varepsilon^2 = \frac{2(1-\rho^2)^2}{n}\ \sigma_\xi^4 + o(\frac{1}{n}).$$

Similarly, the following theorem can be proved (see Theorem 3 in §3.3):

<u>Theorem 4. As $\kappa \to \infty$, the estimators $\hat{m} \sim m$, $\hat{\sigma}_\varepsilon^2 \sim \sigma_\varepsilon^2$ and $\hat{\rho} \sim \rho$ are asymptotically efficient, and the distribution of the random vector</u>

(3.5.54) $$\frac{\hat{m} - m}{\sqrt{\frac{1+\rho}{1-\rho} \cdot \frac{\sigma_\xi^2}{n}}}, \qquad \frac{\hat{\sigma}_\varepsilon^2 - \sigma_\varepsilon^2}{\sqrt{\frac{2(1-\rho^2)^2\,\sigma_\xi^2}{n}}}, \qquad \frac{\hat{\rho} - \rho}{\sqrt{\frac{1-\rho^2}{n}}}$$

<u>tends to the normal distribution with parameters</u>

$$(0,0,0, \begin{Vmatrix} 1 & 0 & 0 \\ 0 & 1 & 0 \\ 0 & 0 & 1 \end{Vmatrix}).$$

The following theorem can be proved in an even easier manner.

<u>Theorem 5. As $n \to \infty$ the estimator $\hat{\sigma}_\varepsilon^2 \sim \sigma_\varepsilon^2$ obtained from (3.5.1) is asymptotically efficient, and the distribution of the ratio</u>

(3.5.55) $$\frac{\hat{\sigma}_\varepsilon^2 - \sigma_\varepsilon^2}{\sigma_\varepsilon^2 \sqrt{\frac{2}{n}}}$$

<u>tends to (0,1) normal distribution.</u>

<u>Proof.</u> We have seen in (3.5.11), that the characteristic function of the random variable

(3.5.56)
$$\zeta_n = -\sqrt{\frac{n}{2}} + \frac{1}{2\sigma_\xi^2}\sqrt{\frac{2}{n}}\{(1-\rho^2)(\xi(1)-m)^2 + \sum_{2}^{n}[\xi(i)-m-\rho(\xi(i-1)-m)]^2\}$$

has the form

$$(1-\frac{2it}{\sqrt{2n}})^{-\frac{n}{2}} \exp(-it\sqrt{\frac{n}{2}}).$$

Hence ζ_n is asymptotically normally distributed as $n \to \infty$. On the other hand, $\hat{\rho} \to \rho$ and $\hat{m} \to m$ in probability, and therefore according to Cramer's theorem the asymptotic distribution of the random variable

$$\zeta_n^* = -\sqrt{\frac{n}{2}} + \frac{1}{2\sigma_\varepsilon^2}\sqrt{\frac{2}{n}}\{(1-\hat{\rho}^2)(\xi(1)-\hat{m}) + \frac{1}{2\sigma_\varepsilon^2}\sqrt{\frac{2}{n}}\sum_{2}^{n}[\xi(i) - \hat{m} -$$

$$-\hat{\rho}(\xi(i-1)-\hat{m})^{2}\} = -\sqrt{\frac{n}{2}} + \frac{1}{2\sigma_\varepsilon^2}\sqrt{\frac{2}{n}}(1-\hat{\rho}^2)\hat{s}_\xi^2 + \frac{(\xi(n)-\hat{m})^2+(\xi(1)-\hat{m})^2}{\sqrt{2n}\cdot\sigma_\varepsilon^2}$$

coincides with that of ζ_n for $n \to \infty$. The solution of the equation $\zeta_n^* = 0$ yields the estimator $\hat{\sigma}_\varepsilon^2$ when one neglects a term of order $0(1/\sqrt{n})$, which completes the proof.

Theorems 1, 2 and 3 in the previous 3.3§ remain true also in the discrete case when parameters m, σ_ε^2 and ρ are considered to be unknown. To verify this we only need to prove that continuous functionals of trajectories of a discrete stationary Gaussian Markov process converge in probability to functionals of trajectories of a process continuous at the corresponding points, and that this convergence is uniform in the parameter space.

Let $\xi_n(t)$ $(0 \leq t \leq T)$ be the polygonal function associated with the process $\xi(t)$, i.e.,

(3.5.57)
$$\xi_n(t) = \xi(\frac{kT}{n}) + \frac{n}{kT}(t-\frac{kT}{n})[\xi(\frac{k+1}{n}T) - \xi(\frac{k}{n}T)],$$

for

$$\frac{kT}{n} < t \leq \frac{k+1}{n}T, \quad k = 0,1,\ldots, n-1.$$

The following lemma is true.

<u>Lemma 1. Let $\xi(t)$ be a stationary Gaussian Markov process. Then</u>

<u>uniformly in</u> $-\infty < m < \infty$, $0 < \lambda \le \lambda_0$ <u>(and</u> $2\sigma_\xi^2 = \sigma_\varepsilon^2$ = <u>constant) the following inequality is valid:</u>

$$(5.5.58) \qquad P\{\sup_{|t'-t''|<\delta} |\xi(t')-\xi(t'')| > \varepsilon\} \le \frac{2\sigma_w^2\,\delta + \sigma_w^2\,\lambda_o\,\delta^2}{\varepsilon^2}\,.$$

<u>Proof.</u> Formula (2.2.1) shows that

$$\xi(t') - \xi(t'') = -\lambda\int_{t''}^{t'} \xi(s)ds + w(t')-w(t''),$$

and therefore

$$(5.5.59) \qquad E\{\sup_{|t'-t''|<\delta} |\xi(t')-\xi(t'')|^2\} \le 2E\{\sup_{|t'-t''|<\delta} |\lambda\int_{t''}^{t'}\xi(s)ds|^2\} +$$

$$+ 2E\{\sup_{|t'-t''|<\delta} |w(t')-w(t'')|^2\} \le$$

$$\le 2\delta\int_{t''}^{t'} E(\lambda\xi(s))^2\,ds + 2\sigma_w^2\cdot\delta \le \sigma_w^2\,\lambda\delta^2 + 2\sigma_w^2\,\delta\,.$$

Using the Čebyshev inequality, we obtain (5.5.58) from (5.5.57). The case of $\lambda \to \infty$ would require spearate discussion; but, as the results in §3.3 show, in this case confidence intervals can be constructed, and therefore we can now leave it aside.

Lemma 1 ensures that the assumptions for the following statement are satisfied (see Gikhman, Skorokhod [1]) this means that the following theorem is valid:

<u>Theorem 6.</u> <u>Let</u> $\xi(t)$ <u>be a stationary Gaussian Markov process, and let</u> $\xi_n(t)$ <u>be the corresponding polygonal function (5.5.57); further, let f(t) and g(t) be continuous functions on the interval</u> $0\le t\le T$ <u>such that</u> $f(0) < \xi(0) < g(0)$.
<u>Then</u>

$$(5.5.60) \qquad \lim_{n\to\infty} P\{f(t) \le \xi_n(t) \le g(t),\ 0\le t\le T\} = P\{f(t)\le\xi(t)\le g(t),\ 0\le t\le T\}$$

<u>uniformly in</u> $-\infty < m < \infty$, $0 < \lambda \le \lambda_0$.

From Theorem 6 immediately follows that for ρ we can construct confidence intervals by the distribution (3.3.5), see Example 2 in §3.3. Further we have:

<u>Corollary.</u> Let $\xi(t)$ be a stationary Gaussian Markov process, let $\overline{h}(\xi(t))$ and $\underline{h}(\xi(t))$ $(0 \le t \le T)$ be continuous functionals and let ϵ be

a positive real number such that

(5.5.61) $P\{\underline{h}(\xi(t)) < m < \overline{h}(\xi(t))\} < 1 - \varepsilon.$

Then for any $\epsilon_1 > 0$ there exists (uniformly for $-\infty < m < \infty$ and $0 < \lambda \leq \lambda_0$) an integer n, which depends on ϵ and ϵ_1 only, such that

(5.5.62) $P\{\underline{h}(\xi_n(t)) < m < \overline{h}(\xi_n(t)) > 1 - \varepsilon - \varepsilon_1.$

This result, considered in conjunction with Theorem 3 of §3.3, means that in the discrete case no finite confidence interval can be constructed.

3.6 The Moments of Estimators and the Asymptotic Theory

As we saw the maximum likelihood estimator $\hat{\lambda}$ of the unknown parameter λ has the form (see (3.3.4))

$$\lambda = \frac{-\frac{1}{2}[\xi^2(0)+\xi^2(T)-T] + \sqrt{\frac{1}{4}[\xi^2(0)+\xi^2(T)-T]^2 + \int_0^T \xi^2(t)\,dt}}{2\int_0^T \xi^2(t)dt} .$$

This estimator is biased and in the following we shall calculate the bias and the higher moments too.

A more simple estimator can be gotten if we assume that $\xi(0)=x_0$ and in this case the Radon-Nikodym derivative has the form

$$\frac{dP_\lambda}{dP_w}(T,\xi) = \exp\{-\frac{\lambda^2}{2}\int_0^T \xi^2(t)\,dt - \lambda\int_0^T \xi(t)\,d\xi(t)\} =$$

$$= \exp\{-\frac{\lambda^2}{2}\int_0^T \xi^2(t)dt - \frac{\lambda}{2}(\xi^2(T) - \xi^2(0) - T)\},$$

and the conditional (under condition $\xi(0) = x_0$) maximum likelihood estimator $\tilde{\lambda}$ has the simple form

$$(3.6.1)\qquad \tilde{\lambda} = -\frac{\int_0^T \xi(t)d\xi(t)}{\int_0^T \xi^2(t)dt} = -\frac{\xi^2(T)-x_0^2 - T}{2\int_0^T \xi^2(t)dt} ,$$

which is used almost in every textbook (see Liptser-Shiryaev [1], Basawa-Prakasa Rao [1] and Bartlett [1]).

For simplicity let us denote the P_w measure by P_0 (this is the case when $\lambda = 0$) and E_λ means the expectation under the measure P_λ.

<u>Lemma 1. The bias of $\hat{\lambda}$ and second moment of $\tilde{\lambda} - \lambda$ are</u>

(3.6.2) $$E_\lambda(\tilde{\lambda}-\lambda) = \frac{\partial}{\partial\lambda} E_\lambda \left(\int_0^T \xi^2(t)dt \right)^{-1},$$

(3.6.3) $$E_\lambda(\tilde{\lambda}-\lambda)^2 = \frac{\partial^2}{\partial\lambda^2} E_\lambda \left(\int_0^T \xi^2(t)dt \right)^{-2} + E_\lambda \left(\int_0^T \xi^2(t)dt \right)^{-1},$$

where the negative moments can be calculated

(3.6.4) $$E_\lambda \left(\int_0^T \xi^2(t)dt \right)^{-k} = \int_0^\infty \mu^{k-1}\psi(\mu,\lambda, T)\, d\mu,$$

and (see (3.3.5) and the definition of $u(T,x)$)

(3.6.5) $$\psi(\mu,\lambda,T) = E_\lambda\, e^{-\mu\int_0^T \xi^2(t)dt} = [\cosh \Lambda T + \frac{\lambda}{\Lambda} \sinh \Lambda T]^{-1/2} \cdot$$

$$\cdot \exp\{\frac{\lambda T}{2} - \mu x_0^2 \ (\Lambda \coth \Lambda T + \lambda)^{-1}\},$$

$$\Lambda = \sqrt{\lambda^2 + 2\mu}.$$

Proof. Using (3.6.1)

$$E_\lambda(\tilde{\lambda}-\lambda) = E_0[(\tilde{\lambda}-\lambda) \exp\{ -\lambda\int_0^T \xi(t)d\xi(t) - \frac{\lambda^2}{2}\int_0^T \xi^2(t)\, dt\}] =$$

$$= E_0[\frac{-\int_0^T \xi(t)d\xi(t) - \int_0^T \xi^2(t)dt}{\int_0^T \xi^2(t)dt} \exp\{ -\lambda\int_0^T \xi(t)d\xi(t) - \frac{\lambda^2}{2}\int_0^T \xi^2(t)dt\}] =$$

$$= E_0 [\frac{\partial}{\partial\lambda} \exp\{ -\lambda\int_0^T \xi(t)\, d\xi(t) - \frac{\lambda^2}{2}\int_0^T \xi^2(t)\, dt\} \cdot \frac{1}{\int_0^T \xi^2(t)dt}]$$

and as differentation and expectation are interchangeable the right side equals to

$$\frac{\partial}{\partial\lambda} E_\lambda \left(\int_0^T \xi^2(t)\, dt \right)^{-1}$$

which gives (3.6.2). (3.6.3) can be proved analogously. (3.6.4) is the consequence of the definition of $\psi(\mu,\lambda, T)$, and Lemma is proved.

In the same way we get that for the maximum likelihood estimator $\hat{\lambda}$

$$(3.6.6)\qquad E_\lambda(\lambda-\lambda) = \frac{\partial}{\partial\lambda} E_\lambda \left(\int_0^T \xi^2(t)\,dt\right)^{-1} + $$

$$+ E_\lambda\left[\frac{-\frac{1}{2}x_o^2 + \sqrt{\frac{1}{4}(x_o^2 + \xi^2(T)-T)^2 + \int_0^T \xi^2(t)dt}}{\int_0^T \xi^2(t)\,dt}\right] .$$

<u>Lemma 2. If $\xi(0) = 0$ and T is fixed the following asymptotic results hold</u>

$$(3.6.7)\qquad E_\lambda(\tilde{\lambda}-\lambda) = \begin{cases} \frac{2}{T} \left(1 - \frac{3}{4\lambda T} + 0\left(\frac{1}{\lambda^2}\right)\right), & \lambda \to \infty, \\ \frac{1.78}{T} \left(1 + 2.34\lambda T + 0(\lambda^2)\right), & \lambda \to 0, \end{cases}$$

$$(3.6.8)\qquad E_\lambda(\tilde{\lambda}-\lambda)^2 = \begin{cases} \frac{2\lambda}{T} \left(1 + \frac{13}{2\lambda T} + 0\left(\frac{1}{\lambda^2}\right)\right), & \lambda \to \infty, \\ \frac{13.3}{T^2} \left(1 + 0.156\lambda T + 0(\lambda^2)\right), & \lambda \to 0. \end{cases}$$

<u>Remark 1.</u> (3.6.7) and (3.6.8) may be compared with Theorem 2 and Theorem 3 in §3.3.

The proof of Lemma 2 can be carried out by direct calculations. Lemma 1 and 2 were proved by A. Novikov (1972).

<u>Remark 2.</u> In Tables 1-3 we give the quantiles of empirical distributions of the estimators

Table 1.

In the table the values z_p are given for which $P_\lambda^{(n)}\{\hat{\lambda}_1 > z_p\} = p$.

λ \ p	0,05	0,1	0,9	0,95	Empirical mean
0,00001	0,04	0,01	-0,02	-0,04	$0{,}5 \cdot 10^{-2}$
0,0001	0,15	0,14	-0,04	-0,08	0,03
0,01	1,32	0,67	-0,25	-0,41	0,22
0,1	4,13	2,45	-0,39	-0,64	0,84
1	7,69	5,69	-0,09	-0,44	2,48
2	9,57	7,27	0,67	0,36	3,59
5	15,16	12,64	2,50	1,97	6,88
10	22,75	18,04	5,71	5,04	11,97

Table 2.

In the table the values z_p are given for which $p^{(n)}\{\tilde{\lambda} > z_p\} = p$.

λ \ p	0,05	0,1	0,9	0,95	Empirical mean
0,00001	$0,73-10^{-2}$	0,0007	$0,33.10^{-5}$	$0,24.10^{-5}$	$0,5.10^{-2}$
0,0001	0,03	0,009	$0,37.10^{-4}$	$0,27.10^{-4}$	0,03
0,01	0,92	0,42	$0,37.10^{-2}$	$0,27.10^{-2}$	0,21
0,1	3,96	2,21	0,04	0,03	0,82
1	6,86	5,21	0,42	0,3	2,44
2	8,64	6,86	0,91	0,74	3,44
5	13,99	11,08	2,65	2,23	6,67
10	20,78	18,45	5,87	5,09	11,69

Table 3.

In the table the values z_p are given for which $p^{(n)}\{\hat{\lambda} > z_p\} = p$.

λ \ p	0,05	0,1	0,9	0,95	Empirical mean
0,00001	0,004	0,0004	$0,18.10^{-5}$	$0,13.10^{-5}$	$0,3.10^{-2}$
0,0001	0,02	0,005	$0,2.10^{-4}$	$0,14.10^{-4}$	0,02
0,01	0,75	0,37	$0,3.10^{-2}$	$0,2.10^{-2}$	0,25
0,1	3,67	1,72	0,03	0,02	0,74
1	6,81	4,98	0,36	0,25	2,21
2	8,34	6,55	1,01	0,71	3,30
5	14,36	11,78	2,40	1,91	6,50
10	21,53	18,16	6,23	5,07	11,47

$\hat{\lambda} \sim$ maximum likelihood estimator,

$$\tilde{\lambda} \sim \frac{1}{2\int_0^1 \xi^2(t)\,dt}$$

$$\hat{\lambda}_1 \sim -N \log \frac{\sum_1^N \xi_i \xi_{i-1}}{\sum_1^N \xi_i^2},$$

$$\xi_i = \xi\left(\frac{i}{N}\right), \quad (i=0,\dots,N).$$

The number of observations in the time interval [0,1], N = 100. The sample size, n, is 1000 (for $\lambda < 10^{-3}$ it is 2000). These tables show that the maximum likelihood estimator $\hat{\lambda}$ is better, for $\lambda << 1$, then the conditional one, $\tilde{\lambda}$.

3.6.1. Sequential Estimation. Let τ be a Markov time with respect to the system $F_t^{\xi}(t \geq 0)$ and $\delta(\tau(\xi),\xi)$ an estimator of the function $g(\lambda)$ on the basis of observations of the trajectory over the time interval $[0,\tau(\xi)]$. $\Delta = (\tau,\delta)$ prescribes a sequential estimation plan. An inequality analogous to Cramer-Rao inequality can be obtained, which we shall call as Cramer-Rao-Wolfowitz inequality, which provides a lower bound for the value $E_\lambda\,[g(\lambda) - \delta(\tau,\xi)]^2$.

The density function is given in the following form

$$(3.6.9) \qquad f(\lambda,\xi) = \frac{dP_\lambda}{dP_0}(\tau(\xi),\xi) = \exp\{-\int_0^{\tau(\xi)} \xi(t)d\xi(t) - \frac{\lambda^2}{2}\cdot \int_0^{\tau(\xi)} \xi^2(t)\,dt\},$$

and we easily verify that the regularity conditions of 3.1 holds (interchangeability of expectation and derivation). We assume that $g(\lambda)$ is differentiable, and so is

$$b(\lambda) = E_0\,\delta(\tau(\xi),\xi)\,f(\lambda,\xi) - g(\lambda)$$

$$\frac{d}{d\lambda}[b(\lambda) + g(\lambda)] = E_0\,\delta(\tau(\xi),\xi)\,\frac{\partial f(\lambda,\xi)}{\partial\lambda}$$

Theorem 1. Let $\Delta = (\tau,\delta)$ be a sequential scheme of estimation with $E_\lambda\,\delta^2(\tau,\xi) < \infty \quad (\lambda \geq 0)$.
Then

$$(3.6.10) \qquad E_\lambda(g(\lambda) - \delta(\tau,\xi))^2 \geq \frac{\left(\frac{d}{d\lambda}[g(\lambda) + b(\lambda)]\right)^2}{E_\lambda \int_0^{\tau(\xi)} \xi^2(t)\,dt} + b^2(\lambda).$$

In the case when Δ is unbiased, i.e., $b(\lambda) = 0$ and $g(\lambda) = \lambda$

$$(3.6.11) \qquad E_\lambda\,[\lambda-\delta(\tau,\xi)]^2 \geq \frac{1}{E_\lambda \int_0^{\tau(\xi)} \xi^2(t)\,dt}.$$

The proof may be found in Liptser-Shiryaev [1] (§7.8). Let H be a non-negative number. Define

$$\tau(H) = \inf\{t: \int_0^t \xi^2(s)\,ds \geq H\},$$

where $P(\tau(H) < \infty) = 1$.

The sequential scheme $\Delta_H = \Delta(\tau(H), \delta_H)$ with

$$\delta_H(\tau,\xi) = -\frac{1}{H}\int_0^{\tau(H,\xi)} \xi(s)d\xi(s) = -\frac{\xi^2(\tau(H)) - \xi^2(0) - \tau(H)}{2\int_0^{\tau_H} \xi^2(s)\,ds}$$

define the sequential likelihood equation estimator of λ.

Theorem 2. The sequential plan $\Delta_H = (\tau(H), \delta(\tau,\xi))$ has the following properties for all $\lambda \geq 0$.

(i) $\delta_H(\tau,\xi)$ is normally distributed with

(3.6.12) $$E_\lambda\,(\delta_H(\tau,\xi)) = \lambda, \qquad D^2\,(\delta_H(\tau,\xi)) = \frac{1}{H},$$

(ii) the plan Δ_H is efficient in the sense that

(3.6.13) $$E_\lambda\,(\delta_H(\tau,\xi) - \lambda)^2 \leq E_\lambda\,(\delta(\xi) - \lambda)^2.$$

Proof. As $d\xi = -\lambda\xi(t)\,dt + dw\,(t)$ we get

$$\delta_H\,(\tau,\xi) = -\frac{\int_0^{\tau(\xi)} \xi(t)d\xi(t)}{H} = -\frac{-\lambda\int_0^{\tau(\xi)} \xi^2(t)dt + \int_0^{\tau(\xi)} \xi(t)dw(t)}{H} =$$

$$= -\frac{\int_0^{\tau} \xi(t)\,dw\,(t)}{H} + \lambda.$$

It is well known (see Appendix B1, Theorem 7) that the random process

$$(H) = \int_0^{\tau(H)} \xi(t)\,dw\,(t)$$

as a function of H is a Wiener-process which proves (i).

From the Cramer-Rao-Wolfowitz inequality (Theorem 1) we get, that for any unbiased scheme $\Delta = \Delta(\tau,\delta)$ with $E_\lambda\,\delta^2 < \infty$,

$$E_\lambda \int_0^{\tau} \xi^2(t)\,dt \leq H,$$

$$E_\lambda\,(\delta(\xi) - \lambda)^2 \geq [E_\lambda \int_0^{\tau} \xi^2(t)dt]^{-1} \geq \frac{1}{H} = E_\lambda[\delta_H(\tau,H) - \lambda]^2,$$

which proves (ii).

<u>Remark.</u> Theorems 1 and 2 remain valid (under some natural conditions on $a(t,\xi)$) if we regard the process

$$d\xi(t) = \lambda\, a(t,\xi(t))dt + dw(t),$$

(see Liptser-Shiryaev [1]§17.5).

A natural question arises whether the advantages of sequential estimators are consequences of a rather long mean observation time $E_\lambda(\tau(H))$? For general $a(t,\xi)$ this question is unsolved. Novikov (1972) examined the moments of $\tau(H,\xi)$. The following statement is true.

<u>Theorem 3. For</u> $\lambda \geq 0$ as $T \to \infty$,

$$(3.6.14) \qquad P_\lambda(\tau(H) \geq T) = 4\left(\frac{H}{\pi T^2}\right)^{1/2} \exp\left\{-\frac{\lambda^2 H}{2} - \frac{T^2}{8H} + \frac{\lambda T}{2}\right\}(1+\sigma(1)),$$

$$(3.6.15) \qquad E_\lambda(\tau(H)) \leq 2\,[\lambda H + 2\sqrt{H}\,] + \sqrt{8(\lambda^2 H^2 + 4\lambda H) + 2H}\,,$$

<u>Further, if $\lambda^2 H \to \infty$, then</u>

$$(3.6.16) \qquad E_\lambda\, \tau(H) = 2\lambda H\left(1 + \frac{3}{4\lambda^2 H} + \sigma\left(\frac{1}{\lambda^2 H}\right)^2\right),$$

<u>and if $\lambda^2 H \to 0$, then</u>

$$(3.6.17) \qquad E_\lambda\, \tau(H) = H^{1/2}\,[2.09 + 0.856\,\lambda H^{1/2} + \sigma(\lambda^2 H)].$$

CHAPTER 4.

THE MULTI-DIMENSIONAL PROCESSES

4.1. The Complex Process.

Suppose that on the interval $0 \leq t \leq T$ we observe the two-dimensional process $\underline{\xi}(t) = (\xi^1(t), \xi^2(t))$ with zero mean $E\xi^1(t) = E\xi^2(t) = 0$ and differential

$$(4.1.1) \qquad d\xi^1(t) = -\lambda\xi^1(t)\,dt - \omega\xi^2(t)\,dt + dw^1(t),$$

$$d\xi^2(t) = -\lambda\xi^2(t)\,dt + \omega\xi^1(t)\,dt + dw^2(t).$$

Here $w^1(t)$, $w^2(t)$ are independent Wiener processes $E\,(dw^i(t))^2 = a\cdot dt$, and they are independent of $\underline{\xi}(0)$. Assuming that

$$\eta(t) = \xi^1(t) + i\xi^2(t), \quad \chi(t) = w^1(t) + iw^2(t), \quad \gamma = \lambda - i\omega,$$

we can rewrite the system (4.1.1) as a single differential equation:

$$(4.1.1) \qquad d\eta(t) = -\gamma\eta(t)\,dt + d\chi(t).$$

The complex correlation function of the process has the form $(\lambda > 0)$

$$(4.1.2) \qquad C(\tau) = A(\tau) + iB(\tau) = E\eta(t)\,\overline{\eta}(t+\tau) = \frac{a}{\lambda}\exp\{-\lambda|\tau| - i\omega\tau\}.$$

Using a realization of $\eta(t)$ on the interval $[0,T]$ the empirical correlation function is

$$(4.1.3) \qquad c(\tau) = a(\tau) + ib(\tau) = \frac{1}{T-\tau}\int_0^{T-\tau} \eta(t)\overline{\eta}(t+\tau)\,dt,$$

which is differentiable from the right at 0 with probability 1, and the following statement is true.

Lemma 1.

$$(4.1.4) \qquad c'(0) = -a - \frac{1}{T}s_1^2 + \frac{1}{T}s_2^2 - ir,$$

where a is a parameter characterizing the intensity of the "white noise" processes $w^i(t)$, while

$$(4.1.5) \qquad s_1^2 = \frac{1}{2}[|\eta(0)|^2 + |\eta(T)|^2], \quad s_2^2 = \frac{1}{T}\int_0^T |\eta(t)|^2\,dt,$$

$$r = \frac{1}{T} \int_0^T |\eta(t)|^2 d\theta.$$

The variable $\theta(t)$ is determined in the last integral by

$$\eta(t) = |\eta(t)| e^{i\theta(t)}$$

It can easily be shown that (if $h \sim 0$)

$$\frac{c(\tau+h) - c(\tau)}{h} = \frac{1}{T-\tau} \int_0^{T-\tau} \eta(t) d\bar{\eta}(t) + \frac{1}{(T-\tau)^2} \int_0^{T-\tau} \eta(t)\bar{\eta}(t+h) dt -$$

$$- \frac{1}{T-\tau} \eta(T-\tau)\bar{\eta}(T) + \sigma(1).$$

On the other hand, uwing the relation

$$\sum_2^n |\eta(t_i) - \eta(t_{i-1})|^2 \to 2aT$$

valid for a complex process, it is easy to see that the value of the integral

$\int_0^T \eta(t) d\eta(t)$, understood as the limit of the sum

$$\sum_2^n \eta(t_{i-1})[\eta(t_i) - \eta(t_{i-1})]$$

is equal to

$$(4.1.6) \qquad -aT + \frac{|\eta(T)|^2 - |\eta(0)|^2}{2} - i \int_0^T |\eta(t)|^2 d\theta.$$

In fact, with simple calculations it can be seen that (see (2.3.20) - (2.3.21)

$$\sum_2^n |\eta(t_i) - \eta(t_{i-1})|^2 = \sum_2^n [\eta(t_{i-1})\bar{\eta}(t_i) - \eta(t_i)\bar{\eta}(t_{i-1})] + |\eta(T)|^2 - |\eta(0)|^2 +$$

$$- 2\sum_2^n \eta(t_{i-1})\overline{[\eta(t_i) - \eta(t_{i-1})]} = - 2 \sum_2^n \eta(t_{i-1})\overline{[\eta(t_i) - \eta(t_{i-1})]} +$$

$$+ \sum_2^n |\eta(t_i)||\eta(t_{i-1})| \, [-e^{i(\theta(t_i) - \theta(t_{i-1}))} + e^{i(\theta(t_{i-1}) - \theta(t_i))}].$$

From (4.1.6) we now get (4.1.4).

In changing to discrete-time case the following relation is very useful

$$(4.1.5') \qquad \int_0^T |\eta(t)|^2 \, d\theta = \int_0^T (\xi^1(t)d\xi^2(t) - \xi^2(t)d\xi^1(t)) ,$$

which can easily be proved starting from the identity

$$\Sigma[\eta(t_i)\bar{\eta}(t_{i-1})-\eta(t_{i-1})\bar{\eta}(t_i)] = 2i\Sigma[\xi^2(t_{i-1})(\xi^1(t_i)-\xi^1(t_{i-1})) -$$

$$-\xi^1(t_{i-1})(\xi^2(t_i)-\xi^2(t_{i-1}))].$$

The diffusion parameter a can be precisely estimated from a single realization, since (see (2.2.29))

$$\sum_1^n |\xi^1(t_i) - \xi^1(t_{i-1})|^2 \to aT, \qquad n \to \infty , \max |t_i-t_{i-1}| \to 0.$$

Therefore the unknown parameters are λ and ω. Let P be the measure on the space of realizations on $[0,T]$ associated with the process $\eta(t)$. Let us introduce on the same space the standard measure $P_0=L\times P_w$, where L is the usual Lebesgue measure in the $\eta(0)$ plane, and let P_w be the two-dimensional Wiener measure on the space of increments $\eta(0)$ with the parameters of the process w(t) (see (4.1.1')). It is known (Ch. 2.§3, (2.3.46)) that

$$(4.1.7) \qquad \frac{dP}{dP_0} = \frac{\lambda}{\pi a^2} \exp\left[-\frac{\lambda^2+\omega^2}{2a} T\, s_2^2 - \frac{\lambda}{a} s_1^2 + \lambda T + \frac{\omega}{a} T\, r\right].$$

According to (4.1.7) the system s_1^2, s_2^2, r is a sufficient system for the problem. Taking derivatives with respect to ω and λ in the formula

$$L = \log \frac{dP}{dP_0} = - \log \pi a^2 + \log\lambda - \frac{\lambda^2+\omega^2}{2a} Ts_2^2 - \frac{\lambda}{a} s_1^2 + \lambda T + \frac{\omega}{a} T\, r$$

we get

$$(4.1.8) \qquad \frac{\partial L}{\partial \omega} = - \frac{\omega}{a} T\, s_2^2 + \frac{T}{a} r = 0,$$

$$(4.1.9) \qquad \frac{\partial L}{\partial \lambda} = \frac{1}{\lambda} - \frac{\lambda}{a} T\, s_2^2 - \frac{s_1^2}{a} + T = 0,$$

from which the maximum likelihood estimators $\hat{\omega}$ and $\hat{\lambda}$ can be obtained. From (4.1.8) we have

(4.1.8′) $$\hat{\omega} = \frac{r}{s_2^2}$$

and, denoting $\lambda T = \kappa$ and $\hat{\lambda} T = \hat{\kappa}$, $\hat{\kappa}$ can be determined from (4.1.9) as a solution of the equation

(4.1.9′) $$\frac{s_2^2}{aT} x^2 + \left(\frac{s_1^2}{aT} - 1\right) x - 1 = 0$$

Lemma 2. For each t the Gaussian vedtor $(\xi^1(t), \xi^2(t))$ has independent components with $E\,(\xi^i(t))^2 = \frac{a}{2\lambda}$, $i = 1,2$, and the density function of $\xi^1(0)$, $\xi^2(0)$ (see (2.3.43)) has the form

(4.1.10) $$f(x_1, x_2) = \frac{\lambda}{\pi a^2} \exp\{- \frac{\lambda}{a} (x_1^2 + x_2^2)\}.$$

Proof. Because of stationarity the eigenvalues of $A = \begin{pmatrix} -\lambda & -\omega \\ \omega & -\lambda \end{pmatrix}$ must lie on the left half-plane we have $\lambda > 0$. Further $B(0) = E\, \underline{\xi}(t)\underline{\xi}^*(t)$ (see (2.2.3)) is the unique solution of $A\,B(0) + B(0)\,A^* = - B_w$, where $B_w = \begin{pmatrix} a & 0 \\ 0 & a \end{pmatrix}$, i.e.

(4.1.11) $$B(0) = \begin{pmatrix} \frac{a}{2\lambda} & 0 \\ 0 & \frac{a}{2\lambda} \end{pmatrix}, \qquad B(t) = e^{-\lambda|t|} \begin{pmatrix} \cos\omega t & \sin\omega t \\ -\sin\omega t & \cos\omega t \end{pmatrix} B(0).$$

Equations (4.1.8′) and (4.1.9′) show that the estimators of the damping parameter λ, and period ω can be treated independently and so we shall do. We expect that $\hat{\lambda}$ (or $\hat{\kappa}$) has a similar distribution as in the one-dimensional case, and it turns out that the characteristic function of s_1^2, Ts_2^2 (see (4.2.16)) is the square of that in one dimensional case (see (3.3.5)).

An interesting feature will be that for estimate $\hat{\omega}$ in (4.1.8′) we get an exact, and not asymptotic, distribution which does not depend upon (λ,ω). We shall prove that

(4.1.12) $$\sqrt{\frac{T\, s_2^2}{a}}\; (\hat{\omega} - \omega)$$

has a Gaussian distribution N(0,1). Heuristically this can easily be explained as

$$(\hat{\omega}-\omega)\sqrt{\frac{Ts_2^2}{a}} = \sqrt{\frac{1}{aT}}\ \frac{\int_0^T |\eta(t)|^2(d\theta-\omega dt)}{(\int_0^T |\eta(t)|^2 dt)^{1/2}},$$

and for "small"

$$(\hat{\omega}-\omega)\sqrt{\frac{T\ s_2^2}{a}} \approx |\eta(T)|\ (d\theta-\omega dt)/\sqrt{adT} = (dw^1\cos\theta - dw^2\ \sin\theta)/\sqrt{adT},$$

which is normally distributed.

4.2 Construction of Confidence Intervals for Parameter λ.

In this section we shall assume that, except of λ, all parameters of a complex stationary Gaussian Markov process $\eta(t)$ are known. For the sake of simplicity let $E\ \xi^1(t) = E\xi^2(t) = 0$ and $E\ (\xi^1(t))^2 = E(\xi^2(t))^2 = 1/2\lambda$. Under these assumptions we shall determine the characteristic function of a sufficient system of statistics and shown a method for obtaining the corresponding confidence intervals.

It can be seen from (4.1.7) that the statistics

$$\chi_1(T) = \frac{1}{2}\{|\eta(0)|^2 + |\eta(T)|^2\}, \tag{4.2.1}$$

$$\chi_2(T) = \int_0^T |\eta(t)|^2 dt = \int_0^T ((\xi^1(t))^2 + (\xi^2(t))^2)dt, \tag{4.2.2}$$

form a sufficient system. To determine the characteristic function of these variables, the following partial differential equation may be considered (the problem of existence and uniqueness in differential equations connected with various functionals are dealt with in Gihman - Skorokhod [1])

$$u(T,x,y) = E\ \{e^{i\alpha_1\chi_1(T)+\alpha_2\chi_2(T)}\ |\xi^1(0)=x,\xi^2(0)=y\}, \tag{4.2.3}$$

i.e. $u(T,x,y)$ is the conditional characteristic function of χ_1 and χ_2 under the condition that $\xi^1(0) = x$ and $\xi^2(0) = y$. Evidently

$$\frac{\partial u}{\partial T} = \frac{1}{2}\left(\frac{\partial^2 u}{\partial x^2} + \frac{\partial^2 u}{\partial y^2}\right) + \frac{\partial u}{\partial x}(-i\alpha_1 x-\lambda x-\omega y)+\frac{\partial u}{\partial y}(-i\alpha_1 y + \tag{4.2.4}$$
$$+\ \omega x-\lambda y)+u[-i\alpha_1+i\alpha_1\lambda(x^2+y^2) - \frac{\alpha_1^2}{2}(x^2+y^2)+i\alpha_2(x^2+y^2)\ ].$$

Let

$$u(T,x,y) = u_1(T,x,y)\, e^{i\alpha_1 \frac{x^2+y^2}{2}} ,$$

hence we have from (2.4)

(4.2.5) $$\frac{\partial u_1}{\partial T} = \frac{1}{2}\left(\frac{\partial^2 u_1}{\partial x^2}+\frac{\partial^2 u_1}{\partial y^2}\right)+ \frac{\partial u_1}{\partial x}(\lambda x+\omega y)+ \frac{\partial u_1}{\partial y}(-\omega x+\lambda y)+u_1 i\alpha_2\cdot$$

$$\cdot(x^2+y^2),$$

or in polar coordinates

(4.2.5′) $$\frac{\partial u_1}{\partial T} = \frac{1}{2}\frac{\partial^2 u_1}{\partial r^2} + \frac{\partial u_1}{\partial r}\left(\frac{1}{2r} + \lambda r\right) + u_1 i\alpha_1 r^2.$$

After the transformation $u_1(T,r) = v(T,r^2) = v(T,\rho)$, the last function satisfies

(4.2.6) $$\frac{\partial v}{\partial T} = 2\rho\frac{\partial^2 v}{\partial \rho^2} + 2(1-\lambda\rho)\frac{\partial v}{\partial \rho} + i\alpha_2 \rho v$$

with the initial condition

(4.3.7) $$v(0,\rho) = e^{\frac{i\alpha_1}{2}\rho}.$$

Let w be the Laplace transform of v, i.e.

$$w(T,\gamma) = \int_0^\infty e^{-\gamma\rho}\, v(T,\rho)\, d\rho .$$

From (4.2.6) and (4.2.7) we have

(4.2.8) $$\frac{\partial w}{\partial T} = \frac{\partial w}{\partial \gamma}(-2\gamma^2 + 2\lambda\gamma - i\alpha_2) + w\cdot(2\lambda - 2\gamma),$$

(4.2.9) $$w(0,\gamma) = \frac{1}{\gamma - \frac{i\alpha_1}{2}}$$

The solution of (4.2.8) can easily be obtained by commonly known methods (see, for example Stepanov, Chapter VIII, §2). Let the solutions of the equation

(4.2.10) $$\gamma^2 - \lambda\gamma + \frac{i\alpha_1}{2} = 0, \qquad \gamma_{1,2} = \frac{\lambda \pm \sqrt{\lambda^2 - 2i\alpha_1}}{2}$$

be denoted by γ_1 and γ_2. The first integrals of (4.2.8) are

(4.2.11) $$c_1 = T - \frac{1}{2(\gamma_1-\gamma_2)} \log \frac{\gamma-\gamma_1}{\gamma-\gamma_2} ,$$

$$c_2 = \log w + \frac{1}{2} \log (\gamma-\gamma_1)(\gamma-\gamma_2) - \frac{\lambda}{2(\gamma_1-\gamma_2)} \log \frac{\gamma-\gamma_1}{\gamma-\gamma_2} .$$

for T = 0 we have

(4.2.12) $$\frac{\gamma-\gamma_1}{\gamma-\gamma_2} = e^{-2(\gamma_1-\gamma_2)c_1} ,$$

$$\gamma = \frac{\gamma_1-\gamma_2 e^{-2(\gamma_1-\gamma_2)c_1}}{1-e^{-2(\gamma_1-\gamma_2)c_1}} ,$$

and therefore

(4.2.13) $$\log w + \frac{1}{2} \log \frac{e^{-2(\gamma_1-\gamma_2)c_1} (\gamma_1-\gamma_2)^2}{(1-e^{-2(\gamma_1-\gamma_2)c_1})^2} + \lambda c_1 = c_2,$$

$$w = \frac{(1-e^{-2(\gamma_1-\gamma_2)c_1})}{(\gamma_1-\gamma_2)\, e^{-c_1(\gamma_1-\gamma_2)}} e^{c_2-\lambda c_1} .$$

From (4.2.9) and (4.2.13) we get

$$\frac{1 - e^{-2c_1(\gamma_1-\gamma_2)}}{e^{-c_1(\gamma_1-\gamma_2)} (\gamma_1-\gamma_2)} e^{c_2-\lambda c_1} - \frac{1}{\dfrac{\gamma_1 -\gamma_2 e^{-2(\gamma_1-\gamma_2)c_1}}{1- e^{-2(\gamma_1-\gamma_2)c_1}} - \dfrac{i\alpha_1}{2}} = 0.$$

Substituting (4.2.11), we obtain

$$\frac{1 - \frac{\gamma-\gamma_1}{\gamma-\gamma_2} e^{-2T(\gamma_1-\gamma_2)}}{e^{-T(\gamma_1-\gamma_2)} \sqrt{\frac{\gamma-\gamma_1}{\gamma-\gamma_2}} (\gamma_1-\gamma_2)} \cdot e^{\log w+1/2\log(\gamma-\gamma_1)(\gamma-\gamma_2)-\frac{\lambda}{2(\gamma_1-\gamma_2)}\log\frac{\gamma-\gamma_1}{\gamma-\gamma_2}} \cdot$$

$$\cdot\, e^{-\lambda T+ \frac{\lambda}{2(\gamma_1-\gamma_2)} \log \frac{\gamma-\gamma_1}{\gamma-\gamma_2}} = \left[\frac{\gamma_1-\gamma_2 \frac{\gamma-\gamma_1}{\gamma-\gamma_2} e^{-2T(\gamma_1-\gamma_2)}}{1 - \frac{\gamma-\gamma_1}{\gamma-\gamma_2} e^{-2T(\gamma_1-\gamma_2)}} - \frac{i\alpha_1}{2} \right]^{-1} ,$$

whence

(4.2.14)

$$w = \frac{(\gamma_1 - \gamma_2)\, e^{\lambda T - T(\gamma_1 - \gamma_2)}}{(\gamma-\gamma_2)(\gamma_1 - \frac{i\alpha_1}{2}) + (\gamma-\gamma_1)(\frac{i\alpha_1}{2} - \gamma_2)\, e^{-2T(\gamma_1-\gamma_2)}} .$$

By means of the substitution $\gamma = \lambda - i\alpha_1/2$ and (4.2.10) the (unconditioned) characteristic function of the random variables $\chi_1(T)$ and $\chi_2(T)$ can be obtained from (4.2.14):

(4.2.15)

$$u_{\chi_1,\chi_2}(T) = E\,(e^{i\alpha_1\chi_1(T)+i\alpha_2\chi_2(T)}) = w(T,\lambda - \frac{i\alpha_1}{2}) =$$

$$= \frac{4\lambda(\lambda^2-2i\alpha_2)^{1/2}\, e^{\lambda T - T\sqrt{\lambda^2-2i\alpha_2}}}{(\lambda-i\alpha_1+\sqrt{\lambda^2-2i\alpha_2})^2 - (\lambda-i\alpha_1-\sqrt{\lambda^2-2i\alpha_2})^2\, e^{-2T\sqrt{\lambda^2-2i\alpha_2}}} .$$

So we proved the following statement.

<u>Theorem 1. Let</u> $\kappa = \lambda T$; <u>the characteristic function of the random variables</u> $\lambda\chi_1$ <u>and</u> $\lambda^2\chi_2$ <u>will be</u>

(4.2.16)

$$\frac{4(1-2i\alpha_2)^{1/2}\, e^{\kappa}}{(1-i\alpha_1+\sqrt{1-2i\alpha_2})^2\, e^{\kappa\sqrt{1-2i\alpha_2}} - (1-i\alpha_1-\sqrt{1-2i\alpha_2})^2\, e^{-\kappa\sqrt{1-2i\alpha_2}}} .$$

As follows from (4.1.9'), the maximum likelihood equation for the unknown parameter λ will take the following form:

$$\frac{1}{\lambda} - (\chi_1 - T) - \lambda\chi_2 = 0.$$

This has the unique positive solution

$$\lambda = \frac{-(\chi_1-T) + \sqrt{(\chi_1-T)^2 + 4\chi_2}}{2\chi_2} . \tag{4.2.17}$$

In order to determine the distribution of the estimator $\hat{\lambda}$ let us consider the relation

$$(4.2.18)\quad P_\lambda(\hat{\lambda}<x) = P_\lambda\{\chi_1 - \frac{1}{x^2} - \frac{1}{x}(T-\chi_1) > 0\} = P_\lambda\{x^2\chi_2 + x\chi_1 > Tx+1\},$$

from which, substituting $x = \lambda y$, $\zeta_y = \lambda y \chi_1 + \lambda^2 y^2 \chi_2$ and $\lambda T = \kappa$, we obtain

$$(4.2.19)\quad P_\lambda\{\hat{\lambda} < \lambda y\} = P_\kappa\{\zeta_y > \kappa y + 1\},$$

where the characteristic function of ζ_y becomes

(4.2.20)

$$\frac{4(1-2iy^2\alpha)^{1/2}\, e^{\kappa}}{(1-iy\alpha+\sqrt{1-2iy^2\alpha})^2\, e^{\kappa\sqrt{1-2iy^2\alpha}} - (1-iy\alpha-\sqrt{1-i2y^2\alpha})^2\, e^{-\kappa\sqrt{1-2iy^2\alpha}}},$$

as follows from (4.2.16). If κ and y are given, the relevant probabilities can be calculated from (4.2.20).

Let $p = 1 - 2iy^2\alpha$; then from (4.2.20) the Laplace transform of the distribution function of the random variable ζ_y can be obtained:

$$\frac{8y^4 e^{\kappa}}{(y-1)^2}\ \frac{p\, e^{-\kappa\sqrt{p}}}{(p-1)}\ \left[\frac{1}{(\sqrt{p}+1)^2} + \frac{1}{(\sqrt{p}+2y-1)^2} - \frac{2}{(\sqrt{p}+1)(\sqrt{p}+2y-1)}\right]\cdot$$

$$\cdot \sum_{k=0}^{\infty} \left(\frac{1-\frac{p-1}{2y}-\sqrt{p}}{1-\frac{p-1}{2y}+\sqrt{p}}\right)^{2k} \cdot e^{-2k\kappa\sqrt{p}},$$

or, substituting $s^2 = p$ and $a = 2y - 1$,

(4.2.21)

$$\frac{8y^4 e^{\kappa}}{(y-1)^2} \sum_{k=0}^{\infty} e^{-\kappa(2k+1)s}\ \frac{s^2}{(s-1)(s+1)(s+a)} \cdot \frac{(s-1)^{2k}}{(s+1)^{2k+1}} \cdot \frac{(s-a)^{2k}}{(s+a)^{2k+1}}\ .$$

The first term in the sum (3.21) has the inverse Laplace transform (using e.g. the tables of Ditkin and Kuznecov [1])

$$\frac{1}{2}\left[1-\Phi\left(\frac{y}{\sqrt{2(y+1)}} - \frac{\sqrt{\kappa y+1}}{y\sqrt{2}}\right)\right] +$$

$$+ [1 - \tilde{\Phi}(\frac{\kappa y}{\sqrt{\kappa y+1}} + \frac{\sqrt{\kappa y+1}}{y\sqrt{2}})] e^{2\kappa} [- \frac{y^2(y^2+4y-2)}{2(y-1)^4} +$$

$$+ \frac{(6y-2)(\kappa y^2+\kappa y+1)}{2(y-1)^3} - \frac{\kappa y+1}{(y-1)^2} - \frac{(\kappa y^2+\kappa y+1)^2}{y^2(y-1)^2}]+$$

(4.2.22)

$$+[1- \tilde{\Phi}(\frac{\kappa y}{\sqrt{2(\kappa y+1)}} + \frac{(2y-1)\sqrt{\kappa y+1}}{y\sqrt{2}})]e^{2\kappa y + \frac{\kappa y+1}{2y^2}[(2y-1)^2-1]} \cdot$$

$$\cdot [\frac{(2y-1)(4y^2-2y+1)}{2(y-1)^4} - \frac{(2y-1)^2}{y(y-1)^3} (\kappa y^2+(2y-1)(\kappa y+1))] +$$

$$+ \sqrt{\frac{2}{\pi}} \frac{\sqrt{\kappa y+1}}{y(y-1)^2} e^{\kappa - \frac{\kappa y+1}{2y^2} \frac{\kappa^2 y^2}{2(\kappa y+1)}} [\frac{7y^2-5y+1}{y(y-1)} y^2+\kappa y^2+\kappa y+1],$$

where

$$\tilde{\Phi}(x) = \frac{2}{\sqrt{\pi}} \int_0^x e^{-u^2} du .$$

This approximation of the distribution of ζ_y is satisfactory with large values of κ. A better approximation can be obtained by recalling that the inverse Laplace transform of

$$\frac{(p-a)^{2k}}{(p+a)^{2k+1}}$$

equals

$$e^{-ax} [\frac{e^{2a x}}{k!} \frac{d^k (2ax)^k e^{-2ax}}{dx^k}] ,$$

consequently the corresponding Laguerre polynomials will appear. Using the theorem on the inverse Laplace transform, for small values of κ the density function of ζ_y becomes

$$f_\zeta(x) = \sum_k e^{s_k x} b(s_k)$$

where s_κ stand for the poles of (2.20) and $b(s_\kappa)$ stand for the corresponding residues. The equation which determines the poles is

$$\frac{(1+ys+\sqrt{1+2y^2 s})^2}{(1+ys-\sqrt{1+2y^2 s})^2} = e^{-2\kappa\sqrt{1+2y^2 s}} .$$

The characteristic function (4.2.20) permits one to determine the distribution function of the random variable ζ_y, and hence the distribution function $F(z,\kappa) = P\{\hat{\kappa} < z\}$ of the random variable $\hat{\kappa}$ as well (the variables ζ_y and $\hat{\kappa}$ are continuously distributed). We have established that, for fixed z, $F(z,\kappa)$ is a monotonically increasing continuous function of κ, $0 < \kappa < \infty$, taking on values from 0 to 1. Therefore, the solution $\kappa_p(\hat{\kappa})$ of the equation

$$(4.2.23) \qquad F(\hat{\kappa},\kappa) = 1-p \ , \quad 0 < p < 1 \ ,$$

(with respect to κ) is a lower confidence limit for the unknown value of κ, the confidence coefficient being equal to p:

$$p = P(\ \kappa_p(\hat{\kappa}) < \hat{\kappa}) = P\{F(\hat{\kappa},\ \kappa_p(\hat{\kappa}) < F(\hat{\kappa},\kappa)\ \}= P(1-p < F(\hat{\kappa},\kappa)).$$

Hence, in particular, it follows that $\kappa_{1-p}(\hat{\kappa})$ is an upper confidence limit for κ with confidence coefficient p. Table 1. of the function $\hat{\kappa}_p(\kappa)$, inverse to $\kappa = \kappa_p(\hat{\kappa})$, is given on page 234 (in other words, $z = \kappa_p(\kappa)$ - the solution of the quation $F(z,\kappa) = 1 - p$ with respect to z - is the (1-p) - quantile of the distribution of the random variable $\hat{\kappa}$ for a given value of the parameter κ). In order to find the confidence limit $\kappa_p(\hat{\kappa})$ we must solve the equation $\hat{\kappa}_p(\kappa) = \hat{\kappa}$ using an inverse interpolation of this table.

<u>Theorem 2.</u> <u>If $\kappa \to 0$</u>

$$(4.2.25) \qquad \lim_{\kappa\to 0} P_\kappa\ \{\hat{\kappa} < y\kappa\ \} = \exp\ \{-\ \frac{1}{y}\}\ ,$$

<u>which means that the ratio $\hat{\kappa}/\kappa$ has a χ^2 distribution with two degrees of freedom.</u>

<u>Proof.</u> Immediately follows from (4.2.19) and (4.2.20) by direct calculations.

<u>Theorem 3.</u> <u>For large κ, i.e., $\kappa \to \infty$ we have</u>

$$(4.2.26) \qquad \lim_{\kappa\to\infty} P_\kappa\ \{\kappa < \hat{\kappa}+ y\sqrt{\hat{\kappa}}\} = \frac{1}{\sqrt{2\pi}} \int_{-\infty}^{y} e^{-t^2/2}\ dt\ ,$$

<u>and $\hat{\kappa}$ is of normal distribution with variance $D^2\ \hat{\kappa} \sim \kappa$.</u> The proof comes again immediately from (4.2.16) and (4.2.20).

Because $\hat{\kappa}$ has a continuous distribution, there exists for any $0 < p < 1$ and $0 < \kappa < \infty$ a κ that $P_\kappa(\hat{\kappa}>k) = p$. From $k = k_p\ (\kappa)$ we get $\kappa = \kappa_p(k)$, so evidently

$$P_\kappa \{\kappa < \kappa_p(\hat{\kappa})\} \equiv p ;$$

and $\kappa_p(\hat{\kappa})$ becomes the limit of the confidence interval for κ. Note that for $\kappa \to \infty$ the characteristic function $\psi(\alpha)$ of random variable $\frac{\hat{\kappa}-\kappa}{\sqrt{\kappa}}$

$$(4.2.26') \qquad |\psi(\alpha) - e^{-\frac{\alpha^2}{2}}| < \frac{c}{\sqrt{\kappa}} ,$$

where c is a constant.

As our calculations show, normal approximation holds only for values of $\kappa > 100$. For comparison exact values of $y = \hat{\kappa}_p(\kappa)/\kappa$, are given, computed from our Table 1 (the first row), as well as values of y obtained with the aid of normal approximation (the second row).

Table 2/a

p	0.001	0.010	0.025	0.050	0.100	0.900	0.950	0.975	0.990
κ=10	1.972	1.734	1.620	1.516	1.403	0.597	0.484	0.380	0.266
	2.575	2.073	1.867	1.701	1.530	0.714	0.641	0.588	0.527
κ=100	1.3090	1.2326	1.1960	1.1646	1.1281	0.8719	0.8355	0.8040	0.7674
	1.3641	1.2654	1.2205	1.1832	1.1413	0.8847	0.8533	0.8269	0.7972

For y = 1, i.e., for probabilities $P_\kappa\{\hat{\kappa} > \kappa\}$, the following values

Table 2/b

κ	100	300	500	1000
p	0.5196	0.5109	0.5073	0.5018

are obtained.

Let F(x) be the distribution function of the random variable ζ_y, then

$$F(x) = \frac{1}{2\pi i} \int_{\sigma-i\infty}^{\sigma+i\infty} e^{px} F^*(p)\, dp ,$$

where (see (4.2.20)) $p = \sigma + is$, $-\infty < s < \infty$,

$$F^*(p) = \frac{4(1+2y^2p)^{1/2} \exp(\kappa-\kappa(1+2y^2p)^{1/2})}{p(1+yp+\sqrt{1+2y^2p})^2 - (1+yp-\sqrt{1+2y^2p})^2 \exp(-2\kappa(1+2y^2p)^{1/2})} ,$$

In order to obtain values of the distribution of the estimate $\hat{\kappa}$ for given probability values, $F(\kappa y+1)$ must be determined. For small values of κ the remainder theorem can be used, while for large values we can use the approximation:

$$F(p) \approx \frac{4(1+2y^2p)^{1/2} \exp(\kappa-\kappa(1+2y^2p)^{1/2})}{p(1+yp+\sqrt{1+2y^2p})^2} \tag{4.2.27}$$

permitting an exact inversion. But the remainder theorem can be used for $\kappa < 10$, and the approximation (4.2.27) for $\kappa \sim 1000$. Therefore, the calculations have been performed directly with the aid of the inversion formula for the Laplace transform. The following formula

$$F(\kappa y+1) = \frac{2e^{\sigma(\kappa y+1)}}{\pi} \int_{-\infty}^{\infty} \frac{\sqrt{r}\,.\,e^{\kappa(1-\sqrt{r}\cos\psi/2)}\{\alpha_1[\sigma\cos\gamma+s\sin\gamma]+\alpha_2[\sigma\sin -s\cos\gamma]}{(\sigma^2+s^2)\ (\alpha_1^2+\alpha_2^2)}\,ds$$

is obtained by simple calculations, where

$$\alpha_1 = \cos(\sqrt{r}\kappa\sin\psi/2)\ [(A_1^2-A_2^2)-(B_1^2-B_2^2)\ e^{-2\kappa\sqrt{r}\cos\psi/2}] -$$

$$- 2\sin(\sqrt{r}\ \kappa\ \sin\psi/2)\ [A_1A_2+B_1B_2\ e^{-2\kappa\sqrt{r}\cos\psi/2}],$$

$$\alpha_2 = \sin(\sqrt{r}\ \kappa\ \sin\psi/2)\ [(A_1^2-A_2^2)+(B_1^2-B_2^2)\ e^{-2\kappa\sqrt{r}\cos\psi/2}] +$$
$$+ 2\cos(\sqrt{r}\kappa\ \sin\psi/2)[A_1A_2-B_1B_2\ \exp(-2\kappa\sqrt{r}\cos\psi/2]$$

$$\gamma = (\kappa y+1)s+\psi/2,\quad A_1 = (1+y\sigma+\sqrt{r}\cos\psi/2),\quad A_2 = (ys+\sqrt{r}\sin\psi/2),$$
$$B_1 = (1+y\sigma-\sqrt{r}\cos\psi/2),\quad B_2 = (ys-\sqrt{r}\sin\psi/2)$$

$$r = [(1+2\sigma^2y^2)^2 + (2y^2s)^2]^{1/2},\qquad \psi = \text{arc tg}\,\frac{2y^2s}{1+2y^2\sigma}\ .$$

From computational considerations it is convenient to choose

$$\sigma = \begin{cases} \kappa & \text{for } \kappa \leq 1, \\ 1/\kappa & \text{for } \kappa > 1. \end{cases}$$

For computing the integral with an accuracy of 10^{-4} we had to perform the basic calculations in the interval $(-30/\kappa y,\ 30/\kappa y)$, and to estimate errors in the intervals $(-60/\kappa y,\ -30/\kappa y)$ and $(30/\kappa y,\ 60/\kappa y)$. The last integrals gave corrections which are smaller than the required accuracy. For a given value of the probability p, the method of successive approximations were used.

Table 1
Quantiles of the distribution of the random variable $\hat{\kappa}$
Values of z are given in the table for which $P_{\kappa}\{\hat{\kappa} > z\} = p$

κ \ p	0.100	0.050	0.025	0.010	0.001
0.05	-	-	-	1.660	3.800
0.1	0.679	1.092	1.620	2.500	5.333
0.2	1.123	1.666	2.350	3.368	6.336
0.3	1.474	2.119	2.860	3.975	7.004
0.4	1.773	2.478	3.270	4.418	7.697
0.5	2.039	2.793	3.622	4.790	8.178
0.6	2.279	3.971	3.938	5.166	8.573
0.7	2.503	3.333	4.225	5.481	8.931
0.8	2.723	3.580	4.496	5.781	9.250
0.9	2.927	3.812	4.748	6.057	9.599
1.0	3.124	4.032	4.980	6.317	9.892
1.5	4.020	5.028	6.068	7.484	11.208
2.0	4.832	5.916	7.020	8.492	12.348
2.5	5.592	6.748	7.903	9.448	13.40
3.0	6.321	7.530	8.733	10.329	14.256
3.5	7.025	8.292	9.524	11.172	15.309
4.0	7.732	9.020	10.316	11.988	16.204
4.5	8.384	9.738	11.052	12.789	17.082
5.0	9.045	10.450	11.750	13.550	17.915
5.5	9.697	11.132	12.507	14.322	18.76
6.0	10.338	11.820	13.236	15.072	19.572
6.5	10.985	12.487	13.942	15.815	20.332
7.0	11.641	13.153	14.631	16.527	21.105
7.5	12.233	13.815	15.322	17.250	21.930
8.0	12.856	14.464	16.000	17.960	22.672
8.5	13.472	15.113	16.677	18.558	23.468
9.0	14.085	15.750	17.343	19.368	24.219
9.5	14.706	16.398	18.003	20.036	24.966
10	15.30	17.01	18.67	20.73	25.75
20	26.98	29.21	31.25	33.74	39.62
30	38.32	40.88	43.62	46.11	52.71
40	49.41	52.28	54.88	58.06	65.16
50	60.38	63.50	66.26	69.74	77.46
60	71.24	74.61	77.66	81.32	89.58
70	82.03	85.62	88.87	92.78	101.24
80	92.78	96.57	100.00	104.12	112.90
90	103.46	107.48	111.06	115.36	124.84
100	114.13	118.32	122.05	126.54	136.41

κ \ p	0.999	0.990	0.975	0.950	0.900
1.0	0.224	0.298	0.354	0.420	0.519
1.5	0.341	0.474	0.567	0.669	0.816
2.0	0.480	0.670	0.802	0.936	1.130
2.5	0.640	0.885	1.050	1.220	1.456
3.0	0.804	1.119	1.317	1.518	1.800
3.5	0.994	1.362	1.593	1.827	2.146
4.0	1.184	1.616	1.880	2.148	2.504
4.5	1.377	1.886	2.183	2.475	2.867
5.0	1.595	2.160	2.485	2.810	3.235
5.5	1.815	2.442	2.800	3.146	3.608
6.0	2.082	2.736	3.120	3.492	3.984
6.5	2.308	3.036	3.445	3.848	4.374
7.0	2.576	3.346	3.780	4.200	4.753
7.5	2.835	3.645	4.118	4.552	5.145
8.0	3.096	3.952	4.456	4.904	5.528
8.5	3.383	4.284	4.794	5.296	5.933
9.0	3.690	4.608	5.157	5.661	6.345
9.5	3.962	4.930	5.510	6.042	6.726
10	4.22	5.27	5.88	6.41	7.14
20	10.56	12.42	13.42	14.35	15.51
30	17.41	20.15	21.49	22.73	24.25
40	24.99	28.22	29.86	31.36	33.17
50	32.60	36.50	38.42	40.13	42.24
60	40.69	44.84	47.09	49.00	51.35
70	48.58	53.49	55.78	57.97	60.56
80	56.74	62.13	64.76	67.05	69.94
90	64.89	70.97	73.68	76.27	79.11
100	73.23	79.72	82.69	85.33	88.47

4.3. Estimation of the Period.

Before formulating the main result of this section with respect to the distribution of the maximum likelihood estimator of the parameter ω in equation (4.1.1) we recall the well known theorem of P. Levy on square integrable continuous martingales (see Appendix B1, Theorems 1 and 2).

Theorem 1. Let $(w(t), F_t)$ *a continuous square integrable martingale, i.e.,*

$$E(w(t)|\, F_s) = w(s), \quad s < t,\ w(0) = 0, \tag{4.3.1}$$

where F_t *is a non-decreasing family of* σ*-algebras. If*

$$E((w(t)-w(s))^2|F_s) = t-s, \tag{4.3.2}$$

then $w(t)$ *is a standard Wiener process* $(E\, w(t) = 0,\ E(w(t))^2 = t)$.

Theorem 2. Let $(\underline{w}^*(t),\ F_t) = (w^1(t),\dots,w^k(t),\ F_t)$ *a k-dimensional continuous martingale*

$$E(w^i(t)|\, F_s) = w^i(s), \quad w^i(0) = 0, \quad s \leq t,\ i = 1,2,\dots,k. \tag{4.3.1'}$$

If

$$E[(\underline{w}(t)-\underline{w}(s))(\underline{w}(t)-\underline{w}(s))^* \,|F_s] = I\cdot(t-s) \tag{4.3.2'}$$

then $\underline{w}(t)$ *is a k-dimensional standard Wiener process with independent components.*

Let $\underline{\xi}^*(t) = (\xi^1(t),\ \xi^2(t))$ denote the two-dimensional elementary Gaussian process, satisfying equation (4.1.1), then Theorem 1 immediately gives that the process

$$\tilde{w}^1(t) = \int_0^t \frac{\xi^1(s)}{\sqrt{\eta(s)}}\, dw^1(s) + \int_0^t \frac{\xi^2(s)}{\sqrt{\eta(s)}}\, dw^2(s), \tag{4.3.3}$$

where $\eta(t) = (\xi^1(t))^2+(\xi^2(t))^2$, is a Wiener process.

Using Ito's formula one can calculate that

$$d\eta(t) = 2\xi^1(t)\, d\xi^1(t) + 2\xi^2(t)\, d\xi^2(t) + 2\, a\, dt =$$

$$= 2\xi^1(t)\, [-\lambda\xi^1(t)dt - \omega\xi^2(t)dt] + 2\xi^2(t)\, [-\lambda\xi^2(t)dt+\omega\xi^1(t)dt] +$$

$$+ 2\, a\, dt + 2\xi^1(t)\, dw^1(t) + 2\xi^2(t)\, dw^2(t) =$$

$$= -2\lambda\, [(\xi^1(t))^2+(\xi^2(t))^2]dt + 2\, a\, dt + 2[\xi^1 dw^1 + \xi^2\, dw^2] =$$

$$= 2\ [1-\lambda\ \eta(t)]dt + 2\sqrt{\eta(t)}\ d\tilde{w}^1(t).$$

i.e.

$$d\eta(t) = 2\ [1-\lambda\ \eta(t)]dt + 2\sqrt{\eta(t)}\ d\tilde{w}^1(t). \tag{4.3.4}$$

Taking the process

$$\tilde{w}^2(t) = -\int_0^t \frac{\xi^2(s)}{\sqrt{\eta(s)}}\, dw^1(s) + \int_0^t \frac{\xi^1(s)}{\sqrt{\eta(s)}}\, dw^2(s) \tag{4.3.5}$$

one can be convinced using Theorem 2 that $\underline{\tilde{w}}^*(t) = (\tilde{w}^1, \tilde{w}^2)$ is a two-dimensional Wiener-process with independent components.

The process $\eta(t) = [(\xi^1(t))^2 + (\xi^2(t))^2]$ is the solution,and the only one (see Jamada - Watanabe (1971), or Liptser - Shiryaev [1]), of equation (4.3.4).

We have the following result.

Lemma 1. *If* $(w^1(t), w^2(t))$ *is a Wiener process with independent components and* $\underline{\xi}(t)$ *is the process in (4.1.1), and*

$$\zeta^1(t) = \frac{\xi^1(t)}{\sqrt{\eta(t)}}, \qquad \zeta^2(t) = \frac{\xi^2(t)}{\sqrt{\eta(t)}}, \tag{4.3.6}$$

then $(\zeta^1(t))^2 + (\zeta^2(t))^2 = 1$ *and the process* $\underline{\tilde{w}}^*(t) = (\tilde{w}^1(t), \tilde{w}^2(t))$ *defined as*

$$d\tilde{w}^1(t) = \zeta^1(t)\ dw^1(t) + \zeta^2(t)\ dw^2(t), \tag{4.3.7}$$

$$d\tilde{w}^2(t) = -\ \zeta^2(t)dw^1(t) + \zeta^1(t)\ dw^2(t),$$

is again a Wiener process with independent components.

Theorem 3. *The characteristic function of variables*

$$\eta(0) + \eta(T) = (\xi^1(0))^2 + (\xi^2(0))^2 + (\xi^1(T))^2 + (\xi^2(T))^2,$$

$$\int_0^T \eta(t)\ dt = \int_0^T [(\xi^1(t))^2 + (\xi^2(t))^2]dt,$$

(see (4.2.15) and (4.2.16)) has the form

$$u(\alpha_1,\alpha_2, T,\lambda) = E \exp[i\alpha_1(\eta(0)+\eta(T))+i\alpha_2 \int_0^T \eta(t)\ dt]= \tag{4.3.8}$$

$$= [\psi(\alpha_1,\alpha_2)]^2$$

where $\psi(\alpha_1,\alpha_2)$ is given in (3.3.5).

Proof. The solution of equation (4.3.4) does not depend on ω and from this it follows that the random variable $\int_0^T \eta(t)\,dt$ has a distribution independent on ω. Putting in (4.1.1) $\omega = 0$ we get a process $\xi(t)$ with independent components, where for $\int_0^T (\xi^1(t))^2\,dt$ we get the characteristic function (3.3.5), and this proves the statement.

We get, that (see (3.6.5))

$$\psi(\mu,\lambda,T) = E_\lambda\{\exp(-\mu\int_0^T \eta(t)dt) \mid \xi^1(0) = x,\ \xi^2(0)=y\} =$$

(4.3.9) $$= [\cosh \Lambda T + \frac{\lambda}{\Lambda}\,\mathrm{sh}\Lambda T]^{-1} \quad \exp\{\frac{\lambda T}{2} - \mu\eta(0)(\Lambda\coth\Lambda T+\lambda)^{-1}\},$$

$$\Lambda = \sqrt{\lambda^2+2\mu}.$$

Corollary 1. Using the same formulas as in (3.6.2) and (3.6.3) we get for $\hat{\omega}$ (see (4.1.8′))

(4.3.10) $$E_{\lambda,\omega}(\hat{\omega}-\omega) = E_{\lambda,\omega}\left(\frac{r}{s_2^2} - \omega\right) = 0,$$

(4.3.11) $$E_{\lambda,\omega}(\hat{\omega}-\omega)^2 = \int_0^\infty \psi(\mu,\lambda,T)\,d\mu.$$

Theorem 4. The random variable (assuming $a = 1$)

(4.3.12) $$(\hat{\omega}-\omega)\left(\int_0^T |\zeta(t)|^2 dt\right)^{1/2}$$

has an exact $N(0,1)$, Gaussian distribution.

Proof. Using (4.1.5′) and (4.1.8′) we get (using notation $\eta(t) = |\xi(t)|^2$)

$$\hat{\omega} - \omega = \frac{r - \omega s_2^2}{s_2^2} = \frac{\int_0^T |\zeta(t)\ ^2 d\theta - \int_0^T \omega|\zeta(t)|^2 dt}{\int_0^T |\zeta(t)|^2\,dt} =$$

$$= \frac{\int_0^T [\xi^1 d\,\xi^2 - \xi^2 d\xi^1 - \omega(\xi^1)^2\,dt - \omega\,(\xi^2)^2\,dt]}{\int_0^T |\xi(t)|^2\,dt} =$$

$$= \frac{\int_0^T [\xi^1(-\lambda\xi^2 dt+\omega\xi^1 dt+dw^2)-\xi^2(-\lambda\xi^1 dt-\omega\xi^2 dt+dw^1)-\omega(\xi^1)^2 dt-\omega(\xi^2)^2 dt]}{\int_0^T |\zeta(t)|^2 dt}$$

$$= \frac{\int_0^T \xi^1(t)\, dw^2(t) - \int_0^T \xi^2(t)\, dw^1(t)}{\int_0^T |\zeta(t)|^2 dt}$$

Now by the help of (4.3.3)

$$(4.3.13) \quad \hat{\omega} - \omega = \frac{\int_0^T \xi^1(t)\, dw^2(t) - \int_0^T \xi^2(t) dw^1(t)}{\int_0^T |\zeta(t)|^2 dt} = \frac{\int_0^T |\zeta(t)|\, d\tilde{w}^2(t)}{\int_0^T |\zeta(t)|^2\, dt}$$

The process $|\zeta(t)|^2 = \eta(t)$ is the solution of (4.3.4) and so it is $F_t^{\tilde{w}^1}$ measurable, and independent of $\tilde{w}^2(t)$. We get that processes $|\zeta(t)|$, $\tilde{w}^2(t)$ are mutually independent. Hence it follows that the conditional distribution of random variable $\int_0^T |\zeta(t)| d\tilde{w}^2(t)$ under the condition $(\eta(t),\ 0 \leq t \leq T)$ is Gaussian with mean 0 and variance $\int_0^T |\zeta(t)|^2\, dt$, which proves the theorem.

Remark 1. We really proved the following statement (which is more than Theorem 4): The conditional distributions

$$P_\omega(\hat{\omega} < x \,|\, |\zeta(t)|^2,\ 0 \leq t \leq T)$$

are Gaussian with parameters ω and $D^2(\hat{\omega}-\omega) = (\int_0^T |\zeta(t)|^2 dt)^{-1}$.

4.4 The unknown mean

4.4.1 The Complex Process. In the one dimensional case (see section 3.4) it was shown, that the unknown mean of the process $\eta(t) = \xi(t) + m$ $(E\xi(t) = 0)$ can be estimated very poorly. In the complex, or two dimensional, case the situation is not the same. Heuristically this can be explained in the following way: the process $\underline{\eta}^*(t) = (\eta^1(t), \eta^2(t))$ travels round its mean position nearly as a circular motion with period $2\pi/\omega$, this mean can be estimated even in the case when a relatively small path is known.

Let us assume that the process $\underline{\eta}(t)$ equals to

(4.4.1) $$\underline{\eta}(t) = \underline{\xi}(t) + \underline{m}$$

where $\underline{\xi}(t)$ is defined by (4.1.1). If both λ and ω are known the maximum likelihood estimators for m_1, m_2 are the following (use (4.1.7))

(4.4.2) $$\hat{m}_1 = \frac{(\lambda^2+\omega^2)\int_0^T \eta^1(t)dt + \lambda(\eta^1(T)+\eta^1(0)) - \omega(\eta^2(T)-\eta^2(0))}{2\lambda+T(\lambda^2+\omega^2)},$$

$$\hat{m}_2 = \frac{(\lambda^2+\omega^2)\int_0^T \eta^2(t)dt + \lambda(\eta^2(T)+\eta^2(0)) + \omega(\eta^1(T)-\eta^1(0))}{2\lambda+T(\lambda^2+\omega^2)}.$$

The estimators $\hat{m}_1$, $\hat{m}_2$ are normally distributed $E\,\hat{m}_1 = m_1$, $E\,\hat{m}_2 = m_2$ and

(4.4.3) $$D^2\left(\int_0^T \eta^i(t)dt\right) = \frac{a}{(\lambda^2+\omega^2)^2}\left[\frac{\lambda^2+\omega^2}{\lambda} e^{-\lambda T}\left(\cos\omega T - \frac{2\lambda\omega}{\lambda^2+\omega^2}\sin\omega T\right) + \right.$$

$$\left. + \frac{\omega^2-\lambda^2}{\lambda}(1+T) + \lambda T(\lambda^2+\omega^2)\right], \qquad i = 1,2,$$

(4.4.4)

$$\mathrm{cov}\left(\int_0^T \eta^i(t)dt, \eta^i(0)\right) = \mathrm{cov}\left(\int_0^T \eta^i(t)dt, \eta^i(T)\right) = \frac{a\lambda}{2(\lambda^2+\omega^2)}.$$

$$\cdot\left[-e^{-\lambda T}\frac{\cos\omega T}{\lambda} + e^{-\lambda T}\frac{\omega}{\lambda^2}\sin\omega T + \frac{1}{\lambda}\right].$$

From (4.4.3) and (4.4.4) one can easily conclude that for fixed ω even in case $\lambda \sim 0$ the variances remain finite and if $T \to \infty$ they tend to 0 in order $\frac{1}{T}$.

4.4.2 Linear Regression. Let $\eta(t)$ be a one dimensional process with representation

(4.4.5) $$\eta(t) = \sum_{i=1}^{N} \theta_i \beta_i(t) + \xi(t),$$

where $\underline{\theta}^* = (\theta_1, \ldots, \theta_N)$ is the vector of unknown parameters and $\underline{\beta}^*(t) = (\beta_1(t), \beta_2(t), \ldots, \beta_N(t))$ is the known vector function. The process $\xi(t)$ is assumed to be Gaussian, stationary with the rational spectral density (see (2.2.34))

$$f_\xi(\lambda) = \frac{1}{2\pi} \frac{|Q(i\lambda)|^2}{|P(i\lambda)|^2}, \quad P(z) = z^n + \sum_{i=1}^{n} a_i z^{n-i}, \tag{4.4.6}$$

$$Q(z) = \sum_{i=0}^{n-1} b_i z^{n-1-i}.$$

We shall find the maximum likelihood estimators of $\underline{\theta}$ from the realization $\xi(t)$, $0 \leq t \leq T$.

According to representation B (see (2.2.40)) $\xi(t)$ is the first component in the system

$$d\xi^1(t) = \sum_{i=1}^{n} d_i \, \xi^i(t) \, dt + b_o \, dw(t), \tag{4.4.7}$$

$$d\xi^i(t) = \lambda_i \, \xi^i(t) \, dt + dw(t), \quad i = 2,3,\ldots,n,$$

and so

$$d\eta(t) = \sum_2^n d_i \, \xi^i(t) \, dt + d_1(\eta(t) - \sum_1^N \theta_j \beta_j(t) dt + b_o dw(t), \tag{4.4.8}$$

$$d\xi^i(t) = \lambda_i \xi^i(t) dt + dw(t), \quad i = 2,3,\ldots,n.$$

In the system of equations (4.4.8) the components $\xi^2(t), \xi^3(t), \ldots, \xi^n(t)$ are unobservable and $\eta(t)$ is observable.

Using theorem 4 in Section 2.3, we have that (see (2.3.49)), $\underline{\eta}^*(t) = (\eta(t) - \sum_1^N \theta_j \beta_j(t), \xi^2(t), \ldots, \xi^n(t))$,

$$\frac{dP_\eta}{dP_w} = \exp\{\int_o^T \underline{\alpha}^*(s,\underline{\eta}) \, d\underline{\eta} - \frac{1}{2} \int_o^T \underline{\alpha}^*(s,\underline{\eta}) \, \underline{\alpha}(s,\underline{\eta}) \, ds \} \tag{4.4.9}$$

The process $\underline{\alpha}(t)$ is given in the following way (see (2.3.64))

$$\underline{\alpha}(t) = e^{\int_o^t [a_1 - \underline{b}_1 \underline{A}_1^* - \gamma(s) \underline{A}_1 \underline{A}_1^*] \, ds} \{ \underline{\alpha}(0) +$$

$$+ \int_o^t e^{\int_o^s \underline{b}_1 \underline{A}_1^* + \gamma(u) \underline{A}_1 \underline{A}_1^* - a_1 \, du} [-(\underline{b}_1 + \gamma(s) \underline{A}_1) d_1 (\eta(s) - \tag{4.4.10}$$

$$- \sum_{j=1}^N \theta_j \beta_j(s)) \, ds + (\underline{b}_1 + \gamma(s) \underline{A}_1)(d\eta - \sum_1^N \theta_j \, \beta_j(s) \, ds)\},$$

and $\gamma(t)$ is explicitely given (see (2.3.63))

$$\gamma(t) = e^{-2[\underline{b}_1\underline{A}_1^* - a_1]t}\left[\gamma^{-1}(0) + \underline{A}_1\underline{A}_1^* \int_0^t e^{-2[\underline{b}_1\underline{A}_1^* - a_1]u} du\right]^{-1}, \tag{4.4.11}$$

where

$$a_1 = \begin{pmatrix} \lambda_2 & 0 & \cdots & 0 \\ 0 & \lambda_3 & \cdots & 0 \\ \vdots & & & \vdots \\ 0 & 0 & & \lambda_n \end{pmatrix}, \quad \underline{b}_1 = \begin{pmatrix} 1 \\ \vdots \\ 1 \end{pmatrix}, \quad \underline{A}_1 = \begin{pmatrix} d_2 \\ \vdots \\ d_n \end{pmatrix}$$

Using the "chain" ru le in differentiation one can obtain

$$\frac{dP_{\eta,\theta}}{dP_{\eta o}} = \frac{dP_{\eta\theta}}{dP_w} \frac{dP_w}{dP_{\eta o}} = c \exp\{ \int_0^t \sum_1^N c_i(s)\ \theta_i\beta_i(s)d\eta(s) + \tag{4.4.12}$$

$$+ \int_0^t \sum_{i,j}^N \tilde{c}_i(s)\theta_i\beta_i(s)\theta_j\beta_j(s)ds + \int_0^t \sum_{i,j} e_{ij}\theta_i\theta_j\beta_i(s)\beta_j'(s) +$$

$$+ \int_0^t \sum_{i,j} \tilde{e}_{ij}(s)\ \theta_i\theta_j\beta_i'(s)\ \beta_j'(s)\},$$

with given $c(s)$, $\tilde{c}(s)$, $e(s)$, $\tilde{e}(s)$. This shows that the system of maximum likelihood equations is a linear system for $\underline{\theta}^* = (\theta_1,\ldots,\theta_N)$

__Remark.__ The maximum likelihood estimators of Example 2 in Section 2.2.2 can be obtained on the basis of calculations in Section 2.3.4 (see formulas (2.3.51) - (2.3.57)), but here we shall not give these long expressions.

4.4.3. __Correct Estimates.__ Let $\xi(t)$ $(0 \leq t \leq T)$ be a real and (strictly) stationary process and suppose that it is continuous with probability 1 and assume that $\theta = E\xi(t)$ is unknown while the covariance function $B(t) = E\ \xi(\ (s+t)\ -\theta) \cdot (\xi(s) - \theta)$ is known.

__DEFINITION.__ The functional $\hat{\theta}(\xi(t))$ is said to be a correct estimate if

$$\hat{\theta}(\xi(t)+c) = \hat{\theta}(\xi(t)) + c$$

for arbitrary $-\infty < c < \infty$.

It is easy to see that the functionals $T^{-1}\int_0^T \xi(t)\,dt$ and $\xi(t_o)$

(t_o fixed) are correct estimates of θ.

Let $\mathfrak{K}$ be the class of correct estimates. If $\hat{\theta}\in\mathfrak{K}$, then

$$(4.4.13) \qquad E_\theta(\hat{\theta}-\theta)^2 = E_\theta(\hat{\theta}(\xi(t))-\theta)^2 = E_o\ (\hat{\theta}(\xi(t)))^2.$$

Pitman introduced the concept of a correct estimate for sequences of independent random variables. We call the estimate

$$(4.4.14) \qquad u = \xi(0) - E_o\ (\xi(0)\,|\ \xi(t)-\xi(0),\ 0 \leq t \leq T\ )$$

of the so-called location parameter θ a Pitman estimate (in case the expected value exists). In the following we denote the σ-algebra generated by the variables $\xi(t) - \xi(0)$ $(0 \leq t \leq T)$ by A_o^T; then

$$u = \xi(0) - E_o(\xi(0)\,|\ A_o^T).$$

Clearly

$$(4.4.15) \qquad u\ (\xi(t)+c) = \xi(0) + c - E_o(\xi(0)\,|A_o^T) = u\ (\xi(t)) + c,$$

that is $u\in\mathfrak{K}$ and

$$E_\theta u = E_\theta\xi(0) - E_\theta(E_o(\xi(0)\,|A_o^T)) = \ .$$

Therefore u is an unbised estimate of θ. If $\theta\in$ we have, on account of the orthogonality of the variables $E_o(\hat{\theta}|A_o^T)$ and $\hat{\theta}-E_o(\hat{\theta}|A_o^T)$,

(4.4.16)

$$E_\theta(\hat{\theta}-\theta)^2 = E_o(\hat{\theta})^2 = E_o(\hat{\theta}- E_o(\hat{\theta}|A_o^T))^2 + E_o(E_o(\hat{\theta}|A_o^T))^2 \geqq$$

$$\geqq E_o[\hat{\theta}-E_o(\hat{\theta}|A_o^T)]^2.$$

If $\hat{\theta}$ is a correct estimate, then $\hat{\theta} - E(\hat{\theta}|A_o^T)$ remains a correct estimate. We show that the equality

$$(4.4.17) \qquad \hat{\theta}_1 - E_o(\hat{\theta}_1|A_o^T) = \hat{\theta}_2-E_o(\hat{\theta}_2|A_o^T)$$

is valid for arbitrary correct estimates $\hat{\theta}_1$, $\hat{\theta}_2$. The equality holds here almost everywhere.

It is clear that

$$(4.4.18)\qquad \hat{\theta}_1(\xi(t))-\hat{\theta}_2(\xi(t)) = \hat{\theta}_1(\xi(t)-\xi(0)) -\hat{\theta}_2(\xi(t)-\xi(0)) =$$

$$= \hat{h}\ (\xi(t) - \xi(0)),$$

so that $\hat{\theta}_1 - \hat{\theta}_2$ is a functional of $\xi(t) - \xi(0)$ which we denote by $\hat{h}$. It follows from a well-known property of conditional expectations that

$$(4.4.19)\qquad E(\hat{\theta}_1 - \hat{\theta}_2|A_o^T) = \hat{h}(\xi(t) -\xi(0)).$$

Formulas (4.4.18) and (4.4.19) yield

$$E(\hat{\theta}_1|A_o^T) = E(\hat{\theta}_2|A_o^T) + \hat{h}\ (\xi(t)-\xi(0)) = E(\hat{\theta}_2|A_o^T) + \hat{\theta}_1-\hat{\theta}_2,$$

so that (4.4.17) is established. The relation

$$\frac{1}{T}\int_o^T \xi(t)\ dt - E_o(\frac{1}{T}\int_o^T \xi(t)|A_o^T) = \xi(0) - E_o(\xi(0)|A_o^T),$$

is a particular case of (4.4.17).

It follows from (4.4.16) and (4.4.17) that

$$E_o(\hat{\theta})^2 \geq E_o(u)^2,$$

for arbitrary $\hat{\theta}\in K$ and we have proved the following statement:
Theorem 1. The estimate u of θ is a minimum variance estimate in the class of correct estimates.

It is easy to show that

$$(4.4.20)\qquad D^2(\xi(0)|A_o^T) = E_o\ [(\xi(0)-E(\xi(0)|A_o^T))^2|A_o^T\ ] =$$

$$= E_o[\xi^2(0)|A_o^T) - E(\xi(0)|A_o^T)^2].$$

Example. Let P_θ be the measure which belong to the one dimensional elementary Gaussian process with parameters (θ,λ) $(E_o\xi(t)^2 = 1/2\lambda)$.

The maximum likelihood estimate

$$\hat{\theta}(\xi(t)) = \frac{\xi(0) +\xi(T) +\lambda \int_o^T \xi(t)\ dt}{2 + \lambda T}$$

is an unbiased estimate of θ and is at the same time a sufficient statistic. It follows from the Blackwell-Kolmogorov-Rao inequality that it is an unbiased minimum variance estimate. Since we deal with an exponential family of distributions $\hat{\theta}$ is, on account of its completeness (see Lehmann [1]), the only optimal unbiased test.

Since $\hat{\theta}$ is also a correct estimate, we see on account of $E(\hat{\theta}(\xi(t))|A_o^T) = 0$ and of (4.4.17) that

$$\xi(0) - E_o(\xi(0)|A_o^T) = \frac{1}{T}\int_o^T \xi(t)dt - E_o\left(\frac{1}{T}\int_o^T \xi(t)dt|A_o^T\right) = \frac{\xi(0)+\xi(T)+\int_o^T \xi(t)dt}{2+\lambda T}$$

hence

$$E(\xi(0)|A_o^T) = -\frac{\lambda\int_o^T (\xi(t)-\xi(0))dt+(\xi(T)-\xi(0))}{2+\lambda T},$$

$$E\left(\frac{1}{T}\int_o^T \xi(t)dt|A_o^T\right) = \frac{2\int_o^T (\xi(t)-\xi(0))dt - T(\xi(T)-\xi(0))}{T(2+\lambda T)}.$$

4.4.4. Pitman's Estimates. If we assume the existence of a density function we can express the Pitman estimate for a sample $\xi_1,\xi_2,\ldots,\xi_n$ of identically distributed but not independent random variables in a form which differs from (4.4.14). Let $p_o(x_1,\ldots,x_n)$ be the joint density function of the random variables $\xi_1, \xi_2,\ldots,\xi_n$. The random variables $\eta_1 = \xi_1$, $\eta_2 = \xi_2 - \xi_1,\ldots,$ $\eta_n = \xi_n - \xi_1$ then have the joint density $p_o(y_1, y_2,\ldots,y_n) =$

$= p_o(y_1, y_2 + y_1,\ldots,y_n + y_1)$. Therefore

$$E_o(\xi_1|\xi_2-\xi_1,\ldots,\xi_n-\xi_1) = E_o(\xi_1|\eta_2,\ldots,\eta_n) =$$

$$= \frac{\int tp_o(t,\eta_2+t,\ldots,\eta_n+t)\,dt}{\int p_o(t,\eta_2+t,\ldots,\eta_n+t)\,dt},$$

hence, substituting $t = \xi_1 - x$, we obtain

$$E_o(\xi_1|A_2^n) = \xi_1 - \frac{\int xp_o(\xi_1-x, \xi_2-x,\ldots,\xi_n-x)\,dx}{\int p_o(\xi_1-x_1\ \xi_2-x,\ldots, \xi_n-x)\,dx}.$$

We combine this with (4.14) and get the estimate

$$\text{(4.4.21)}\qquad \hat{\theta} = \frac{\int xp_o(\xi_1-x, \xi_2-x,\ldots,\xi_n-x)\,dx}{\int p_o(\xi_1-x, \xi_2-x,\ldots, \xi_n-x)\,dx}.$$

This estimate exists even if the expected value of the ξ_i does not exist (for the case of the Cauchy distribution). Similarly one can see that

$$E_o(\xi_1^2|A_o^T) = \frac{\int t^2 p_o(t,\eta_2+t,\dots,\eta_n+t)\,dt}{\int p_o(t,\eta_2+t,\dots,\eta_n+t)\,dt} .$$

Using (4.4.20) one gets

$$(4.4.22) \qquad D^2(\xi_1|A_2^T) = \frac{\int x^2 p_o(\xi_1-x,\xi_2-x,\dots,\xi_n-x)\,dx}{\int p_o(\xi_1-x,\dots,\xi_n-x)\,dx} - \left(\frac{\int x p_o(\xi_1-x,\dots,\xi_n-x)\,dx}{\int p_o(\xi_1-x,\dots,\xi_n-x)\,dx}\right)^2 .$$

Stein proved that the estimate (4.4.21) is an admissible estimate of θ provided that (i) the expression (4.4.22) raised to the power 3/2 has a finite expected value and (ii) the observations are independent. (The loss function which occurs in the definition of admissibility is here and in the following the expected value of the squared deviation from the parameters.) Clearly, the theorem is also valid for stationary sequences.

Example 2. Let $\xi(n)$ be a stationary Gaussian sequence which satisfies the difference equation

$$\xi(n+k) + a_{k-1}\xi(n+k-1) + \dots + a_o\xi(n) = \varepsilon(n+k).$$

Here $\varepsilon(n)$ ($E\varepsilon(n) = 0$, $\mathrm{Var}\,\varepsilon(n) = 1$) is an independent Gaussian sequence. The Pitman estimate (4.4.21) agrees with the maximum likelihood estimate; this can be seen by a simple but tedious computation.

In particular, for $k = 1$ one has

$$\xi(n+1) = \rho\xi(n)+\varepsilon(n+1),$$

and obtains

$$\xi_1 - E_o(\xi_1|A_2^N) = \frac{\xi_1+\xi_N + (1-\rho)\sum_2^{N-1}\xi_i}{2+(1-\rho)(N-2)}$$

and

$$E_o(\xi_1|A_2^N) = -\frac{(\xi_N-\xi_1) + (1-\rho)\sum_2^{N-1}(\xi_i-\xi_1)}{2+(1-\rho)(N-2)} .$$

We define the Pitman estimate for a time-continuous stationary process whose expectation does not exist in the following way. Let the measure P_θ which belongs to the process $\xi(t)$ be absolutely continuous with respect to a measure Q and let

$$\frac{dP_\theta}{dQ}(\xi(t)) = f_\theta(\xi(t))$$

be its Radon-Nikodym derivative.

Define the Pitman estimate of θ by

$$\hat{\theta} = \frac{\int x f_o(\xi(t)-x)\,dx}{\int f_o(\xi(t)-x)\,dx} \tag{4.4.23}$$

The estimates (4.4.23) and (4.4.14) agree if the expected value of $\xi(t)$ exists.

4.4.5. Admissible Estimates. In case of one-dimensional, elementary Gaussian process one can directly show that the maximum likelihood estimate, respectively the Pitman estimate, is admissible. The extension of Stein's proof to stochastic processes remains an open question, but it seems that the extension could be carried out. To prove the admissibility one needs the method of Hodges and Lehmann which we repeat here for the sake of completeness.

Suppose that the stationary Gaussian process is n - 1 times differentiable and that it satisfies the differential equation (see (2.2.31))

$$d\xi^{(n-1)}(t) + [a_1\xi^{(n-1)}(t)+\ldots+ a_n(\xi(t)-\theta)]dt = dw(t). \tag{4.4.24}$$

Here $w(t)$ is the Wiener process with the parameters $Edw(t) = 0$, $E(dw)^2 = dt$. It is known (see (2.2.33))

$$f_\xi(\lambda) = \frac{1}{2\pi} \cdot \frac{1}{|(i\lambda)^n+a_1(i\lambda)^{n-1}+\ldots+a_n|^2}.$$

We denote the measure belonging to the process $\xi(t)$ by P_θ; this emphasizes that $E_\theta\xi(t) = \theta$. The following theorem is valid.

Theorem 2. The maximum likelihood estimate

$$m^* = \frac{\sum_{o}^{n-1} a_{n-k}[\xi^{(k)}(T)+(-1)^k\xi^{(k)}(0)] + a_n\int_o^T\xi(t)dt}{2a_{n-1} + a_n T} \tag{4.4.25}$$

is also a Pitman estimate and is an admissible estimate of the parameter $E_\theta\xi(t) = \quad (-\infty < \theta < \infty)$.

The proof of the theorem is carried out in several steps.

Lemma 1. If A has the form (2.2.31) then

$$(4.4.26) \qquad B^{-1}(0) = 2 \begin{pmatrix} a_n a_{n-1} & 0 & a_n a_{n-3} & 0 \ldots & & a_n a_o \\ 0 & a_{n-1}a_{n-2} - a_n a_{n-3} & 0 & a_{n-1}a_{n-4} - a_n a_{n-5} \ldots & & 0 \\ \cdot & & & & & \cdot \\ \cdot & & & & & \cdot \\ \cdot & & & & & \cdot \\ a_n a_o & & & & & a_1 a_o \end{pmatrix}$$

if n is odd. Here $a_o = 1$ *and* $a_i = 0$ *if either* $i < 0$ *or* $i > n$, *while the elements of* $B_{ij}^{-1}(0)$ *are*

$$b_{ij}^{-1} = b_{ji}^{-1} = \begin{cases} 0, & i \equiv j+1 \pmod 2 \\ 2\sum_{o}^{n}(-1)^k a_{i-k}\, a_{j+1+k}, & i \equiv j \pmod 2,\ i < j. \end{cases}$$

Substitution into (2.2.3) shows that the lemma is correct. We turn now to the derivation of the Radon-Nikodym derivatives which belong to the measures P_o and P_θ belonging to the processes $\xi(t)$ $[E_o(\xi(t))=0]$ and $\xi(t)+\theta$ $[E_\theta(\xi(t)) = \theta]$respectively.

Lemma 2. The following relation holds:

$$(4.4.27) \qquad \frac{dP_\theta}{dP_o}(\xi(t)) = \exp\{-\frac{1}{2}a_n\theta^2(a_nT+2a_{n-1}) + a_n^2\theta\int_o^T \xi(t)\,dt + a_n\theta\sum_o^{n-1} a_{n-k}[\xi^{(k)}(T) + (-1)^k\xi^k(0)]\}.$$

Proof: According to a well-known factorization theorem (see Lehmann [1], p. 75)

$$a_n\int_o^T \xi(t)\,dt + \sum_o^{n-1} a_{n-k}\,[\xi^{(k)}(T) + (-1)^k\,\xi^{(k)}(0)]$$

is a sufficient statistic and the maximum likelihood estimate (4.4.25) is, according to the Blackwell-Kolmogorov-Rao inequality, a minimum variance unbiased estimate. It follows from the Lehmann--Scheffe theorem (Lehmann [1], p. 183) that the sufficient statistic is complete and also that the maximum likelihood estimate (4.4.25) is the only minimum variance unbiased estimate. We do not go into the explicit computation here of the Pitman estimate on the basis of (4.4.26).

It is easy to show that

$$(4.4.28) \qquad D^2(m^*) = \frac{1}{a_n(2a_{n-1} + a_n T)}$$

Let $\hat{\theta}(\xi(t))$ be an estimate of θ and $E_\theta(\hat{\theta}) = \theta + b_{\hat{\theta}}(\theta)$. According to the Cramer-Rao inequality

$$(4.4.29) \qquad D^2_\theta(\hat{\theta}) = E_\theta(\hat{\theta}-\theta)^2 - b^2_{\hat{\theta}}(\theta) \geq \frac{(1+b'_{\hat{\theta}}(\theta))^2}{E_\theta\left(\frac{\partial}{\partial\theta}\log\frac{dP_\theta}{dP_o}(\xi(t))\right)^2}$$

Since m^* is a sufficient statistic, one has

$$E_\theta\left(\frac{\partial}{\partial\theta}\log\frac{dP_\theta}{dP_o}\right)^2 = (D^2(m^*))^{-1} = a_n\ (2\ a_{n-1} + a_n\ T).$$

(this can also be seen by means of a simple computation).

<u>Lemma 3. If the estimate $\hat{\theta}$ satisfies the inequality</u>

$$b^2_{\hat{\theta}}(\theta) + \frac{[1+b'_{\hat{\theta}}(\theta)]^2}{a_n(2a_{n-1}+a_nT} \leq \frac{1}{a_n(2a_{n-1} + a_nT)}$$

for all θ, then $b_{\hat{\theta}}(\theta) \equiv 0$ $(-\infty < \theta < \infty)$.

<u>Proof.</u> It follows from the assumption that $|b_{\hat{\theta}}(\theta)|$ is bounded $(-\infty < \theta < \infty)$ and that $b'_{\hat{\theta}}(\theta)$ is nonpositive. Therefore $b_{\hat{\theta}}(\theta)$ is a monotone decreasing function of θ; since it is bounded there exist sequences θ'_n and θ_n $(\theta'_n \to \infty\ -\infty,\ \theta_n \to \infty)$ such that $b'(\theta_n)$ and $b'(\theta'_n)$ converge to zero. The statement of the lemma follows.

Finally we need a lemma due to Hodges and Lehmann:

<u>Lemma 4. If the estimate θ^* satisfies the Cramer-Rao inequality for every θ and if the validity of the inequality</u>

$$(4.4.30) \qquad b^2_{\hat{\theta}}(\theta) + \frac{(1+b'_{\hat{\theta}}(\theta))^2}{E\left(\frac{\partial}{\partial\theta}\log\frac{dP_\theta}{dP_o}\right)^2} \leq b^2_{\theta^*} + D^2_\theta(\theta^*)$$

<u>for an arbitrary estimate $\hat{\theta}$ implies that</u> $b_{\hat{\theta}}(\theta) \equiv b_{\theta^*}(\theta)$, <u>then $\hat{\theta}$ is an admissible estimate of θ.</u>

<u>Proof.</u> We prove that the validity of the indequality $E_\theta(\hat{\theta}-\theta)^2 \leq E_\theta(\theta^*-\theta)^2$ (for all θ) implies that $\hat{\theta} = \theta^*$; that is θ^* is admissible. If

$$E_\theta(\hat{\theta}-\theta)^2 \le E_\theta(\theta^*-\theta)^2 = b^2_{\theta^*}(\theta) + \frac{[1+b'_{\theta^*}(\theta)]^2}{E\left(\frac{\partial}{\partial\theta}\log\frac{dP_\theta}{dP_o}\right)^2},$$

then according to (4.4.24), the inequality

$$b^2_{\hat{\theta}}(\theta) + \frac{(1+b'_{\hat{\theta}}(\theta))^2}{E\left(\frac{\partial}{\partial\theta}\log\frac{dP_\theta}{dP_o}\right)^2} \le b^2_{\hat{\theta}}(\theta) + D^2(\hat{\theta}) = E_\theta(\hat{\theta}-\theta)^2 \le$$

$$\le b^2_{\theta^*}(\theta) + \frac{(1+b'_{\theta^*}(\theta))^2}{E_\theta\left(\frac{\partial}{\partial\theta}\log\frac{dP_\theta}{dP_o}\right)^2},$$

is also satisfied. But this means that $b_{\hat{\theta}}(\theta) = b_{\theta^*}(\theta)$ and also $b'_{\hat{\theta}}(\theta) = b'_{\theta^*}(\theta)$. Therefore $\mathrm{Var}_\theta(\hat{\theta}) = \mathrm{Var}_\theta(\theta^*)$ and $E_\theta(\hat{\theta}-\theta)^2 = E_\theta(\theta^*-\theta)^2$, that is $\theta = \theta^*$; q.e.d.

Tehorem 2 follows immediately from Lemmas 3 and 4.

The proof indicates that the method cannot be extended to the estimation of the expectation of multivariate, stationary Gaussian processes. It is known that admissibility is not true for independent observations in case $n \geq 3$ (n is the number of unknown mean values). It is an open question whether the admissibility is true for the two-dimensional stationary Gaussian process.

4.4.6. Minimax Weights in Trend Detection. Let F_M be a class of real functions $f(t)$, $-\infty < t < \infty$, which can be written in the form

$$f(t) = a_o + g(t)\cdot t,$$

where a_o is a constant, and $\sup_t |g(t)| \le M$, where M is a known constant. It is required to estimate optimally the value $a_o = f(0)$ from a realization

$$\xi(t) = f(t) + \eta(t), \quad -\infty < t < \infty,$$

where $f(t) \in F_M$, while $\eta(t)$ is an elementary Gaussian process.

As an estimate of $a_o = f(0)$ we shall consider the linear functionals

$$\hat{a}_o = \int_{-\infty}^{\infty} \ell(t)\xi(t)\,dt, \quad E\hat{a}_o = \int_{-\infty}^{\infty} \ell(t)\,f(t)\,dt,$$

characterized by the weight function $\ell(t) \geq 0$ in the class

(4.4.31) $$L = \{ \ell(t): \int_{-\infty}^{\infty} \ell^2(t)\, dt < \infty \,,\ \int_{-\infty}^{\infty} |t\ell(t)|dt < \infty,$$

$$\int_{-\infty}^{\infty} \ell(t)\, dt = 1\}.$$

Let $f(t) \in F_M$ and $\ell(t) \varepsilon L$ and

$$\Delta(\ell, f) = E(f(0) - \hat{a}_o)^2,$$

then

(4.4.32) $$\Delta(\ell, f) = E(f(0)-\hat{a}_o)^2 = E(f(0) - \int_{-\infty}^{\infty} \ell(t)\xi(t)\, dt)^2 =$$

$$= (f(0)-E\hat{a}_o)^2 + E(\hat{a}_o-E\hat{a}_o)^2 =$$

$$= (f(0)-E\hat{a}_o)^2 + E(\int_{-\infty}^{\infty} \ell(t)\eta(t))^2 =$$

$$= (\int_{-\infty}^{\infty} t\,\ell(t)\ g(t)dt)^2 + \sigma_\eta^2 \iint_{-\infty}^{\infty} \ell(t)\ell(s)e^{-\lambda|t-s|}dsdt =$$

$$= (\int_{-\infty}^{\infty} t\ell(t)\ g(t)dt)^2 + d^2$$

where

$$d^2 = \sigma_\eta^2 \iint_{-\infty}^{\infty}\ell(t)\ell(s)\ e^{-\lambda|t-s|}\ dsdt,$$

and $\Delta(\ell,f) < \infty$. The weight function $f^*(t)\ \varepsilon L$ is said to be a <u>minimax</u> if

(4.4.33) $$\sup_{f\in F_M} \Delta(\ell^*;f) = \inf_{\ell\in L} \sup_{f\in F_M} \Delta(\ell, f).$$

Since

$$\inf_{\ell\in L} \sup_{f\in F_M} \Delta(\ell,f) = \inf_{\sigma>0} \inf_{\ell\in L^\sigma} \sup_{f\in F_M} \Delta(\ell, f),$$

where

$$L^\sigma = L \cap \{\ell: \int_{-\infty}^{\infty} \ell^2(s)\, ds = \sigma^2 \},$$

to find minimax weights in L it suffices to find minimax weight functions only in the classes L^σ, $\sigma > 0$.

For functions $\ell(t)\ \ \varepsilon L^\sigma$

(4.4.34) $$\Delta(\ell, f) =[\int_{-\infty}^{\infty} t\ell(t)g(t)dt]^2+d^2 \leq [M \int_{-\infty}^{\infty} t\ell(t)dt]^2 + d^2 .$$

Thus seeking a minimax weight function in the class L^σ is equivalent to finding a function at which

$$\inf_{\ell} \{ \int_{-\infty}^{\infty} |t\ell(t)|dt + d^2\}. \quad (4.4.35)$$

is attained under the restrictions

$$\int_{-\infty}^{\infty} \ell(t)dt = 1, \quad \int_{-\infty}^{\infty} \ell^2(t)\, dt = \sigma^2. \quad (4.4.36)$$

The solution of this variational problem can be gotten in the case when $\eta(t)$ is a "white noise" process with intensity a (see Legostayeva, Shiryaev (1971)), when

$$d^2 = a^2 \int_{-\infty}^{\infty} \ell^2(t)dt = a^2\sigma^2. \quad (4.4.37)$$

In this case the minimax weight function $\ell^\sigma(t)$ has the form

$$\ell^\sigma(t) = \max\{ 0, \frac{2/3\ \sigma^{-2} - |t|}{(2/3\ \sigma^{-2})^2}\} .$$

By (4.4.34) - (4.4.35)

$$\inf_{\sigma>0} \{ M \int_{-\infty}^{\infty} |t\ \ell^\sigma(t)|dt]^2 + a^2\sigma^2 \} = \inf_{\sigma>0}\{[\frac{2}{9} M\ \sigma^{-2}]^2 + a^2\sigma^2\},$$

and the inf is attained for

$$\sigma_o = 2\cdot 3^{-4/3}(Ma^{-1})^{2/3}, \quad (4.4.38)$$

which gives

$$\ell^*(t) = \ell^{\sigma_o}(t) = \max\{ 0, \frac{(3a^2M^{-2})^{1/3} - |t|}{(3a^2\ M^{-2})^{2/3}} \}, \quad (4.4.39)$$

and

$$\Delta(\ell^*; f) = \inf_{\sigma>0}\{ [M \int_{-\infty}^{\infty} |t\ \ell^\sigma(t)|dt]^2 + a^2\sigma^2\} = \frac{1}{\sqrt{3}} (Ma^2)^{2/3} . \quad (4.4.40)$$

From the above result it follows that the most "difficult" functions for estimation of a_o are functions of the form $\tilde{f}(t) = a_o + Mt\cdot$ sign t, for them

$$\Delta(\ell^*; \tilde{f}) = \frac{1}{\sqrt{3}} (Ma^2)^{2/3}.$$

The "easiest" function for estimation of a_o are contants $\tilde{f} = a_o$, for them

$$\Delta(\ell^*; \tilde{f}) = \frac{2}{3} \cdot \frac{1}{\sqrt{3}} (Ma^2)^{2/3}$$

Further generalizations of this problem the reader can find in

Legostayeva, Shiryaev (1971).

4.5. Real Roots and Other Special Cases

In the earlier sections of this chapter we supposed that the matrix A in equation

$$d\underline{\xi}(t) = A\underline{\xi}(t)\,dt + d\underline{w}(t),$$

has complex eigenvalues $\lambda_1 = -\lambda+i\omega$, $\lambda_2 = -\lambda-i\omega$, $(\lambda > 0)$.
If A has different real eigenvalues $-\lambda_1$, $-\lambda_2$ $(\lambda_i > 0)$, then by the linear transformation $\underline{\eta} = B\underline{\xi}$ we have equation

$$\text{(4.5.1)}\qquad \begin{aligned} d\eta^1(t) &= -\lambda_1\eta^1(t)\,dt + d\tilde{w}_1(t), \\ d\eta^2(t) &= -\lambda_2\,\eta^2(t)\,dt + d\tilde{w}_2(t), \end{aligned}$$

where $(\tilde{w}^1(t), \tilde{w}^2(t))$ is a Wiener process (not necessarily with independent components). λ_1 and λ_2 can be estimated as the parameters of one-dimensional processes (see Chapter 3.).

If $-\lambda$ is an eigenvalue with multiplicity 2 then by linear transformation we get the Jordan decomposition (see Appendix A. Th. 1)

$$\text{(4.5.2)}\qquad \begin{aligned} d\eta^1(t) &= -\lambda\,\eta^1(t)\,dt + \eta^2(t)\,dt + dw^1(t) \\ d\eta^2(t) &= -\lambda\,\eta^2(t)\,dt + dw^2(t). \end{aligned}$$

The following lemma holds.

Lemma 1. Let P_λ denote the measure generated by the process $\underline{\eta}(t)$, given in (4.5.2), where $B_w = \begin{pmatrix} \sigma_w^2 & 0 \\ 0 & 1 \end{pmatrix}$ and P_o the measure when $\lambda=0$.

Then

$$\text{(4.5.3)}\qquad \frac{dP_\lambda}{dP_o}(\eta,t) = f_\lambda(\underline{\eta}(0))\exp\{-\frac{\lambda}{\sigma_w^2}\int_o^t \eta^1(s)d\eta^1(s) + \frac{1}{\sigma_w^2}\int_o^t\eta^2(s)d\eta^1(s) +$$

$$-\lambda\int_o^t\eta^2(s)d\eta^2(s) - \frac{1}{2}\int_o^t[\frac{\lambda^2}{\sigma_w^2}(\eta^1(s))^2 + \frac{1}{\sigma_w^2}(\eta^2(s))^2 - \frac{2\lambda}{\sigma_w^2}\eta^1(s)\eta^2(s) +$$

$$+\left(\frac{1}{\sigma_w^2}+\lambda^2\right)(\eta^2(s))^2]ds\} =$$

$$= \frac{(2\lambda)^2}{2\pi\sqrt{1+(\sigma_w\,2\lambda)^2}}\exp\{-\frac{1}{2}\int_o^t[\frac{\lambda^2}{\sigma_w^2}(\eta^1(s))^2 - \frac{2\lambda}{\sigma_w^2}\eta^1(s)\eta^2(s) +$$

$$+(\lambda^2+\frac{1}{\sigma_w^2})(\eta^2(s))^2]ds+t\lambda+\frac{1}{\sigma_w^2}\int_0^t\eta^2(s)d\eta^1(s)-\frac{\lambda}{2\sigma_w^2}(\eta^1(t))^2-\frac{\lambda}{2}(\eta^2(t))^2-$$

$$-[\frac{\lambda}{2\sigma_w^2}+\frac{(2\lambda)^3}{1+(2\lambda\sigma_w)^2}](\eta^1(0))^2-[\frac{\lambda}{2}+\lambda(1+\frac{1}{1+(2\lambda\sigma_w)^2}](\eta^2(0))^2+$$

$$+\frac{(2\lambda)^2}{1+(2\lambda\sigma_w)^2}\eta^1(0)\ \eta^2(0)\}.$$

The proof is a straightforward consequence of Theorem 1 in § 2.3.3, where the unique solution of

$$AB(0) + B(0)\ A^* = -B_w$$

is

$$(4.5.4)\qquad B(0) = E\underline{\eta}(t)\underline{\eta}^*(t) = \begin{pmatrix} \frac{\sigma_w^2}{2\lambda}+\frac{1}{\lambda(2\lambda)^2} & \frac{1}{(2\lambda)^2} \\ \frac{1}{(2\lambda)^2} & \frac{1}{2\lambda} \end{pmatrix},$$

$$\det B = \frac{(2\lambda)^2}{\sqrt{1+(2\lambda\sigma_w)^2}}$$

Note that in (4.5.3) the exponent contains terms not depending on λ.

To show that in Section 4.3 we really used the symmetry of the complex process let us examine the second order ARMA process $\xi(t)$ as a two dimensional one. If $\xi(t)$ is defined as the solution of the following equation

$$(4.5.5)\qquad d\,\xi'(t)+[a_1\,\xi'(t)+a_2\xi(t)]dt = dw(t),\ E(dw(t))^2 = \sigma_w^2\cdot dt,$$

with complex roots of the characteristic equation, i.e., $-\lambda + i\omega$ and $-\lambda - i\omega$, where

$$\lambda = \frac{a_1}{2},\quad \omega = \sqrt{a_2-\frac{a_1^2}{4}},\quad (a_1 > 0,\ a_2 > \frac{a_1^2}{4}),$$

then (4.5.5) can be written in the form ($\xi(t) = \xi^1(t)$)

$$d\xi^1(t) = \xi^2(t)\,dt,$$

$$(4.5.5')\qquad d\xi^2(t) = -[a_1\,\xi^2(t) + a_2\,\xi^1(t)]dt + dw(t).$$

<u>Lemma 2. Let P_{a_1,a_2} denote the measure generated by the process</u>

$\xi(t)$ <u>and</u> P_o <u>the standard Wiener - Lebesgue measure. Then</u>

(4.5.6)

$$\frac{dP_{a_1 a_2}}{dP_o}(t,\xi) = \frac{a_1\sqrt{a_2}}{\pi\sigma_w^2}\exp\{-\frac{a_2^2}{2\sigma_w^2}\int_0^t(\xi(s))^2\,ds - \frac{a_1^2-2a_2}{2\sigma_w^2}\int_0^t(\xi'(s))^2\,ds +$$

$$+\frac{a_1 t}{2} - \frac{a_2 a_1}{2\sigma_w^2}[\xi^2(t)+\xi^2(0)] - \frac{a_1}{2\sigma_w^2}[(\xi'(t))^2-(\xi'(0))^2] - \frac{a_2}{\sigma_w^2}[\xi(s)\xi'(s)]_0^t\}.$$

First of all note that in this case the covariance function has the form

(4.5.7) $\quad B(t) = E\underline{\xi}(t+s)\,\underline{\xi}^*(s) = E^{A|t|}.\ B(0) =$

$$= \sigma_w^2\, e^{-\lambda|t|}\begin{pmatrix} \cos\omega t + \frac{\lambda}{\omega}\sin\omega|t| & \frac{1}{\omega}\sin\omega|t| \\ -\frac{\lambda^2+\omega^2}{\omega}\sin\omega|t| & \omega\cos\omega t + \lambda\sin\omega|t| \end{pmatrix}\begin{pmatrix} \frac{1}{4\lambda(\lambda^2+\omega^2)} & 0 \\ 0 & \frac{1}{4\lambda} \end{pmatrix}$$

The proof of Lemma 2. immediately follows from Theorem 1 in § 2.3.3.

By transformation $C\underline{\xi}(t) = \underline{\eta}(t)$, where

$$C = \begin{pmatrix} -a_2\frac{\lambda}{\omega} & \frac{\lambda}{\omega} \\ -a_2\frac{\lambda}{\omega}+a_1 & -1 \end{pmatrix},\quad C^{-1} = \begin{pmatrix} -1 & -\frac{\lambda}{\omega} \\ a_2\frac{\lambda}{\omega}-a_1 & -a_2\frac{\lambda}{\omega} \end{pmatrix}\cdot\frac{1}{a_2(\frac{\lambda}{\omega}+(\frac{\lambda}{\omega})^2)-a_1\frac{\lambda}{\omega}}$$

we can write (4.5.5') in the following equivalent from

(4.5.5'') $\quad d\eta^1(t) = (-\lambda\eta^1(t)dt - \omega\eta^2(t)dt) + \rho d\tilde{w}(t),\quad \rho = \frac{\lambda}{\omega}\sigma_w,$

$\quad d\eta^2(t) = (\omega\eta^1(t)dt - \lambda\eta^2(t)dt) - d\tilde{w}(t),\quad E(d\tilde{w}) = dt.$

For (4.5.5') we have, solving equation $AB_\xi(0) + B_\xi(0)A = -\begin{pmatrix} 0 & 0 \\ 0 & \sigma_w^2 \end{pmatrix}$

(4.5.8) $\quad B_\xi(0) = \sigma_w^2\begin{pmatrix} \frac{1}{2a_1a_2} & 0 \\ 0 & \frac{1}{2a_1} \end{pmatrix},$

while for (4.5.5'')

$$(4.5.8')\qquad B_\eta(0) = \begin{pmatrix} \dfrac{\rho^2}{2\lambda} & \dfrac{1+\frac{2\lambda}{\omega}(\rho+\frac{\rho^2}{2})}{2[\frac{2\lambda}{\omega}(\omega-\lambda)-\omega]} \\ \dfrac{1+\frac{2\lambda}{\omega}(\rho+\frac{\rho^2}{2})}{2[\frac{2\lambda}{\omega}(\omega-\lambda)-\omega]} & \dfrac{\omega-\lambda}{\omega}\cdot\dfrac{1+\frac{2\lambda}{\omega}(\rho+\frac{\rho^2}{2})}{2[\frac{2\lambda}{\omega}(\omega-\lambda)-\omega]} - \dfrac{1}{\omega}(\rho+\dfrac{\rho^2}{2}) \end{pmatrix}$$

Note that in (4.5.5'') we have only one Wiener process $\tilde{w}(t)$ and so to find e.g. the Radon-Nikodym derivative is rather more complicated than in case (4.5.6) (compare with the calculations for rational spectral density in Section 4.3.4). Because of this antisymmetry it is impossible to use results of §4.3. E.g. for

$$|\eta(t)|^2 = (\eta^1(t))^2 + (\eta^2(t))^2 \quad \text{we get}$$

$$(4.5.9)\qquad d|\eta(t)|^2 = [(\rho-1)-2\lambda|\eta(t)|^2]dt+2\eta(t)\,\frac{\rho\eta^1(t)-\eta^2(t)}{|\eta(t)|}\,d\tilde{w}\,,$$

where $\dfrac{\rho\eta^1(t)-\eta^2(t)}{|\eta(t)|}\,d\tilde{w}$ depends on ω.

Using the estimator

$$(4.5.10)\qquad \hat{\omega} = \frac{\int_0^t[\eta^1 d\eta^2 - \eta^2 d\eta^1]}{\int_0^t|\eta(s)|^2\,ds}$$

we get that

$$(4.5.11)\qquad \hat{\omega} - \omega = \frac{\int_0^t [\eta^1 d\tilde{w} + \rho\eta^2 d\tilde{w}\,]}{\int_0^t |\eta(s)|^2\,ds}\,,$$

where $\eta^1 + \rho\eta^2$ depends on $\tilde{w}$ and may be has not a normal distribution. The maximum likelihood estimators from (4.5.6) $\hat{a}_1$, $\hat{a}_2$ can be used for λ and ω

$$(4.5.12)\qquad \hat{\lambda} = \frac{\hat{a}_1}{2} \qquad \hat{\omega} = \sqrt{\hat{a}_2 - \frac{\hat{a}_1^2}{4}}\,,$$

for which the distributions in §3.3 and §4.3 respectively can be used as approximations.

For (4.5.11) the method of sequential estimation (see §3.6) can be used.

Remark 1. Note that in this special case the "natural" parameters of the process $\xi(t)$ in (4.5.5) are the damping (or decay) parameter $\lambda = \frac{a_1}{2}$ and the "resonance" frequency $\omega_o = \sqrt{a_2}$, which is always greater than ω $(\omega_o^2 = \omega^2 + \lambda^2)$. The process $\xi(t)$ has a period $2\pi/\omega_o = T_o$ (where $T_o < T = 2\pi/\omega$), which in many cases can be estimated e.g. by the number of peaks of a long realization. This means, that the process $\xi(t)$ has shorter periods than its covariance function (4.5.7).

Remark 2. The spectral density of the process $\xi(t)$ is the following

$$(4.5.13) \qquad f_\xi(\lambda) = \frac{\sigma_w^2}{2\pi} \frac{1}{|(i\lambda)^2 + a_1(i\lambda) + a_2|^2} .$$

The discrete time process satisfies the following equation ($\underline{\xi}_\Delta(n) = \underline{\xi}(n\Delta)$)

$$(4.5.14) \qquad \underline{\xi}_\Delta(n+1) = Q\underline{\xi}_\Delta(n) + B_\varepsilon \underline{\varepsilon}(n+1),$$

where (see Theorem 2 in Section 2.2.1)

$$(4.5.15) \qquad Q = e^{A\Delta} = e^{-\lambda\Delta} \begin{pmatrix} \cos\omega\Delta + \frac{\lambda}{\omega}\sin\omega\Delta & \frac{\lambda}{\omega}\sin\omega\Delta \\ -\frac{\lambda^2+\omega^2}{\omega}\sin\omega\Delta & \omega\cos\omega\Delta + \lambda\sin\omega\Delta \end{pmatrix}$$

$$(4.5.16) \qquad B_\varepsilon = B(0) - QB(0)Q^* , \qquad B(0) = \sigma_w^2 \begin{pmatrix} \frac{1}{4\lambda(\lambda^2+\omega^2)} & 0 \\ 0 & \frac{1}{4\lambda} \end{pmatrix},$$

and $\underline{\xi}_\Delta(n)$ has the covariance function

$$(4.5.17) \qquad B(n) = e^{nA\Delta} . B(0) = Q^n . B(0).$$

The one-dimensional process $\xi^1(n\Delta) = \tilde{\xi}(n)$ has the covariance function

$$(4.5.18) \qquad \tilde{B}(n) = e^{-\lambda\Delta n}(\cos n\omega\Delta + \frac{\lambda}{\omega}\sin n\omega\Delta) \frac{\sigma_w^2}{4\lambda(\lambda^2+\omega^2)} =$$

$$= e^{-\lambda\Delta n} \frac{\cos(n\omega\Delta + \psi)}{\cos\psi} \cdot \frac{\sigma_w^2}{4\lambda(\lambda^2+\omega^2)} ,$$

where

$$\mathrm{tg}\psi = \frac{\lambda}{\omega} , \quad \tilde{\rho}(n) = e^{-\lambda\Delta n} \frac{\cos(n\omega\Delta + \psi)}{\cos\psi}$$

Using the notations

$$(4.5.19) \qquad a = -2\, e^{-\lambda\Delta} \cos \omega\Delta, \qquad a = -\frac{\tilde\rho(1)\ (1-\tilde\rho(2))}{1-(\tilde\rho(1))^2},$$

$$b = e^{-2\lambda\Delta}, \qquad b = -\frac{\tilde\rho(2) - \tilde\rho(1)}{1-(\tilde\rho(1))^2},$$

we obtain

$$\tilde B(n) = (b)^{n/2} \frac{\cos(n\omega\Delta+\psi)}{\cos\psi} \cdot \frac{\sigma_w^2}{4\lambda(\lambda^2+\omega^2)},$$

$$(4.5.18') \qquad \tilde\rho(1) = -\frac{a}{1+b}, \qquad \tilde\rho(2) = -b + \frac{a^2}{1+b},$$

and $\tilde B(n)$ is the colution of equation

$$(4.5.20) \qquad \tilde B(n+2) + a\tilde B(n+1) + b\tilde B(n) = 0.$$

Equation (4.5.20) suggests that

$$(4.5.21) \qquad \tilde\xi(n+2) + a\tilde\xi(n+1) + b\tilde\xi(n) \sim \tilde\varepsilon(n),$$

with the spectral density (see (4.5.16))

$$(4.5.22) \qquad f_{\tilde\xi}(\lambda) = \frac{\sigma_{\tilde\varepsilon}^2}{2\pi} \frac{1}{|e^{2i\lambda} + a\, e^{i\lambda} + b|^2}, \qquad \sigma_{\tilde\varepsilon}^2 = \frac{\sigma_w^2}{4\lambda(\lambda^2+\omega^2)} [1 - e^{-\lambda\Delta} \frac{\cos(\omega\Delta+\psi)}{\cos\psi}],$$

with a peak at

$$-\frac{a(1+b)}{4b}.$$

4.6. Multidimensional Case, Asymptotic Theory

Suppose we have the stochastic differential equation (2.2.1), i.e.,

$$(*) \qquad d\,\underline{\xi}(t) = A\,\underline{\xi}(t)\,dt + B_w^{1/2}\,d\underline{w}(t),$$

where $\underline{\xi}(t)^*$ is a k-dimensional vector process, $k > 2$. $\underline{w}(t)$ is a k-dimensional standard Wiener process with independent components and $B_w = (\sigma_{ij})$ $(B_w^{-1} = (\sigma_{ij}^{-1}))$ is positive definite.

Let $\underline{\xi}(0) = \underline{x}$ then we can state the following theorems

Theorem 1. *The conditional (under condition $\underline{\xi}(0) = \underline{x}$) likelihood equation estimator of A is*

$$(4.6.1) \qquad \hat A_T = \int_0^T \{d\underline{\xi}(t)\ \underline{\xi}^*(t)\} \left\{\int_0^T \underline{\xi}(t)\ \underline{\xi}^*(t)\,dt\right\}^{-1},$$

$$(4.6.1') \qquad \hat{A}_T = \frac{1}{2}\{\ \underline{\xi}(T)\underline{\xi}^*(T) - \underline{x}\ \underline{x}^* - B_w T\}\ \{\int_0^T \underline{\xi}(t)\ \underline{\xi}^*(t)\ dt\}^{-1}\ .$$

<u>Theorem 2.</u> <u>If the roots</u> $\lambda_i, |\ \lambda I - A| = 0$ <u>has negative real parts, i.e., A is a real stability matrix in the Lyapunov's sense, then</u> A_T <u>is consistent and the vector variable</u>

$$(4.6.2) \qquad T^{1/2}(\underline{\hat{A}}_T - \underline{A})$$

<u>is asymptotically normally distributed, as</u> $T \to \infty$<u>, with parameters</u> $(\underline{0}, B^{-1})$, where $B^{-1}_{k^2 x k^2} = (\sigma_{rp}\ b_{qs})^{-1}$, and $B(0) = (b_{ij})$ is the <u>solution of equation</u>

<u>(see(2.2.3))</u> $\qquad A\ B(0) + B(0)A^* = -B_w.$

<u>Remark 1.</u> Note that the asymptotic normality is not true uniformly (only in the case $\mathrm{Re}\ (\lambda_i) < -\varepsilon,\ \varepsilon > 0$ is prescribed, we have uniform asymptotic normality).

<u>Proof of Theorem 1.</u> Using formulas (2.3.25′) and (2.3.35) we get for the conditional log-likelihood function

$$(4.6.3)\ L_A = \log \frac{dP_{\xi x}}{dP_w} = \mathrm{Sp}[B_w^{-1}A \int_0^T \underline{\xi}(t)d\ \underline{\xi}^*(t)] - \frac{1}{2}\mathrm{Sp}[A\ B_w^{-1}A\int_0^T \underline{\xi}(s)\underline{\xi}^*(s)ds] =$$

$$= \int_0^T (B_w^{-1}\ A\underline{\xi}(t), d\underline{\xi}(t)) - \frac{1}{2}\int_0^T (A\underline{\xi}(t), B_w^{-1}\ A\underline{\xi}(t))\ dt =$$

$$= \frac{1}{2}\int_0^T (B_w^{-1}A\underline{\xi}(t),\ \underline{\xi}(t))dt + \int_0^T (B_w^{-1}\ A\underline{\xi}(t),\ d\underline{w}(t)).$$

or

$$(4.6.4) \quad L_A = -\frac{1}{2}\int_0^T (A\underline{\xi}(t), B_w^{-1}\ A\underline{\xi}(t))dt + \frac{1}{2}\ [(B_w^{-1}\ A\underline{\xi}(T), \underline{\xi}(T)) -$$

$$- (B_w^{-1}\ A\underline{x}, \underline{x})] - \frac{1}{2}T \cdot \mathrm{Sp}\ A =$$

$$= -\frac{1}{2}\ \mathrm{Sp}[A^*B_w^{-1}A \int_0^T \underline{\xi}(t)\ \underline{\xi}^*(t)dt] + \frac{1}{2}\mathrm{Sp}[B_w^{-1}A(\underline{\xi}(T)\underline{\xi}^*(T) -$$

$$- \underline{x}\ \underline{x}^*) - T\ A].$$

The system of conditional likelihood equations can be obtained by

(4.6.5) $$dL_A = Sp[dA \int_0^T \underline{\xi}(t)d\,\underline{\xi}^*(t)\,B_w^{-1} - dA\int_0^T \xi(t)\,\xi^*(t)dt\,A\,B_w^{-1}] = 0$$

or

(4.6.6) $$dL_A = Sp[\tfrac{1}{2}\,dA[\underline{\xi}(T)\underline{\xi}^*(T) - \underline{x}\,\underline{x}^* - B_w.T]B_w^{-1} - dA\int_0^T \underline{\xi}(t)\underline{\xi}^*(t)dtA^*B_w^{-1}] = 0,$$

which prove the statement of Theorem 1. That A minimizes L_A, and maximizes the likelihood, immediately follows from the last equation of (4.6.3).

<u>Proof of Theorem 2.</u> Equation (*) (or (2.2.1)) gives $(\underline{\tilde{w}}(t) = B_w^{1/2}\,\underline{w}(t))$,

(4.6.7) $$\int_0^T \xi^q(s) \sum_{j=1}^k \sigma_{pj}^{-1}(d\xi^j(s) - \sum_i a_{ji}\,\xi^i(s)ds) = \int_0^T \xi^q(s)\sum_j \sigma_{pj}^{-1}\,d\tilde{w}^j(s),$$

$p,\ q = 1,2,\ldots,k.$

Further, from (4.6.5)

(4.6.8) $$\frac{\partial}{\partial a_{pq}}\{\int_0^T (B_w^{-1}A\underline{\xi}(s), d\underline{\xi}(s)) - \frac{1}{2}\int_0^T (A\underline{\xi}(s), B_w^{-1}A\underline{\xi}(s))\,ds\} =$$

$$= \int_0^T \xi^q(s)\sum_j \sigma_{pj}^{-1}\,(d\xi^j(s) - \sum_i \hat{a}_{ji}\,\xi^i(s)\,ds) = 0,$$

$p,\ q = 1,2,\ldots,k.$

By substraction from (4.6.7) and (4.6.8) one can get

(4.6.9) $$\frac{1}{T}\int_0^T \xi^q(s)\sum_j \sigma_{pj}^{-1}\sum_i \sqrt{T}(\hat{a}_{ji} - a_{ji})\xi^i(s)\,ds = \frac{1}{\sqrt{T}}\,\cdot$$

$$\cdot\int_0^T \xi^q(s)\sum_j \sigma_{pj}^{-1}\,d\tilde{w}^j(s) = \eta_{pq}(T),$$

$p,\ q = 1,2,\ldots,k,$

where

(4.6.10)
$$E\eta_{pq}(T) = 0,$$

$$E\eta_{pq}(T)\,\eta_{rs}(T) = \frac{1}{T}\int_0^T E\xi^q(s)\xi^s(t)\sum_{j_1,j_2}\sigma_{pj_1}^{-1}\,\sigma_{rj_2}^{-1}\,\delta_{j_1j_2}\,dt =$$

$$= \sigma_{rp}^{-1}\,b_{qs}, \qquad p,q = 1,2,\ldots,k.$$

The ergodic theorem (see Appendix B2 th. 3) gives that the left side of (4.6.9) is asymptotically

$$(4.6.11)\quad \sum_{i,j} \sqrt{T}\,(\hat{a}_{ji}-a_{ji})\,\frac{1}{T}\int_0^T \sigma_{pj}^{-1}\,\xi^q(s)\xi^i(s)ds \sim \sum_{j,k}\sqrt{T}(\hat{a}_{ji}-a_{ji})\,\sigma_{pj}^{-1}b_{qi}$$

i.e.,

$$(4.6.11')\quad B.\sqrt{T}\,(\hat{A}-A) \sim \underline{\eta}(T)$$

where

$$(4.6.12)\quad E\,\underline{\eta}(T) = \underline{0}\,,\qquad E\,\underline{\eta}(T)\,\underline{\eta}^*(T) = B.$$

Now using that $\underline{\xi}(t)$ and $\underline{w}(t)\cdot\underline{\xi}(t)$ are completely regular on the basis of Theorem 4 in Appendix B.2 we get, that $\underline{\eta}(t)$ is asymptotically normally distributed with parameters $(\underline{0}, B)$.

As (see (4.6.11'))

$$(4.6.13)\quad \sqrt{T}\,(\underline{\hat{A}}-\underline{A}) \sim B^{-1}\underline{\eta}(T),\quad E(B^{-1}\underline{\eta})\,(B^{-1}\underline{\eta})^* = B^{-1}.$$

The theorem is proved.

<u>Remark 2.</u> If $\xi(t)$ is a one dimensional AR (see (2.2.30) process

$$d\xi^{(k-1)}(t) + [a_1\,\xi^{(k-1)}(t)+\ldots+a_k\xi(t)]dt = dw\,(t),$$

then the solutions of conditional likelihood equation for $(a_1,\ldots,a_k)$ are asymptotically normally distributed with $\underline{0}$ mean and covariance matrix

$$\frac{1}{\sqrt{T}}\,\sigma_w^2\,B^{-1}(0),\ \text{where (see (4.4.26))}$$

$$(4.6.14)\quad B^{-1}(0) = (b_{ij}^{-1}),\quad b_{ij}^{-1} = \begin{cases} 0 \\ \dfrac{2}{\sigma_w^2}\cdot\sum_{\ell}(-1)^{\ell}a_{i-\ell}\,a_{j+1+\ell} \end{cases}$$

<u>Remark 3.</u> For simplicity let $\underline{\tilde{w}}(t) = B_w^{1/2}\,\underline{w}(t)$ have independent components, i.e., $\sigma_{pq} = 0$ if $p \neq q$, then using the notations

$$\underline{A}_p = \begin{pmatrix} a_{p1} \\ \cdot \\ \cdot \\ \cdot \\ a_{pk} \end{pmatrix},\quad B_p = (\sigma_{pp}^{-1}\,b_{qr}),$$

we get that the random variable

$$(4.6.15)\quad T.(\underline{\hat{A}}_p-\underline{A}_p)^*\,B_p\,(\underline{\hat{A}}_p-\underline{A}_p)$$

has an asymptotic χ_k^2 distribution if $T \to \infty$. The same is true for

(4.6.16) $$T(\hat{\underline{A}}_p-\underline{A}_p)^* B_p (\hat{\underline{A}}_p-\underline{A}_p), \quad \hat{B}_p = \left(\frac{1}{T}\int_0^T \sigma_{pp}^{-1}\, \xi^q(t)\, \xi^r(t)\, dt\right).$$

The last remark gives asymptotic confidence bounds for A.

<u>Theorem 3. The solution $\hat{A}_T$ of the conditional likelihood equation (4.6.4) is an efficient estimator and the convergence rate to the normal distribution is $(|\lambda|\cdot T)^{-1/2}$, where λ means the maximal real part of eigenvalues λ_i of A</u> ($\mathrm{Re}\lambda_i < 0$).

<u>Proof.</u> For simplicity let $\tilde{\underline{w}}(t)$ a Wiener process with independent components, and $\tilde{\underline{A}} = B^{-1}\underline{\eta}(T)$. Then $\tilde{\underline{A}}$ is asymptotically unbiased (see (4.6.12)). Further, if $S = E\,(\hat{\underline{A}}_p-\underline{A}_p)(\hat{\underline{A}}_p-\underline{A}_p)^*$ we get by direct calculations from (4.6.5) that

(4.6.17) $$S \geq \left(E\,\frac{\partial L_A}{\partial a_{pr}} \cdot \frac{\partial L_A}{\partial a_{pq}}\right)^{-1}_{r,q=\overline{1,k}}$$

and as $T \to \infty$

(4.6.18) $$S \geq \sigma_{pp}\, B^{-1}(0)$$

which proves the efficiency.

Let λ_i ($i=1,2,\ldots,k$) denote the roots of characteristic equation A, $|A-\lambda_i I| = 0$, and for simplicity we assume that λ_i are simple (with multiplicity 1). Then A is similar to the diagonal matrix (see (A.1.5))

$$P\,A\,P^{-1} = \begin{pmatrix} \lambda_1 & 0 & \ldots & 0 \\ 0 & \lambda_2 & \ldots & 0 \\ \vdots & & & \vdots \\ 0 & & & \lambda_k \end{pmatrix}.$$

Let $\underline{\xi}\,(t) = PA\underline{P}^{-1}\,\underline{\xi}(t)$, then the parameters $\lambda_i(i=1,2,\ldots,k)$ can be estimated by $\underline{\xi}(t)$ and on the basis of Theorem 3 in Section 3.3 (see (3.3.15)), Theorem 3 in section 4.2 (see (4.2.26')) and Theorems 3, 4 in Section 4.3 (see (4.3.8) and (4.3.12)) we get the statement of our theorem.

APPENDIX A

Linear Differential Equations With Constant Coefficients

Systems of ordinary differential equations with constant coefficients constitute an important class of ordinary differential equations which may be solved completely with the aid of elementary functions. As linear equations with constant coefficients have numerous engineering applications (the performance of many technical devices is described in an adequate manner by these equations), on one side and they are the equations of some mean values of elementary Gaussian processes (the equations of covariance functions) on the other side, we recall here the most important results and notations. The fact that the solution of linear equations with constant coefficients does not present any difficulties, in principle, they have a great interest for the same linear stochastic differential equations with constant coefficients, which are not simple generalizations of nonhomogeneous linear equations.

We have to underline here, that the statistical problems of linear stochastic equations with constant coefficients are not solved until now. E.g., to find the reduction of the A matrix to the Jordan form when a realization of the process $\underline{\xi}(t)$, $0 \leq t \leq T$, is given is not solved. In linear algebra when all eigenvalues of A are distinct, the task of reducing it to Jordan form, i.e., to diagonal form, is quite elementary. In the general case, however, the problem of reducing A to the Jordan form is one of the most complicated ones in linear algebra, and all the more so in statistical problems of elementary Gaussian processes.

1. Preliminary Definitions and Notations, Matrices

$A = (a_{ij})_{nxm}$ denotes a matrix with nxm entries, i.e., with n rows and m columns, where the complex number a_{ij} means the general element, $1 \leq i \leq n$, $1 \leq j \leq m$. A nx1 matrix $\underline{z}$ is termed vector (or column vector) and vectors are underlined. Multiplication of matrices is defined as row-column multiplication. A_{ij} is the cofactor of a_{ij}, and (A_{ij}) is called the adjoint matrix of A.

The zero matrix will be denoted by 0, the zero vector by $\underline{0}$, and the unit matrix by I. If there is danger of confusion concerning the dimension (size), these n x n matrices will be denoted by 0_n and I_n, respectively.

The complex conjugate matrix of $A = (a_{ij})$, denoted by $\overline{A}$, is defined by $\bar{A} = (\bar{a}_{ij})$, here $\bar{a}_{ij}$ is the complex conjugate of a_{ij}. The transposed matrix of A, denoted by A^*, is defined by $A^* = (a_{ji})$. The conjugate transposed matrix of A is $\bar{A}^*$. Note that $(\overline{AB})^* = \bar{B}^* \bar{A}^*$, $(AB)^* = B^*A^*$.

The determinant of A is denoted by det A, or $|A|$. The scalar product of two vectors $\underline{a}$, $\underline{b}$ is defined as $\underline{a}^* \underline{b}$, or $(\underline{a},\underline{b})$. The reduced cofactor; $\tilde{A}_{ij} = \dfrac{A_{ij}}{\det A}$.

If det A = 0, then A is said to be singular. A nonsingular matrix A possesses an inverse, A^{-1}, which satisfies

$$AA^{-1} = A^{-1}A = I, \quad (A^{-1} = (\tilde{A}_{ji})). \tag{A1.1}$$

The polynomial in λ of degree n, $\det (I\lambda - A) = 0$, is called the characteristic polynomial of A, and its roots $\lambda_1, \ldots, \lambda_n$ are the characteristic roots of A. Clearly

$$\det(\lambda I - A) = \prod_{i=1}^{n} (\lambda - \lambda_i).$$

The nxn matrices A and B are said to be similar if there exists a nonsingular P such that

$$B = PAP^{-1},$$

in this case they have the same characteristic polynomial

$$\det (\lambda I - B) = \det (\lambda I - A) \tag{A1.2}$$

for

$$\det(\lambda I - B) = \det(P(\lambda I - A)P^{-1}) = \det P \det (\lambda I - A) \det P^{-1} = \det (\lambda I - A).$$

Under similarity transformations the most important invariants are det A and trace of A, denoted by $\mathrm{Sp}\, A = \sum_{i=1}^{n} a_{ii}$.

The following statement, called Jordan decomposition, concerning the canonical form of a matrix is assumed known.

<u>Theorem 1. Every matrix A is similar to a matrix of the form</u>

$$J = \begin{pmatrix} J_0 & 0 & 0 & \cdots & 0 \\ 0 & J_1 & 0 & \cdots & 0 \\ \cdot & & & & \\ \cdot & & & & \\ \cdot & & & & \\ 0 & 0 & 0 & \cdots & J_p \end{pmatrix} \tag{A1.3}$$

<u>where</u> J_0 <u>is a diagonal with entries</u> $\lambda_1, \ldots, \lambda_q$, <u>and</u>

$$
(A1.4) \qquad J_i = \begin{pmatrix} \lambda_{q+i} & 1 & 0 & \dots & 0 & 0 \\ 0 & \lambda_{q+i} & 1 & \dots & 0 & 0 \\ \cdot & & & & & \\ \cdot & & & & & \\ \cdot & & & & & \\ 0 & 0 & 0 & \dots & \lambda_{1+i} & 1 \\ 0 & 0 & 0 & \dots & 0 & \lambda_{1+i} \end{pmatrix}
$$

$$(i = 1,2,\dots,p).$$

J_j is a Jordan block of the r_j-th order. The λ_j, $(j=1,\dots,n)$, are the characteristic roots of A, which need not all be distinct. If λ_j is a simple root, then it occurs in J_0, and therefore, if all the roots are distinct, A is similar to the diagonal matrix

$$
(A1.5) \qquad J = \begin{pmatrix} \lambda_1 & 0 & 0 & \dots & 0 \\ 0 & \lambda_2 & 0 & \dots & 0 \\ \cdot & & & & \\ \cdot & & & & \\ \cdot & & & & \\ 0 & 0 & 0 & \dots & \lambda_n \end{pmatrix}
$$

<u>Remark 1.</u> Theorem 1 is a theorem on the reduction of a matrix to Jordan form. We say that the sequence of the vectors $\underline{h}_1$, $\underline{h}_2,\dots,\underline{h}_p$ of the space R^n is a <u>basis set</u> or a series with eigenvalue λ for the transformation defined by A if the relations

$$
(A1.6) \qquad A\underline{h}_1 = \lambda \underline{h}_1, \quad A\underline{h}_2 = \lambda\underline{h}_2 + \underline{h}_1, \dots, A\underline{h}_p = \lambda\underline{h}_p + h_{p-1}
$$

$\underline{h}_1 \neq 0$, are fulfilled. If matrix A is real, then the sequence

$$\bar{\underline{h}}_1,\dots,\bar{\underline{h}}_p$$

forms a series with eigenvalue $\bar{\lambda}$.

Theorem 1 states that there exists a basis of the space R^n consisting of all the vectors of one or more series for the transformation A. If matrix A is real, then the series constituting the basis can be chosen in such a way that a series with real eigenvalues is real and series with complex eigenvalues are pairwise conjugate.

Remark 2. It follows immediately that

(A1.7) $$\det A = \prod_{1}^{q+p} \lambda_i^{r_i}, \quad \mathrm{Sp}\, A = \sum_{1}^{q+p} \lambda_i r_i$$

The norm of A, denoted by $||A||$, we define by the spectral norm

(A1.8) $$||A|| = \sup_{x} \frac{||A\underline{x}||}{||\underline{x}||} \overset{1)}{=} \max_{i} |\lambda_i|, \quad ||\underline{x}|| = \sqrt{x_1^2 + \ldots + x_n^2},$$ but we

may use other norms too, as for example

$$||A||_\infty = \max_{1\le i\le n} \sum_{j=1}^{n} |a_{ij}| \quad \text{or} \quad ||A||_1 = \max_{1\le j\le n} \sum_{i=1}^{n} |a_{ij}|.$$ The

distance between A and B is defined by $||A-B||$. The sequence A_n is said to converge to A (or to have a limit A) if $||A_m - A|| \to 0$. The exponential of a matrix A is defined by

(A1.9) $$e^A = I + \sum_{n=1}^{\infty} \frac{A^n}{n!},$$

where A^n is the nth power of A. The series is convergent for all A. Also we have

(A1.10) $$||e^A||_\infty \le (n-1) + e^{||A||_\infty}, \quad ||e^A|| \overset{1)}{=} \max_{i} e^{\lambda_i},$$

$$\det e^A = e^{\mathrm{Sp}\, A}$$

and hence e^A is nonsingular for all A. Since $-A$ commutes with A, $e^{-A} = (e^A)^{-1}$. If B is nonsingular A is the logarithm of B when $e^A = B$ (A is not unique).

Let

(A1.11) $$f(z) = a_0 + a_1 z + a_2 z^2 + \ldots$$

be an analytic function of the complex variable z with radius of convergence ρ, so that for $|z| < \rho$ the series converges, and for $|z| > \rho$ it diverges. If all eigenvalues of the matrix A lie inside the circle of convergence of (A1.11), i.e., $|\lambda_i| < \rho$, $i=1,\ldots,r$, then the matrix series

(A1.11') $$f(A) = a_o I + a_1 A + a_2 A^2 + \ldots$$

1) If A is symmetric.

converges, so that $f(A)$ is defined. The numbers $f(\lambda_i)$, $i=1,\dots,r$ ($r\le n$), which need not be distinct, comprise the set of all eigenvalues of the matrix $f(A)$.

It is both practical and theoretical interest to determine if the non-singularity condition can be removed so that a single set of arguments can be used for the singular as well as the nonsingular situations. Consider a matrix equation

(A1.12) $$A X A = A.$$

If A is an nxn nonsingular matrix, then the unique solution of (A1.12) is $X = A^{-1}$. If the matrix A is singular, or even rectangular, then a solution of Equation (A1.12), even if it exists, cannot be defined uniquely.
A matrix A^+ of order nxm is called the pseudo inverse (or generalized inverse) with respect to the matrix A of order mxn, if the following conditions are satisfied:

(A1.13) $$AA^+ A = A,$$
(A1.14) $$A^+ = UA^* = A^*V,$$
where U and V are matrices. (A1.14) shows that rows and columns of the matrix A^+ are, respectively, linear combinations of rows and columns of the matrix A^*. The following statement is true.

Theorem 2. The matrix A^+ satisfying (A1.13) exists and is unique. The following properties of pseudo inverses to be used further on:

1. $AA^+ A = A$, $A^+AA^+ = A^+$;
2. $(A^*)^+ = (A^+)^*$;
3. $(A^+)^+ = A$;
4. $(A^+A)^2 = A^+A$, the idempotent property, $(A^+A)^* = A^+A$, the symmetric property, $(AA^+)^2 = AA^+$, $(AA^+)^* = AA^+$;
5. $(A^*A)^+ = A^+(A^*)^+ = A^+(A^+)^*$;
6. $A^+ = (A^*A)^+ A^* = A^*(AA^*)^+$;
7. $A^+AA^* = A^*AA^+ = A^*$;
8. If S is an orthogonal matrix, then $(SAS^*)^+ = SA^+S^*$;
9. If A is a symmetric nxn nonnegative definite matrix of rank r<n, then

(A1.15) $$A^+ = T^*(TT^*)^{-2}T,$$
where T, rxn, matrix of rank r is defined by the decomposition
(A1.16) $$A = T^*T$$

10. If the matrix A is nonsingular, then $A^+ = A^{-1}$.

The decomposition in property 9. $A = T^*T$ is not unique. Later we shall use the pseudo inverses to prove a theorem on normal correlation.

As an application of the pseudo inverse let us take the normal equation for a linear least quares estimator

$$A_{nxq}\,\underline{\hat{b}}_{qx1} = B_{nxp}\,\underline{y}_{px1} \tag{A1.17}$$

This system will have a solution if and only if $B\underline{y}$ lies in the column space of A. This solution will be of the form

$$\underline{\hat{b}} = A^+B\underline{y} + \underline{z}, \tag{A1.18}$$

where $\underline{z}$ is a vector othogonal to the column space of A^*

$$A\underline{z} = 0. \tag{A1.19}$$

Let $\phi(t)$ and $\psi(t)$ be matrix functions of t, $0 \leq t \leq T$. $\phi'(t)$ denotes $\frac{d\phi(t)}{dt}$. Note that

$$(\Phi(t)\psi(t))' = \Phi'\psi + \Phi\psi'. \tag{A1.20}$$

If $\phi'(t)$ exists and $\phi(t)$ is nonsingular at t, then $\Phi^{-1}(t)$ is differentiable:

$$(\Phi^{-1}(t))' = -\,\Phi^{-1}\,\phi'\phi^{-1}, (\det \phi \neq 0). \tag{A1.21}$$

If $A(t)$ is continuous on $0 \leq t \leq T$ and

$$\Phi'(t) = A(t)\Phi(t), \tag{A1.22}$$

then

$$(\det \phi(t))' = (\mathrm{Sp}A(t))(\det \phi(t)) \tag{A1.23}$$

which is a first order equation with the solution

$$\det \Phi(t) = \det \Phi(\tau) \exp \int_\tau^t \mathrm{Sp}\, A(s)\, ds \quad , \quad 0 \leq \tau < t \leq T. \tag{A1.24}$$

2. Linear Systems With Constant Coefficients

Let A be an nxn matrix, and consider the corresponding homogeneous system

$$\underline{x}'(t) = A\underline{x}(t). \tag{A2.1}$$

If n=1 the solution assuming value x_0 at τ is given by $x_0 \cdot e^{A(t-\tau)}$.

The solution is the same when $\underline{x}(t)$, $\underline{x}_0$ are vectors of arbitrary finite dimension n. The matrix $\phi(t)$ is called fundamental for this homogeneous system (A2.1) if its n columns are the n linearly independent solutions on $0 \leq t \leq T$. A fundamental matrix for (A2.1) is given by

(A2.2) $$\phi(t) = e^{tA} \qquad (-\infty < t < \infty),$$

and the solution of (A2.1) satisfying $\underline{x}(\tau) = \underline{x}_0$ is given by

(A2.3) $$\underline{x}(t) = e^{A(t-\tau)}\,\underline{x}_0.$$

The fundamental matrix can be expressed by the help of Jordan reduction. By Theorem 1.1 let J be the canonical form of A, and suppose P is a nonsingular constant matrix such that

(A.2.4) $$P^{-1}AP = J.$$

then (using that for any matrix M $Pe^{M}\,P^{-1} = e^{PMP^{-1}}$),

(A2.5) $$e^{tA} = e^{tPJP^{-1}} = Pe^{tJ}\,P^{-1},$$

and

(A2.6) $$e^{tJ} = \begin{pmatrix} e^{tJ_o} & 0 & \cdots & 0 \\ 0 & e^{tJ_1} & \cdots & 0 \\ \cdot & & & \cdot \\ \cdot & & & \cdot \\ \cdot & & & \cdot \\ 0 & 0 & & e^{tJ_p} \end{pmatrix},$$

where

(A2.7) $$e^{tJ_o} = \begin{pmatrix} e^{t\lambda_1} & 0 & \cdots & 0 \\ 0 & e^{t\ 2} & \cdots & 0 \\ \cdot & & & \cdot \\ \cdot & & & \cdot \\ \cdot & & & \cdot \\ 0 & 0 & & e^{t\lambda_q} \end{pmatrix},$$

$$e^{tJ_i} = e^{t\lambda_{q+i}} \begin{pmatrix} 1 & t & \frac{t^2}{2!} & \cdots & \frac{t^{r_i-1}}{(r_i-1)!} \\ 0 & 1 & t & \cdots & \frac{t^{r_i-2}}{(r_i-2)!} \\ \vdots & & & & \vdots \\ 0 & 0 & 0 & \cdots & 1 \end{pmatrix}$$

where J_i is an $r_i \times r_i$ matrix $(n = q+r_1+\ldots+r_p)$.

Another fundamental matrix of (A2.1) is given by

$$e^{tA}\, P = Pe^{tJ}. \tag{A2.8}$$

The nonhomogeneous system

$$\underline{x}'(t) = A\underline{x}(t) + \underline{b}(t) \tag{A2.9}$$

has the general solution

$$\underline{x}(t) = e^{(t-\tau)A} \cdot \underline{x}_0 + \int_\tau^t e^{(t-s)A} \cdot \underline{b}(s)ds \tag{A2.10}$$

The linear differential equation of order n with constant coefficients has the form

$$Lx(t) = x^{(n)}(t) + a_1 x^{(n-1)}(t) + \ldots + a_n x(t) = 0 \tag{A2.11}$$

in the homogeneous case. Its associated vector system

$$\underline{x}'(t) = A\underline{x}(t)$$

where A is the constant matrix

$$A = \begin{pmatrix} 0 & 1 & 0 & \cdots & 0 \\ 0 & 0 & 1 & \cdots & 0 \\ \vdots & & & & \\ 0 & 0 & & \cdots & 1 \\ -a_n & -a_{n-1} & -a_{n-2} & \cdots & -a_1 \end{pmatrix}.$$

A fundamental set of solutions can be exhibited, and their form depends on the characteristic polynomial det $(\lambda I-A) = 0$, which is given by

$$\lambda^n + a_1 \lambda^{n-1} + \dots + a_n = 0. \tag{A2.12}$$

Let $\lambda_1,\dots,\lambda_p$ be the distinct roots of this equation and suppose λ_i has multiplicity m_i. Then a fundamental set of solutions is given by

$$t^k e^{t\lambda_i} \quad (k=0,1,\dots, m_i-1; \; i=1,\dots,p). \tag{A2.13}$$

With the formal operator L the adjoint of L, L^+ is associated, given by

$$L^+ = (-1)^n \frac{d^n}{dt^n} + (-1)^{n-1} \frac{d^{n-1}}{dt^{n-1}} (\bar{a}_1) + \dots + \bar{a}_n. \tag{A2.14}$$

The equation

$$L^+ x(t) = 0 \tag{A2.15}$$

can be rewritten in the form

$$\underline{x}'(t) = -\bar{A}^*\underline{x}(t),$$

where

$$-\bar{A}^* = \begin{pmatrix} 0 & 0 & \dots & 0 & \bar{a}_n \\ -1 & 0 & \dots & 0 & \bar{a}_{n-1} \\ 0 & -1 & \dots & 0 & \bar{a}_{n-2} \\ \cdot & & & & \\ \cdot & & & & \\ \cdot & & & & \\ 0 & 0 & \dots & 0 & \bar{a}_2 \\ 0 & 0 & \dots & -1 & \bar{a}_1 \end{pmatrix}$$

All the proofs of the above mentioned statements can be found in Pontryagin [1], Coddington-Levinson [1], Gantmakher [1] and Bodewig [1].

APPENDIX B

Probability Background

1. Gaussian Systems

We recall that the random vector variable $\underline{\xi}^* = (\xi_1,\ldots,\xi_n)$ is called Gaussian, or normal, if for any vector $\underline{a}$, $\underline{a}^*\underline{\xi}$ is a Gaussian random variable. The characteristic function has the form

(B1.1) $E\exp(i\underline{\alpha}^*\underline{\xi}) = \varphi_\xi(\underline{\alpha}) = \exp(i\underline{\alpha}^*\underline{m} - \frac{1}{2}\underline{\alpha}^*R\underline{\alpha})$, where $\underline{m}^* = (m_1,\ldots,m_n)$ is the mean value vector $\underline{m} = E\underline{\xi}$, and the nonnegative definite symmetric matrix $R = (R_{ij})_{nxn}$ is the covariance matrix

(B1.2) $$R = R_{\xi\xi} = \mathrm{cov}\,(\underline{\xi},\underline{\xi}) = E(\underline{\xi}-\underline{m})(\underline{\xi}-\underline{m})^*.$$

The following simple properties of Gaussian vectors are often used.

__Lemma 1. If $\xi^* = (\xi_1,\ldots,\xi_n)$ is a Gaussian vector, A is an mxn matrix, $\underline{a}^* = (a_1,\ldots,a_m)$, then $\underline{\eta} = A\underline{\xi} + \underline{a}$ is Gaussian with__

(B1.3) $$\varphi_\eta(\underline{\alpha}) = \exp\{i\underline{\alpha}^*(\underline{a}+A\underline{m}) - \frac{1}{2}\underline{\alpha}^*(AR_{\xi\xi}A^*)\,\underline{\alpha}\},$$

(B1.4) $$E\underline{\eta} = \underline{a} + A\underline{m}, \quad R_{\eta\eta} = \mathrm{cov}(\underline{\eta},\underline{\eta}) = AR_{\xi\xi}A^*.$$

__Lemma 2. Let $(\underline{\xi}^1,\underline{\xi}^2)^* = (\xi_1^1,\ldots,\xi_k^1,\xi_1^2,\ldots\xi_\ell^2)$ be a Gaussian vector with__

$$\underline{m}_{\xi^1} = E\,\underline{\xi}^1, \qquad R_{\xi^1\xi^1} = \mathrm{cov}(\underline{\xi}^1,\underline{\xi}^1),$$

$$\underline{m}_{\xi^2} = E\,\underline{\xi}^2, \qquad R_{\xi^2\xi^2} = \mathrm{cov}\,(\underline{\xi}^2,\underline{\xi}^2),$$

$$R_{\xi^1\xi^2} = \mathrm{cov}\,(\underline{\xi}^1,\underline{\xi}^2)$$

__If $R_{\xi^1\xi^2} = 0$, then $\underline{\xi}^1$, $\underline{\xi}^2$ are independent and__

(B1.5) $$\varphi_{\xi^1}(\underline{\alpha}_1)\,\varphi_{\xi^2}(\underline{\alpha}_2) = \varphi_{\xi^1\xi^2}(\underline{\alpha}_1, \underline{\alpha}_2)$$

__Lemma 3. Let $\underline{\xi}^* = (\xi_1,\ldots,\xi_n)$ be a Gaussian vector with $\underline{m} = E\underline{\xi}$, $R_{\xi\xi} = \mathrm{cov}(\underline{\xi},\underline{\xi})$. Then there exists a Gaussian vector $\underline{\varepsilon}^* = (\varepsilon_1,\ldots,\varepsilon_n)$ with independent components, $E\underline{\varepsilon} = \underline{0}$, $\mathrm{cov}\,(\underline{\varepsilon},\underline{\varepsilon}) = I$, such that__

(B1.6) $$\underline{\xi} = R_{\xi\xi}^{1/2}\,\underline{\varepsilon} + \underline{m}\,.$$

To prove (B1.6) let $T = R_{\xi\xi}^{1/2}$ and $\underline{\zeta}^* = (\zeta_1,\ldots,\zeta_n)$ Gaussian and independent of $\underline{\xi}$, $E\underline{\zeta} = \underline{0}$, cov $(\underline{\zeta},\underline{\zeta}) = I$. The vector variable

(B1.7) $$\underline{\varepsilon} = (T^+)^*(\underline{\xi} - \underline{m}) + (I-T\,T^+)\underline{\xi}$$

is Gaussian, since $\underline{\xi}$ and $\underline{\zeta}$ are independent. We have

$$E\underline{\varepsilon} = \underline{0},$$

(B1.8) $$\text{cov}\,(\underline{\varepsilon},\underline{\varepsilon}) = (T^+)^*R_{\xi\xi}T^+ + (I-TT^+)(I-TT^+)^* = I,$$

because by property 4 of pseudoinverse matrices

$$(I-TT^+)^* = I-TT^+,\quad (I-TT^+)^2 = I-TT^+,$$

and

$$(T^+)^*\,R_{\xi\xi}T^+ = (T^+)^*\,T^*TT^+ = [(T^+)^*T^*][TT^+] = TT^+.$$

(B1.8) shows the independence of components $\underline{\varepsilon}$.
From (B1.7) we obtain

(B1.9) $$T^*\underline{\varepsilon} = T^*(T^+)^*(\underline{\xi}-\underline{m}) + (T^*-T^*TT^+)\underline{\zeta} =$$

$$= (\underline{\xi}-\underline{m}) - (I-T^*(T^+)^*)(\underline{\xi}-\underline{m}) + (T^*-T^*TT^+)\underline{\xi}\ .$$

But $T^* = T^*TT^+$ (from property 7), $T^*(T^+)^* = (T^+T)^* = T^+T$ (from property 4) and

$$(I-T^+T)\;R_{\xi\xi}(I-TT^+)^* = (I-T^+T)(T^*T)(I-T^+T)^* = 0$$

which proves (B1.6).

<u>Lemma 4.</u> <u>Let</u> $\underline{\xi}_n$, n=1,2,..., <u>be a sequence of Gaussian vector variables converging in probability to a vector</u> $\underline{\xi}$. <u>Then</u> $\underline{\xi}$ <u>is also a Gaussian vector variable.</u>

<u>Proof.</u> Let $\underline{m}_n = E\underline{\xi}_n$, $R_n = \text{cov}(\underline{\xi}_n,\underline{\xi}_n)$. Then, since lim $\underline{\xi}_n = \underline{\xi}$ in probability and $|\exp(i\underline{\alpha}^*\underline{\xi}_n)| \le 1$, by the Lebesgue bounded convergence theorem

$$\lim_{n\to\infty}\exp(i\underline{\alpha}^*\underline{m}_n - 1/2\ \underline{\alpha}^*\,R_n\underline{\alpha}) = \lim_{n\to\infty} E\exp(i\underline{\alpha}^*\underline{\xi}_n) = E\exp(i\underline{\alpha}^*\underline{\xi}).$$

As $\underline{\alpha}$ was arbitrary there exists a vector $\underline{m}$ and a nonnegative matrix

R such that

$$\underline{m} = \lim_{n\to\infty} \underline{m}_n, \quad R = \lim_{n\to\infty} R_n,$$

and

$$E \exp(i\, \underline{\alpha}^* \underline{\xi}) = \exp(i \underline{\alpha}^* \underline{m} - 1/2\, \underline{\alpha}^* R \underline{\alpha}),$$

which proves the statement.

Theorem 1. Let $(\underline{\xi}^1, \underline{\xi}^2) = (\xi_1^1, \ldots \xi_k^1, \xi_1^2, \ldots, \xi_\ell^2)$ be a Gaussian vector with

$$\underline{m}_{\xi^1} = E\, \underline{\xi}^1, \quad \underline{m}_{\xi^2} = E\underline{\xi}^2,$$

$$R_{\xi^1\xi^1} = \mathrm{cov}\, (\underline{\xi}^1, \underline{\xi}^1),\ R_{\xi^2\xi^2} = \mathrm{cov}(\underline{\xi}^2, \underline{\xi}^2),\ R_{\xi^1\xi^2} = \mathrm{cov}(\underline{\xi}^1, \underline{\xi}^2).$$

Then the conditional expectation and conditional covariance are given by the formulas

$$\text{(B1.10)} \quad E(\underline{\xi}^1 | \underline{\xi}^2) = \underline{m}_{\xi^1} + R_{\xi^1\xi^2} R^+_{\xi^2\xi^2} (\underline{\xi}^2 - \underline{m}_{\xi^2}),$$

$$\text{(B1.11)} \quad \mathrm{cov}(\underline{\xi}^1, \underline{\xi}^1 | \underline{\xi}^2) = E(\,[\underline{\xi}^1 - E(\underline{\xi}^1 |\, \underline{\xi}^2)][\underline{\xi}^1 - E(\underline{\xi}^1 | \underline{\xi}^2)]^* | \underline{\xi}^2) =$$

$$= R_{\xi^1\xi^1} - R_{\xi^1\xi^2} R^+_{\xi^2\xi^2} (R_{\xi^1\xi^2})^*,$$

and the conditional distribution is also Gaussian.

Proof. Consider the covariance matrix

$$\text{(B1.12)} \quad \mathrm{cov}(\underline{\xi}^1 - \underline{m}_{\xi^1} - R_{\xi^1\xi^2} R^+_{\xi^2\xi^2} (\underline{\xi}^2 - \underline{m}_{\xi^2}), \underline{\xi}^2 - \underline{m}_{\xi^2}) =$$

$$= \mathrm{cov}(\underline{\xi}^1 - \underline{m}_{\xi^1}, \underline{\xi}^2 - \underline{m}_{\xi^2}) - R_{\xi^1\xi^2} R^+_{\xi^2\xi^2} \mathrm{cov}(\underline{\xi}^2 - \underline{m}_{\xi^2}, \underline{\xi}^2 - \underline{m}_{\xi^2}) =$$

$$= R_{\xi^1\xi^2} - R_{\xi^1\xi^2} R^+_{\xi^2\xi^2} R_{\xi^2\xi^2} = 0$$

since (see property 9 for pseudoinverses)

$$R^+_{\xi^2\xi^2} = (TT)^+ = T^+T^+,\ T(TT)^+TT = TT^+T^+\ TT = TT^+(T^+T)^*T =$$

$$= (TT^+)^2 T = TT^+T = T,$$

and

$$\text{(B1.13)} \quad \mathrm{cov}(\underline{\xi}^1 - \underline{m}_{\xi^1} - R_{\xi^1\xi^2} R^+_{\xi^2\xi^2} (\underline{\xi}^2 - \underline{m}_{\xi^2}),\ \underline{\xi}^1 - \underline{m}_{\xi^1} - R_{\xi^1\xi^2} R^+_{\xi^2\xi^2} (\underline{\xi}^2 - \underline{m}_{\xi^2})) =$$

$$= R_{\xi^1\xi^1} - R_{\xi^1\xi^2} R^+_{\xi^2\xi^2} R^*_{\xi^1\xi^2} - R_{\xi^1\xi^2} (R_{\xi^1\xi^2} R^+_{\xi^2\xi^2})^* +$$

$$+ R_{\xi^1\xi^2} R^+_{\xi^2\xi^2} R_{\xi^2\xi^2} (R_{\xi^1\xi^2} R^+_{\xi^2\xi^2})^* = R_{\xi^1\xi^1} - R_{\xi^1\xi^2} R^+_{\xi^2\xi^2} R^*_{\xi^1\xi^2} -$$

$$- R_{\xi^1\xi^2} [(R_{\xi^1\xi^2} R^+_{\xi^2\xi^2})^* - R^+_{\xi^2\xi^2} \cdot R_{\xi^2\xi^2} (R_{\xi^1\xi^2} R^*_{\xi^1\xi^2})^*] =$$

$$= R_{\xi^1\xi^1} - R_{\xi^1\xi^2} R^+_{\xi^2\xi^2} R^*_{\xi^1\xi^2} .$$

Since $(\underline{\xi}^1, \underline{\xi}^2)$ is Gaussian so is $(\underline{\eta}, \underline{\xi}^2)$, where

(B1.14) $$\underline{\eta} = \underline{\xi}^1 - \underline{m}_{\xi^1} - R_{\xi^1\xi^2} R^+_{\xi^2\xi^2} (\underline{\xi}^2 - \underline{m}_{\xi^2}).$$

But from (B1.12) $\mathrm{cov}(\underline{\eta}, \underline{\xi}^2) = 0$, which means $\underline{\eta}$ and $\underline{\xi}^2$ are independent (Lemma 2) and the conditional distribution of $\underline{\eta}$ is also Gaussian with

(B1.15) $$E(\underline{\eta}) = E(\underline{\eta}|\underline{\xi}^2) = \underline{0}, \quad \mathrm{cov}(\underline{\eta}, \underline{\eta}) = R_{\xi^1\xi^1} - R_{\xi^1\xi^2} R^+_{\xi^2\xi^2} R^*_{\xi^1\xi^2},$$

which proofs the Theorem as

(B1.16) $$\underline{\eta} = \underline{\xi}^1 - E(\underline{\xi}^1 | \underline{\xi}^2),$$

$$\mathrm{cov}(\underline{\xi}^1, \underline{\xi}^1 | \underline{\xi}^2) = \mathrm{cov}(\underline{\eta}, \underline{\eta} | \underline{\xi}^2) = \mathrm{cov}(\underline{\eta}, \underline{\eta}).$$

Remark 1. If ξ_1 and ξ_2 are real random variables

$$E(\xi_1|\xi_2) = E\xi_1 + \rho \frac{\sigma_{\xi_1}}{\sigma_{\xi_2}} (\xi_2 - E\xi_2),$$

$$D^2(\xi_1|\xi_2) = E([\xi_1 - E(\xi_1|\xi_2)]^2 | \xi_2) = \sigma^2_{\xi_1} (1-\rho^2),$$

where

$$\sigma_\xi = (D^2\xi)^{1/2}, \qquad \rho = \frac{\mathrm{cov}(\xi_1, \xi_2)}{\sigma_{\xi_1} \sigma_{\xi_2}} .$$

Remark 2. Let A_1 $k \times k$, A_2 $k \times \ell$, B_1 $\ell \times k$, B_2 $\ell \times \ell$ be matrices and

(B1.17)
$$A \circ B = A_1 B_1^* + A_2 B_2^*,$$
$$A \circ A = A_1 A_1^* + A_2 A_2^*,$$
$$B \circ B = B_1 B_1^* + B_2 B_2^*.$$

Then the symmetrix matrix

(B1.18) $$A \circ A - (A \circ B)(B \circ B)^+ (A \circ B)^*$$

is nonnegative definite.

The proof is immediate. Set

$$\underline{\xi}^1 = A_1\underline{\zeta}^1 + A_2\underline{\zeta}^2 \tag{B1.19}$$

$$\underline{\xi}^2 = B_1\underline{\zeta}^1 + B_2\underline{\zeta}^2$$

where $\underline{\zeta}^1$ and $\underline{\zeta}^2$ are independent Gaussian vectors with independent components ($E\zeta_i = 0$, $D^2\ \zeta_i = 1$). Then according to (B1.11)

$$\operatorname{cov}(\underline{\xi}^1,\underline{\xi}^1 | \underline{\xi}^2) = A \circ A - (A \circ B)\ (B \circ B)^+(A \circ B)^*,$$

which proves the statement.

In the following we recall some fundamental results in the theory of stochastic processes and stochastic differential equations, the proofs of which the reader may find in most of the textbooks (see e.g., Gikhman-Skorokhod [1], [2]). We are using them in more sophisticated discussions, but not in the construction of elementary Gaussian processes.

<u>Theorem 2. (P. Levy) Let</u> $(w(t),F_t)$ <u>a continuous square integrable martingale</u>

$$E(w(t)|F_s) = w(s),\quad s \le t,\ w(0) = 0,$$

<u>where</u> F_t <u>is a nondecreasing family of σ-algebras. If</u>

$$E([w(t)-w(s)]^2|F_s) = t-s, \tag{B1.20}$$

<u>then w(t) is a standard Wiener process</u> ($Ew(t) = 0$, $Ew^2(t) = 1$).

<u>Theorem 3. Let</u> $\underline{w}(t)^* = (w^1(t),\ldots w^n(t))$ <u>be an n-dimensional continuous martingale</u> *)

$$E(w^i(t)|F_s) = w^i(s),\ w^i(0) = 0,\ s \le t,\ i = 1,2,\ldots,n,$$

<u>and</u>

$$E[(\underline{w}(t)-\underline{w}(s))(\underline{w}(t)-\underline{w}(s))^*|F_s] = (t-s)I.$$

<u>Then</u> $\underline{w}(t)$ <u>is an n-dimensional standard Wiener process with independent components</u> ($Ew^i(t) = 0$, $E(w^i(t))^2 = t$).

<u>Lemma 5. Let</u> $0 \le t_o^{(n)} < t_1^{(n)} <\ldots < t_n^{(n)} = T$ <u>be a subdivision of</u>

*)The components of stochastic processes we shall designate by upper supperscripte $w^i(t)$.

the interval $[0,T]$, with $\max_i [t_{i+1}^{(n)} - t_i^{(n)}] \to 0$, $n \to \infty$. Then

$$\lim_{n\to\infty} \sum_{i=1}^{n} [w(t_i^{(n)}) - w(t_{i-1}^{(n)})]^2 = T, \tag{B1.21}$$

in mean square and with probability 1.

Theorem 4. (Ito's formula) Let the function $f(t,x)$ be continuous and have the continuous partial derivatives f_t, f_x, f_{xx}. Assume that the random process $\xi(t)$, $0 \le t \le T$, has the stochastic differential

$$d\xi(t) = a(t,\omega)\, dt + b(t,\omega)\, dw(t), \tag{B1.22}$$

where $w(t)$ is a standard Wiener process and $a(t,\omega)$, $b(t,\omega)$ are nonanticipative functions (i.e., F_t measurable functions). Then the process $f(t,\xi(t))$ also has a stochastic differential and

$$df(t,\xi(t)) = [f_t(t,\xi(t)) + f_x(t,\xi(t))a(t,\omega) + \tfrac{1}{2}f_{xx}(t,\xi(t))b^2(t,\omega)] \cdot dt + f_x(t,\xi(t))b(t,\omega)\, dw(t). \tag{B1.23}$$

The multidimensional variant of Ito's formula states:

Theorem 5. Let the function $f(t,x_1,\ldots,x_n)$ be continuous and have the continuous derivatives f_t, f_{x_i}, $f_{x_i x_j}$. Assume that the random vector process $\underline{\xi}^*(t) = (\xi^1(t),\ldots\xi^n(t))$ has the stochastic differential

$$d\xi^i(t) = a^i(t,\omega)\, dt + \sum_{j=1}^{n} b_{ij}(t,\omega)\, dw^j(t), \quad i = 1,2,\ldots,n, \tag{B1.24}$$

where $\underline{w}^*(t) = (w^1(t),\ldots,w^n(t))$ is a Wiener process. Then the process $f(t,\xi^1(t),\ldots,\xi^n(t))$ has the stochastic differential

$$\begin{aligned} d\, f(t,\xi^1(t),\ldots,\xi^n(t)) &= \\ = [f_t(t,\xi^1(t),\ldots,\xi^n(t)) &+ \sum_{i=1}^{n} f_{x_i}(t,\xi^1(t),\ldots,\xi^n(t))a^i(t,\omega) + \\ &+ \frac{1}{2}\sum_{i,j=1}^{n} f_{x_i x_j}(t,\xi^1(t),\ldots,\xi^n(t) \sum_{k=1}^{n} b_{ik}(t,\omega)b_{jk}(t,\omega)]dt + \\ &+ \sum_{i,j=1}^{n} f_{x_i}(t,\xi^1(t),\ldots,\xi^n(t))b_{ij}(t,\omega)b_{ij}(t,\omega)dw^j(t). \end{aligned} \tag{B.1.25}$$

<u>Definition 1.</u> The continuous random process $(\underline{\xi}(t), F_t)$ is called Ito process relative to the Wiener process $(\underline{w}(t), F_t)$ if there exists two nonanticipative processes $(\underline{a}(t), F_t)$ and $(B(t), F_t)$ such that ($\underline{a}$ is a vector, B is a matrix)

$$P\{\int_0^T ||\underline{a}(t)||^2 dt < \infty\} = 1,$$

$$P\{\int_0^T ||B(t)||^2 dt < \infty\} = 1,$$

and with probability 1 for $0 \le t \le T$,

$$\text{(B1.22')} \qquad \underline{\xi}(t) = \underline{\xi}(0) + \int_0^t \underline{a}(s,\omega)\, ds + \int_0^t B(s,\omega)\, d\underline{w}(s),$$

or, for brevity

$$\text{(B1.22")} \qquad d\underline{\xi}(t) = \underline{a}(t,\omega)\, dt + B(t,\omega)\, d\underline{w}(t).$$

<u>Definition 2.</u> The Ito process $(\xi(t), F_t)$ is called a process of diffusion type, relative to the Wiener process $(\underline{w}(t), F_t)$, if the functionals $\underline{a}(t,\omega)$ and $B(t,\omega)$ being part in the above formulas are F_t^{ξ} - measurable.
The continuous stochastic process $\xi(t)$ is called a <u>strong</u> solution of the stochastic differential equation

$$\text{(B1.22')} \qquad d\xi(t) = a(t,\xi)\, dt + b(t,\xi)\, dw(t),$$

where $(w(t), F_t)$ is a Wiener process, with F_0 measurable initial condition $w(0)$, if for each t $(0 \le t \le T)$ w(t) is F_t measurable and with probability 1 for each t

$$\text{(B1.26)} \qquad \xi(t) = \xi(0) + \int_0^t a(s\xi)\, ds + \int_0^t b(s,\xi)\, dw(s),$$

and

$$\text{(B1.27)} \qquad P(\int_0^T |a(t,\xi)\, dt < \infty) = 1,$$

$$\text{(B1.28)} \qquad P(\int_0^T b^2(t,\xi)\, dt < \infty) = 1.$$

We say that equation (B1.22) has a weak solution if there may be given a probability space (Ω, F, P), a non-decreasing family of σ-algebras F_t, $t \le T$, a continuous process $(\xi(t), F_t)$ and a Wiener process $(w(t), F_t)$, such that (B1.26) - (B1.28) are satisfied and $\xi(0)$ has the prescribed distribution.

The difference between strong and weak solutions is that in case of strong solution the probability space, the F_t system and Wiener process are given ($F_t = F_t^w$ and $F_t^\xi \subseteq F_t^w$), while in case of weak solution we can construct them.

Theorem 6. Let the elements of the vector function $\underline{a}^*(t) = (a_1(t),\ldots,a_n(t))$ and matrices $A(t) = (a_{ij}(t)), B(t) = (b_{ij}(t))$ $i=1,2,\ldots,n$, be deterministic functions satisfying the conditions

$$\int_0^T |a_1(t)|dt < \infty, \quad \int_0^T |a_{ij}(t)|\, dt < \infty, \quad \int_0^T |b_{ij}(t)|^2 dt < \infty,$$

then the vector stochastic differential equation

$$\text{(B1.29)} \qquad d\underline{\xi}(t) = [\underline{a}(t) + A(t)\underline{\xi}(t)]dt + B(t)d\underline{w}(t), \qquad \underline{\xi}(0) = \underline{\eta}\,,$$

has a unique strong solution, defined by the formula

$$\text{(B1.30)} \qquad \underline{\xi}(t) = \Phi(t)[\underline{\eta} + \int_0^t \Phi^{-1}(s)\, \underline{a}(s)\, ds + \int_0^t \Phi^{-1}(s)\, B(s)\, d\underline{w}(s)],$$

where $\Phi(t)$ is the n x n fundamental matrix

$$\text{(B1.31)} \qquad \Phi(t) = I_n + \int_0^t A(s)\Phi(s)\, ds.$$

Theorem 7. Let $(w(t), F_t)$, $t > 0$, be a Wiener process and $\xi(t)$, $t \geq 0$, be such that

$$\text{(B1.32)} \qquad P\left(\int_0^T \xi^2(t)\, dt < \infty\right) = 1, \qquad 0 \leq T < \infty,$$

$$\text{(B1.33)} \qquad P\left(\int_0^\infty \xi^2(t)\, dt = \infty\right) = 1.$$

Then the random process $(\zeta(s), F_{\tau_s})$, $s \leq 0$, with

$$\text{(B1.34)} \qquad \zeta(s) = \int_0^{\tau_s} \xi(t)\, dw(t),$$

where $\tau_s = \inf\left(t: \int_0^t \xi^2(u)\, du \geq s\right)$, is a Wiener process

and

$$\text{(B1.35)} \qquad P\left(\lim_{t\to\infty} \frac{\int_0^t \xi(u)\, dw(u)}{\int_0^t \xi^2(u)\, du} = 0\right) = 1.$$

2. Some Basic Concepts in Probability Theory

The primary object of probability theory is the probability space (Ω, F, P), where the set Ω consists of elementary events ω, the system F forms a σ-algebra, and P denotes a probability measure ($P(\Omega) = 1$) defined on sets in F. We assume that the Kolmogorov's axiomatics is known. Let (Ω, F) and (E,B) be two measurable spaces. The function $\xi(\omega)$ defined on (Ω,F) with values on E, is called F/B-measurable if the set $\{\omega:\xi(\omega)\in B\}\in F$ for any $B\in B$. Such functions are called random functions with values in E. If $E = R^1$ $\xi=\xi(\omega)$ is called random variable, if $E = R^n$ $\underline{\xi}^*=(\xi_1, \ldots, \xi_n)$ is called random vector variable. The mathematical expectation (denoted by $E\xi$) of a random variable is the Lebesgue integral

(B2.1) $$E\xi = \int_\Omega \xi(\omega)\ P(d\omega) = \int_\Omega \xi(\omega)\ dP.$$

$E\xi$ exists if $E|\xi| < \infty$.

Let F_1 be a sub σ-algebra of F (i.e., $F_1 \subseteq F$) and let $\xi(\omega)$ be a random variable. The conditional mathematical expectation (denoted $E(\xi|F_1)$) is an F_1 measurable function such that for any $\Lambda\varepsilon F_1$

(B2.2) $$\int_\Lambda \xi(\omega)\ P(d\omega) = \int_\Lambda E(\xi|F_1)P(d\omega).$$

If two probability measures P_1 and P_2 are given on (Ω,F) we say that P_1 is absolutely continuous with respect to measure P_2 (denoted $P_1 << P_2$) if $P_1(A) = 0$ for any $A\varepsilon F$ for which $P_2(A) = 0$. The Radon-Nikodym theorem states the following. If $P_1 << P_2$, then there exists a nonnegative random variable $f(\omega)$, which is called the density of one measure P_1 with respect to the other (P_2) or Radon-Nikodym derivative, such that for any $A\varepsilon F$

(B2.3) $$P_1(A) = \int_A f(\omega)\ dP_2.$$

The function $f(\omega)$ is unique within stochastic equivalence (with P_2 probability 1).

The notation

(B2.4) $$\frac{dP_1}{dP_2}(\omega) = f(\omega)$$

is used.

We say that the sequence of random variables $\xi_n, n = 1,2,\ldots,$

converges in probability to a random variable ξ, if for any $\varepsilon > 0$

(B2.5) $$\lim_{n\to\infty} P\{|\xi_n - \xi| > \varepsilon\} = 0.$$

If

(B2.6) $$P\{\omega: \lim_{n\to\infty} \xi_n = \xi\} = 1$$

we say the sequence ξ_n converges to ξ with probability 1 (or almost surely). The sequence of random variables ξ_n, $n = 1,2,\ldots,n$, is called convergent in mean square to ξ, if $E\xi_n^2 < \infty$, $E\xi^2 < \infty$

(B2.7) $$E\,(\xi_n - \xi)^2 \to 0 \qquad \text{as } n \to \infty .$$

We shall assume that such statements as Fatou's lemma, the Lebesgue dominated convergence theorem, Kolmogorov's three series theorem are known.

Let (Ω,F,P) be a probability space and $0 \leq t < \infty$. The family of random variables $\xi(t) = \xi(t,\omega)$ is called a random (stochastic) process with continuous time. When the time parameter is confined to the set $\{0,\pm1,\pm2,\ldots,\}$ $\xi(t)$, or $\xi(n)$, is called a random sequence or stochastic process with discrete time (time series). The time function $\xi(t,\omega)$ with fixed ω is called a trajectory or re alization corresponding to an elementary event ω.

The σ-algebras $F_t^\xi = \sigma\{\xi(s), s\leq t\}$ being the smallest σ-algebras with respect to which the random variables $\xi(s)$, $s\leq t$, are measurable, are naturally associated with any stochastic process $\xi(t)$. For the conditional mathematical expectation $E(\eta|F_t^\xi)$ we may use the notation:

$$E(\eta|F_t) = E(\eta|F^\xi_{o,t}) = E(\eta|\xi(s),\ 0 \leq s \leq t).$$

Let F_t, $0 < t < \infty$, be a nondecreasing family of σ-algebras, $F_s \subseteq F_t$, $s \leq t$. A measurable $\xi(t,\omega)$ process, $0 \leq t < \infty$ is adapted to a family of σ-algebras $\{F_t\}$, $0 \leq t < \infty$, if for any $t > 0$ the random variables $\xi(t)$ are F_t-measurable. Such a stochastic process will be denoted $(\xi(t),F_t)$ and called nonanticipative.

Two random processes $\xi_1(t)$, $\xi_2(t)$, $0 \leq t < \infty$, are called stochastically equivalent if $P(\xi_1(t) \neq \xi_2(t)) = 0$ for all $t \geq 0$, in this case $\xi_2(t)$ is called the modification of $\xi_1(t)$.

$\xi(t)$ is called stochastically continuous in $[a,b]$ if for any $\varepsilon > 0$

(B2.8) $$P\{|\xi(t) - \xi(t_o)| > \varepsilon\} \to 0 \qquad \text{as } t \to t_o,$$

for all $t_0 \in [a,b]$. The stochastic process $\xi(t)$ is called continuous on $[a,b]$ if it is continuous with probability 1. The following theorem is known as Kolmogorov's criterion.

Theorem 1. The stochastic process $\xi(t)$ ($a \le t \le b$) permits a continuous modification if there are constants $a > 0$, $\varepsilon > 0$ and C such that

$$\text{(B2.9)} \qquad E|\xi(t+\Delta t) - \xi(t)|^a \le C|\Delta t|^\varepsilon ,$$

for all t, $t + \Delta t \in [a,b]$.

$\xi(t)$ is called continuous in the mean square on $[a,b]$ if

$$\text{(B2.10)} \qquad E|\xi(t) - \xi(t_o)|^2 \to 0 \qquad \text{as } t \to t_o,$$

for all the points $t_0 \in [a,b]$.

The stochastic process $\xi(t)$ is called stationary in the strict sense if for any real h the finite dimensional distributions do not change with the shift on h:

$$\text{(B2.11)} \quad P\{\xi(t_1+h)<x_1,\ldots,\xi(t_n+h)<x_n\} = P\{\xi(t_1)<x_1,\ldots,\xi(t_n)<x_n\},$$

for any $t_1,\ldots,t_n$ and $x_1,x_2,\ldots,x_n$.

The random process $\xi(t)$ is said to be stationary in the wide sense if

$$\text{(B2.12)} \qquad E\xi(t) = \text{const.}, \quad E\xi^2(t) < \infty, \quad -\infty < t < \infty,$$
$$\operatorname{cov}(\xi(t), \xi(s)) = r(t-s),$$

i.e., if the first two moments do not change with the shift.

The stochastic process $(\xi(t), F_t)$, $0 \le t$, given on (Ω, F, P) is called Markov with respect to the nondecreasing system of σ-algebras $\{F_t\}$ if with probability 1

$$\text{(B2.13)} \qquad P(A \cap B|\xi(t)) = P(A|\xi(t))\, P(B|\xi(t))$$

for any $t \ge 0$, $A \in F_t$, $B \in F_{[t,\infty]}$.

Theorem 2. In order that the stochastic process $\xi(t)$, $t \ge 0$, be Markov, it is necessary and sufficient that for each measurable function $f(x)$ with $\sup_x|f(x)| < \infty$ and any collection t_n with

$0 \le t_1 \le t_2 \le \ldots \le t_n \le t$

(B2.14) $E(f\ (\xi(t))|\xi(t_1),\dots,\xi(t_n)) = E(f\ (\xi(t))|\xi(t_n))$.

Processes with independent increments, i.e., for any $0 < t_1<\dots<t_n$ the increments $\xi(t_n) - \xi(t_{n-1}),\dots,\xi(t_2) - \xi(t_1)$ are independent, are a special case of Markov processes. A process with independent increments is said to be a process with stationary independent increment (or homogeneous) if the distribution of $\xi(t) - \xi(s)$ depends only on $t - s$.

The stochastic process $\{\xi(t),F_t\}$, $t \geq 0$, is called a martingale with respect to the system $\{F_t\}$, $t \geq 0$, if $E|\xi(t)|< \infty$, $t \geq 0$, and with probability 1

(B2.15) $E(\xi(t)|F_s) = \xi(s), \quad s \leq t$.

The random variable $\tau= \tau(\omega)$, on $[0,\infty]$ is called a Markov time relative to the system $\{F_t\}$, $t \geq 0$, if for all $t \geq 0$

$$\{\omega: \tau(\omega) \leq t\} \in F_t .$$

If $P(\tau<\infty) = 1$ the Markov time is called a stopping time. We shall use the following properties

a) If $\tau_1,\tau_2,\dots$ is a sequence of Markov times, then $\sup \tau_n$ is also a Markov time. If, furher, the family $\{F_t\}$ is right continuous $\inf \tau_n$, $\liminf \tau_n$, $\limsup \tau_n$ are also Markov times.

b) Any Markov time τ is F_τ measurable. If τ and σ are two Markov times and $\tau \leq \sigma$ (with probability 1), then $F_\tau \subseteq F_\sigma$.

c) If $\xi(t)$ is a right continuous stochastic process, then $\xi(\tau)$ is F_τ measurable.

d) If $\zeta(\omega)$ has finite expectation $E|\zeta| < \infty$ and τ is a Markov time, then on $\{\tau= t\}$

$$E(\zeta|F_t) = E(\zeta|F_\tau) , (\{\tau=t\}).$$

The stochastic process $w(t)$, $0 \leq t < \infty$, is called a Brownian motion process (or Wiener process) on (Ω, F, P), if

$P(w(0) = 0) = 1$,

$w(t)$ is a process with stationary independent increments,

$w(t) - w(s)$ is normally distributed with

$$E(w(t) - w(s)) = m(t-s),\ D^2(w(t)-w(s)) = \sigma^2|t-s|,$$

$w(t)$ is continuous.

In the case $m = 0$, $\sigma^2 = 1$ the process $w(t)$ is called the standard Brownian motion process (standard Wiener process). The standard Brownian motion process provides the following properties:

a) It is a martingale with respect to F_t^w, i.e.,

(B2.16) $$E(w(t)|F_s^w) = w(s), \quad s \le t,$$

and

(B2.17) $$E((w(t) - w(s))^2|F_s^w) = t-s, \quad s \le t.$$

b) The expectation and distribution,

(B2.18) $$Ew(t) = 0, \quad \operatorname{cov}(w(t), w(s)) = \min(t,s),$$

(B2.19) $$E|w(t)| = \sqrt{\frac{2t}{\pi}}, \quad P(w(t) < x) = \frac{1}{\sqrt{2\pi t}} \int_{-\infty}^{x} e^{-\frac{u^2}{2t}} du.$$

c) $w(t)$ is Markovian, i.e.,

(B2.20) $$E(f(w(t))|F_s^w) = E(f(w(t))|w(s)), \quad s \le t,$$

for any measurable function $f(x)$ with $\sup_x |f(x)| < \infty$.

d) $w(t)$ has the strong Markov property: for any Markov time τ, relative to F_t, the following relation holds

(B2.21) $$E(f(w(t+\tau))|F_{\tau+}^w) = E(f(w(t+\tau))|F_\tau^w),$$

where $P(\tau \le T) = 1$ and $P(\tau+t \le T) = 1$. The strong Markov property can be given in the following way:

$$\tilde{w}(t) = w(\tau+t) - w(\tau), \quad t \ge 0,$$

is also a Brownian motion process independent of $F_{\tau+}^w$, for any Markov stopping time τ.

Let $P(s,x,t,y) = P\{w(t)<y|w(s) = x\}$, $t > s$, denote the conditional distribution, and let

$$\frac{\partial}{\partial y} P(s,x;t,y) = p(s,x;t,y)$$

denote the conditional density. Then for the standard Brownian motion $w(t)$

(B2.22) $$p(s,x;\ t,y) = \frac{1}{\sqrt{2\pi(t-s)}}\ e^{-\frac{(x-y)^2}{2(t-s)}},$$

where $p(s,x;t,y)$ satisfies the Kolmogorov's backward and forward equation (the second is also called Focker-Planck equation)

(B2.23) $$\frac{\partial p}{\partial s} = -\frac{1}{2}\frac{\partial^2 p}{\partial x^2}, \qquad s < t,$$

(B2.24) $$\frac{\partial p}{\partial t} = \frac{1}{2}\frac{\partial^2 p}{\partial x^2}, \quad s < t.$$

From the strong Markov behavior we get the relation (reflection principle of Desire Andre).

(B2.25) $$P\{\sup_{0 \le s \le T} w(s) \ge x\} = 2.P\{w(T) \ge x\} = \frac{2}{\sqrt{2\pi T}} \int_x^\infty e^{-y^2/2T}\,dy.$$

The first crossing of the level $a \ge 0$ by $w(t)$

$$\tau = \inf\{t : w(t) \geq a\}$$

is a Markov time and by definition

$$P(\tau \le t) = P(\sup_{0 \le s \le t} w(s) \ge a),$$

because of (B2.25)

(B2.26) $$P(\tau \le t) = \frac{2}{\sqrt{2\pi t}} \int_a^\infty e^{-y^2/2t}\,dy = \sqrt{\frac{2}{\pi}} \int_{a/\sqrt{t}}^\infty e^{-y^2/2}\,dy.$$

From (B2.26) we find for the density

(B2.27) $$p_\tau(t) = \frac{dP(\tau < t)}{dt} = \frac{a}{\sqrt{2\pi}\ t^{3/2}}\ e^{-\frac{a^2}{2t}},$$

which gives, for $a > 0$, $E\tau = \infty$.
If

$$\tilde{\tau} = \inf\{t : w(t) = a - bt\}, \quad a > 0,\ 0 \le b < \infty, \quad t \ge 0,$$

then

(B2.28) $$p_{\tilde{\tau}}(t) = \frac{dP(\tilde{\tau} < t)}{dt} = \frac{a}{\sqrt{2\pi}\ t^{3/2}}\ e^{-\frac{(bt-a)^2}{2t}}.$$

By the maximal correlation coefficient between the systems of random variables $\{\xi_1(t),\ t \in T\}$ and $\{\xi_2(t),\ t \in T\}$ we mean

(B2.29) $$r(\xi_1,\xi_2) = \sup E\, \eta_1\eta_2,$$

where $\eta_1 \in F_t^{\xi_1}$, $\eta_2 \in F_T^{\xi_2}$

$$E\,\eta_1 = E\eta_2 = 0,\ E|\eta_1|^2 = E|\eta_2|^2 = 1.$$

$r(\xi_1,\xi_2)$ is the cosine of the minimal angle between the Hilbert subspaces H_1 and H_2 of $F_T^{\xi_1}$measurable random variables η_1, $E|\eta_1|^2 < \infty$ and $F_T^{\xi_2}$ measurable random variables η_2, $E|\eta_2|^2 < \infty$, respectively.

The stationary process $\xi(t)$ is completely regular if

(B2.30) $$r(\tau) = r\{(\xi(s),\ s \le t);\ (\xi(n),\ n \ge t+\tau)\} \to 0,\ \tau\to\infty.$$

Note that for Gaussian processes regularity and linear regularity are the same. This is also true for complete regularity.

Processes with rational spectral density functions are completely regular processes (see Ibragimov, Rozanov [1], Chapters V, VI, Theorem 1) with

$$r(\tau) = \sigma(e^{-c\tau}),\ c > 0,\ \tau\to\infty.$$

A strictly stationary process $\xi(t)$ is said to be metrically transitive if every set which is invariant relative to the shift transformation has probability either 0 or 1. For the metric transitivity a necessary and sufficient condition is that for every random variable η, measurable with respect to F^{ξ}, and $E|\eta| < \infty$

(B2.31) $$\lim_{T\to\infty} \frac{1}{T}\int_0^T \eta(s)\, ds = E\,(\eta)$$

holds, i.e., the process

$$\eta(t) = \int_{-\infty}^{\infty} e^{i\lambda t}\Phi_{\xi}(d\lambda)$$

is ergodic, where $\Phi_{\xi}(dt)$is the random spectral measure of the process $\xi(t)$.

The following statement is well known (see Rozanov [1], Chapter 6, Example 6.2).

Theorem 3. If $\xi(t)$ is Gaussian stationary process with $E\xi(t) = 0$, then a necessary and sufficient condition for the metric transitivity of $\xi(t)$ is that the spectral measure

$$F(d\lambda) = E\, \underline{\Phi}_{\xi}\, (d\lambda)\, \underline{\Phi}^{*}(d\lambda)$$

be continuous; that is, that the spectral measure of each point λ *be zero.*

Let $\underline{H}(T)$ be defined by

$$\underline{H}(T) = \int_{0}^{T} \underline{\xi}(n)\, dn, \tag{B2.32}$$

where $\underline{\xi}(t)$ be an ergodic stationary process. We say that the central limit theorem is applicable to the multidimensional stationary process $\underline{\xi}(t)$ if the limit

$$\lim_{T\to\infty} \frac{\underline{H}(T) - E\underline{H}(T)}{\sqrt{T}} \; \frac{(\underline{H}(T)-E\underline{H}(T))^{*}}{\sqrt{T}} = B \tag{B2.33}$$

exists and the limit distribution of random vector variable

$$\frac{\underline{H}(T)-E\underline{H}(T)}{\sqrt{T}} \tag{B2.34}$$

exists and is a Gaussian $N(\underline{0},B)$.

The following theorem holds (see Rozanov [1], Theorem 11.2).

Theorem 4. Suppose that $\underline{\xi}(t)$ *is completely regular and*

$$r(\tau) = \sigma(\tau^{-1-\varepsilon}), \text{ for some } \varepsilon > 0, \tag{B2.35}$$

further for $\delta > 4/\varepsilon$

$$E\|\, \underline{\xi}(t)\, \|^{2+\delta} < \infty\, . \tag{B2.36}$$

Suppose also that the spectral density $f_{\xi}(\lambda)$ *is bounded, and also continuous and nondegenerate at zero.*

Then the central limit theorem is applicable to $\underline{\xi}(t)$, *and, moreover*

$$B = 2\pi f_{\xi}(0). \tag{B2.37}$$

GENERAL BIBLIOGRAPHY

Books

Anderson, T. W., [1], /1958/, An Introduction to Multivariate Statistical Analysis, John Wiley, New York.

[2], /1970/, Time Series Analysis, John Wiley, New York.

Arató, M., Benczúr, A., Krámli, A., and Pergel, J., /1974/, Statistical Problems of the Elementary Gaussian Processes. MTA SZTAKI Stochastic Processes, Part I.

Arnold, L., /1974/, "Stochastic Differential Equations. Theory and Applications", Wiley-Interscience, New York.

Balakrishnan, A. V., [1], /1971/, Introduction to Optimization Theory in a Hilbert Space, Springer, Lecture Notes 42, New York.

[2], /1973/, Stochastic Differential Systems 1, in Lecture Notes in Economics and Math. 84, Springer, Berlin.

[3], /1976/ Applied Functional Analysis, Springer, Berlin.

[4], /1981/, Stochastic Filtering and Control, Optimization Software, Los Angeles.

Bartlett, M. S., /1966/, An Introduction to Stochastic Processes with Special Reference to Methods and Applications, Second Edition, Cambridge University Press, Cambridge.

Basawa, I. W., and Prakasa Rao, B., /1980/, Statistical Inference for Stochastic Processes. Academic Press, London.

Bendat, J. S., and Piersol, A., /1966/, Measurement and Analysis of Random Data. New York: Wiley.

Billingsley, P., /1968/, Convergence of Probability Measures. New York: Wiley.

Blackman, R. B., /1965/, Linear Data Smoothing and Prediction in Theory and Practice. Reading, Mass.: Addison-Wesley.

Blackman, R. B., and Tukey, J. W., /1959/, The Measurement of Power Spectra from the Point of View of Communications Engineering, Dover Publications, New York, /originally printed in 1958, Bell System Techn. J. 37, 185-282/.

Blanc-Lapierre, A., and Fortet, R., [1], /1953/, Théorie des Fonctions Aléatoires. Paris: Masson.

[2], /1965/, Theory of Random Functions. New York: Gordon and Breach. Translation of 1953 French edition.

Bodewig, E., /1959/, Matrix Calculus, 2nd Ed., North Holland, Amsterdam.

Box, G.E.P., and Jenkins, G. M., /1970/, Time Series Analysis. Forecasting and Control, Holden Day, San Francisco.

Breiman, L., /1968/, Probability . Addison-Wesley: Reading.

Brillinger. D. R., /1975/, Time Series Data Analysis and Theory, New York, Holt.

Brunner, W., /1930/, Astronomische Mitteilungen, cxxiv, p. 77.

Coddington, E., and Levinson, N., /1955/, Theory of Ordinary Differential Equations, McGraw-Hill, New York.

Cox, D. R., and Miller, H. D. /1965/, The Theory of Stochastic Processes . Methuen, London.

Cox, D. R., and Hinkley, D. V., /1974/, Theoretical Statistics . Chapman and Hall, London.

Cramér, H., /1946/, Mathematical Methods of Statistics . Princeton University Press.

Cramér, H., and Leadbetter, M. R., /1967/, Stationary and Related Stochastic Processes. New York: Wiley.

Davenport, W., and Root, W., /1958/, An Introduction to the Theory of Random Signals and Noise. McGraw-Hill, New York.

Ditkin, V., and Kuznecov, P., /1951/, Handbook of Operational Calculus. GITTL, Moscow, /in Russian/.

Doob, J. L., /1953/, Stochastic Processes. New York: Wiley.

Draper, N. R., and Smith, H., /1966/, Applied Regression Analysis, New York: Wiley.

Džaparidze, K., /1976/, Lectures on Statistics of Random Processes /in Russian/. University of Jena.

Einstein, A., /1956/, Investigations on the Theory of the Brownien Movement. New York, Dover, /contains translations of Einstein's 1905 papers/.

Ezekiel, M. A., and Fox, C. A., /1959/, Methods of Correlation and Regression Analysis, New York: Wiley.

Feller, W., [1], /1968/, An Introduction to Probability Theory and its Applications, Vol. I, Third Edition, John Wiley, New York.

[2], /1966/, Introduction to Probability Theory and its Applications. Vol. II, John Wiley, New York.

Friedman, A., /1975/, Stochastic Differential Equations and Applications , Vol. 1, Academic Press, New York and London.

Fritz, H., /1873/, Verzeichniss Beobachteter Polarlichter, Wien.

Gantmakher, F. R., /1946/, Theory of Matrices, /in Russian/, Nauka, Moskva.

Gikhman, I. I., and Skorokhod, A. V., [1], /1969/, Introduction to the Theory of Random Processes , W. B. Saunders, Philadelphia.

[2], /1972/, Stochastic Differential Equations , Springer Verlag, Berlin.

[3], /1974/, The Theory of Stochastic Processes , Vol. 1-3, Springer Verlag, Berlin.

Granger, C.W.J., and Hatanaka, M., /1964/, Spectral Analysis of Economic Time Series, Princeton University Press, Princeton, N.S.

Grenander, U., /1950/, Stochastic processes and statistical inference , Ark. Mat. 1:195-277, /Russian ed. 1961/.

Grenander, U., and Rosenblatt, M., /1957/, Statistical Analysis of Stationary Time Series, New York: Wiley.

Grenander, U., and Szegõ, G., /1958/, Toeplitz Forms and Their Applications, Berkeley: Univ. of Cal. Press.

Hajek, J., and Sidak, Z., /1967/, Theory of Rank Tests , Academia, Prague.

Hannan, E. J., [1], /1960/, Time Series Analysis, London: Methuen.

[2], /1970/, Multiple Time Series, New York: Wiley.

Henrici, P., /1962/, Discrete variable methods in ordinary differential equations, J. Wiley, New York.

Hida, T., /1980/, Brownian Motion, Springer, New York - Berlin.

Ibragimov, I. A., and Linnik, Yu, V., /1971/, Independent and Stationary Sequences of Random Variables , Wolters-Noordhoff, Groningen.

Ibragimov, I. A., and Rozanov, Yu, A., /1970/, Gaussian Random Processes, "Nauka", Moscow.

Ito, K., and McKean, G., /1968/, Diffusion processes and their trajectories, /Russian translation/, "MIR", Moscow.

Jenkins, G. M., and Watts, D. G., /1968/, Spectrum Analysis and Its Applications, San Francisco: Holden-Day.

Kagan, A. M., Linnik, Yu, V., and Rao, C. R., /1973/, Characterization Problems in Mathematical Statistics , Wiley, New York.

Kailath, Th. /1980/, Linear Systems, Prentice Hall, Englewood Cliffs.

Kallianpur, G., /1980/, Stochastic Filtering Theory, Springer, Berlin.

Kashyap, A., and Ramachandra, Rao, /1976/, Dynamic Stochastic Models from Empirical Data, Academic Press, New York.

Kendall, M., /1946/, Contributions to the Study of Oscillatory Time Series, Cambridge: Cambridge University Press.

Kendall, M. G., and Stuart, A., [1], /1961/, The Advanced Theory of Statistics, Vol. II, Charles Griffin and Hafner, New York, London.

[2], /1966/, The Advanced Theory of Statistics, Vol. III, Charles Griffin and Hafner, New York.

Koenig, H., Tokad, Y., Kesevan, H., and Hedges, H., /1967/, Analysis of Discrete Physical Systems, McGraw-Hill, New York.

Kolmogorov, A. N., [1], /1933/, Grundbegriffe der Wahrscheinlichkeitsrechnung, Berlin, /reprinted in 1946 by Chelsea, New York/.

[2], /1941a/, Interpolation und Extrapolation von stationären zufalligen Folgen , Bull. Acad. Sci. de l'U.R.S.S., 5:3-14.

[3], /1941b/, Stationary sequences in Hilbert space , /in Russian/, Bull. Moscow State U. Math. 2:1-40. /Reprinted in Spanish in Trab. Estad., 4:55-73, 243-270/.

Kutojanc, J. A., /1980/, Parameter estimation of random processes /in Russian/, A.N., Jerevan.

Lambeck, K., /1980/, The Earth's Variable Rotation, Cambridge University Press, Cambridge.

Lee, Y. W., /1960/, Statistical Theory of Communication, New York: Wiley.

Lehman, E. L., /1959/, Testing Statistical Hypotheses , Wiley, New York.

Levy, P., /1972/, Stochastic Processes and Brownian Motion, "Nauka", Moscow.

Liptser, R. S., and Sirjaev, A. N., [1], /1974/, Statistics of Random Processes , /Russian/, Nauka, Moscow.

[2], /1977/, Statistics of Random Processes I General Theory , Springer Verlag, Berlin.

McKean, H. P., /1696/, Stochastic Integrals , Academic Press, New York.

Medgyessy, P., /1961/, Decomposition of Superpositions of Distribution Functions. Budapest: Hungary Acad. Sci.

Melchior, P., and Yumi, S., /1972/, Rotation of the Earth, Reidel, Holland.

Middleton, D., /1960/, Statistical Communication Theory, New York: McGraw-Hill.

Munk, W. H., and MacDonald, G.J.F., /1960/, The Rotation of the Earth, Cambridge: Cambridge University Press, /second ed. 1978/.

Newton, H. W., /1958/, The Face of the Sun, London: Penguin.

Orlov, A. J., /1958/, Service des latitudes, Izdat Akad Nauk SSSR, Moscow.

Parzen, E., /1967a/, Time Series Analysis Papers, San Francisco: Holden-Day.

Perrin, F., /1916/, Atoms, London, Constable.

Pontryagin, L. S., /1962/, Ordinary differential equations, Addison-Wesley, Reading.

Prohorov, Yu, V., and Rozanov, Yu, A., /1969/, Probability Theory, Springer Verlag, Berlin.

Quenouille, M. H., /1957/, The Analysis of Multiple Time Series, London Griffin.

Rao, C. R., /1965/, Linear Statistical Inference and Its Applications, New York: Wiley.

Reid, W., /1972/, Riccati Differential Equations, Academic Press, N. Y.

Riesz, F., and Sz-Nagy, B., /1955/, Lessons in Functional Analysis, New York: Ungar.

Robinson, E. A., /1967a/, Multichannel Time Series Analysis with Digital Computer Programs, San Francisco: Holden-Day.

Rozanov, Ju. A., [1], /1967/, Stationary Random Processes, San Francisco: Holden-Day.

[2], /1969/, Infinite dimensional Gaussian distributions, Nauka, Moscow.

Scheffé, H., /1959/, The Analysis of Variance, New York: Wiley.

Schuss, Z, /1980/, Theory and Applications of Stochastic Differential Equations, Wiley, New York.

Shiryaev, A. N., [1], /1969/, Statistical sequential analysis /in Russian/, Nauka, Moscow.

[2], /1980/, Probability /in Russian/, Nauka, Moscow.

Skorokhod, A. V., /1965/, Studies in the Theory of Random Processes , Addison-Wesley, Reading, Mass.

Soong, T.T., /1973/, Random Differential Equations in Science and Engineering, Academic Press, London.

Srinivasan, S. K., and Vasudevan, R., /1971/, Introduction to random differential equations and their applications, Elsevier, New York.

Stepanov, V.V. /1959/ The Theory of Differential Equations /in Russian/ Moscow

Stroock, D.W., Varadhan, S.R., /1979/, Multidimensional Diffusion Processes, Springer, Berlin

Ventcel, A.D., /1975/, The Theory of Stochastic Processes, Nauka, Moscow, /in Russian/.

Vitinsky, Ju, V., /1973/, Ciklicnosty i prognozi solnecnoi aktivnosty, Nauka, Moscow.

Wald, A., /1974/, Sequential Analysis , Wiley, New York.

Waldmeir, M., /1961/, The Sunspot Activity in the Years 1610-1960. Zurich: Schulthess.

Whittle, P., [1], /1951/, Hypothesis Testing in Time Series Analysis, Uppsala: Almqvist.

[2], /1963a/, Prediction and Regulation, London: English Universities Press.

Wiener, N.,[1], /1933/, The Fourier Integral and Certain of its Applications, Cambridge: Cambridge University Press.

[2], /1949/, The Extrapolation, Interpolation and Smoothing of Stationary Time Series with Engineering Applications, New York: Wiley.

Wold, H.O.A., /1965/, Bibliography on Time Series and Stochastic Processes, London: Oliver and Boyd.

Wong, E., /1971/, Stochastic Processes in Information and Dynamical Systems, McGraw-Hill, N. Y.

Wonham, W. M., /1969/, Random differential equations in control theory, Academic Press, New York.

Yaglom, A. M., /1962/, An Introduction to the Theory of Stationary Random Functions, Englewood Cliffs: Prentice-Hall.

References

Albert, A., /1962/. Estimating the infinitesimal generator of a continuous time finite state Markov process. Ann. Math. Statist. 38, 727-753.

Andel, J. /1971/. On the multiple autoregressive series, Ann. Math. Statist. 42, No. 2, 755-759.

/1972/. Symmetric and reversed multiple stationary autoregressive series, Ann. Math. Statist. 43, No. 4, 1197-1203.

/1981/. An autoregressive representation of ARMA Processes, /Preprint/, Pannonia Symposium, Austria.

T.W. Anderson - A.M. Walker, /1964/. On the asymptotic distribution of the autocorrelations of a sample from a linear stochastic process, Ann. Math. Statist. 35 /1964/, 1296-1303.

Akaike, H. /1962/. Undamped oscillation of the sample autocovariance function and the effect of prewhitening operation. Ann. Inst. Statist. Math. 13, 127-144.

/1965/. On the statistical estimation of the frequency response function of a system having multiple input. Ann. Inst. Statist. Math. 17, 185-210.

/1966/. On the use of a non-Gaussian process in the identification of linear dynamic system. Ann. Inst. Statist. Math. 18, 269-276.

/1969a/. A method of statistical investigation of discrete time parameter linear systems. Ann. Inst. Statist. Math. 21, 225-242.

/1969b/. Fitting autoregressive models for prediction. Ann. Inst. Statist. Math. 21, 243-247.

Arató M. /1961/. Some remarks on absolute continuity /in Russian/. MTA Mat. Kut. Int. Közleményei VI /1961/ 123-126.

/1961/. Sufficient statistics for stationary Gaussian processes, Teor. Verojatnost. i Primenen. 6 /1961/, 216-218 = Theor. Probability Appl. 6 /1961/, 199-200.

/1962/. On statistical problems concerning stationary Markov processes, Dissertation, Moscov State Univ., Moscow, 1962. /Russian/.

/1962/. Estimation of the parameters of a stationary Gaussian Markov process, Dokl. Akad. Nauk SSSR 145 /1962/, 13-14 = Soviet Math. Dokl. 3 /1962/, 905-909.

/1964/. On the statistical examination of continuous state Markov processes. I, Magyar Tud. Akad. Mat. Fiz. Oszt. Közl. 14 /1964/, 13-34; English transl., Selected Transl. Math. Statist. and Probability, vol. 14, Amer. Math. Soc., Providence, R.I., 1978, pp. 203-225.

Arató, M. /1964/. On the statistical examination of Continuous state Markov processes. II, Magyar Tud. Akad. Mat. Fiz. Oszt. Közl. 14 /1964/, 137-159; English transl., Selected Transl. Math. Statist. and Probability, vol. 14, Amer. Math. Soc., Providence, R.I., 1978, pp. 227-251.

/1964/. On the statistical examination of continuous state Markov processes. III, Magyar Tud. Akad. Mat. Fiz. Oszt. Közl. 14 /1964/, 317-330; English transl., Selected Transl. Math. Statist. and Probability, vol. 14, Amer. Math. Soc., Providence, R.I., 1978, pp. 253-267.

/1965/. On the statistical examination of continuous state Markov processes. IV, Magyar Tud. Akad. Mat. Fiz. Oszt. Közl. 15 /1965/, 107-124; English Transl., Selected Transl. Math. Statist. and Probability, vol. 14, Amer. Math. Soc., Providence, R.I., 1978, pp. 269-288.

/1968/. Confidence limits for the parameter λ of a complex stationary Gaussian Markovian process, /in Russian/, Teorija Verojatn. i Primenen., 13 /1968/, 326-333.

/1968/. Unbiased parameter estimation for complex stationary Gaussian Markovian processes; approximation of the distribution function, /in Russian/, Studia Sci. Math. Hungar., 3 /1968/, 153-158.

/1970/. Exact formulas for density measure of elementary Gaussian processes, /in Russian/, Studia Sci. Math. Hungar., 5 /1970/, 17-27.

/1970/. On the parameter estimation of processes satisfying a linear stochastic differential equation, /in Russian/, Studia Sci. Math. Hungar, 5 /1970/, 11-16.

/1971/. The admissible estimation of the unknown mean of a stationary process with rational spectral density. Selected Transl. in Math. Stat. and Probability 10 /1971/ 211-223. /Original: MTA III. Oszt. Közl. 19 /1969/, 89-99/.

/1973/. On similar tests and admissible estimators for a Stationary Gaussian Markov process. Selected Transl. in Math. Stat. and Probability 13 /1973/, 235-243.

/1975/. Diffusion approximation for multiprogrammed computer systems. Computers and Mathematics with applications. 315-326.

/1976/. Statistical sequential methods in performance evaluation of computer systems. 2nd International Workshop on Modelling and Performance Evaluation of Computer Systems. Stresa, Italy, /1976/, Oct. North Holland, pp. 1-10.

/1976/. A note on optimal performance of page storage. Acta Cybernetica Tom. 3. /1976/, No. 1. 25-30.

/1981/. On optimal stopping times in operating systems. In "Stochastic Systems", Editors: Arató, M., Balakrishman, A., Vermes, D., Springer, /1981/, 1-12.

Arató, M. /1981/. On failure processes in computer systems. /1981/. III. Hungarian Computer Science Conference I, 19-30. In "Mathematical Models in Computer Systems", Akadémiai Kiadó /1981/, Budapest.

/1982/. On sufficient statistics of Gaussian processes with rational spectral density. Analysis Math. /to appear/.

/1982/. Round-off error propagation in the integration of ordinary differential equations by one step method. /Acta Mathematica, Szeged, to appear./

/1982/. Probability bounds and asymptotic properties of error propagation. /Computers and Mathematics with Applications, to appear./

/1982/. Sufficient statistics for confidence interval construction and run length control in simulation and performance evaluations. /Performance Evaluation, to appear./

/1982/. Run length control in simulations and performance evaluation and the elementary Gaussian Processes. /Acta Cybernetica, to appear./

/1982/. Radon-Nikodym derivatives in case of rational spectral densities. Lecture Notes in Control and Information Sciences, Springer 2nd Bad Honnef Workskop on Stochastic Differential Systems /to appear/.

Arató, M. - A. Benczúr. /1970/. Distribution function of the damping parameter of stationary Gaussian processes, /in Russian/, Studia Sci. Math. Hungar., 5 /1970/, 445-456.

/1972/. Some new results in the statistical investigation of elementary Gaussian processes. European meeting of statisticians, Budapest, /1972/. 69-83.

/1981/. Dynamic placement of records and the classical occupancy problem. Computers of Mathematics with Applications 7 /1981/, 173-185.

A General Treatment of Rearrangement Problems in a linear Storage. Performance Evaluation, /in print/.

Arató, M., A. Benczúr, A. Krámli. On the solution of optimal performance of page storage hierarchies with independent reference string. Banach Center Publications. Volume 6. Mathematical Statistics. /1980/, 9-15.

Arató, M., A. Benczúr, A. Krámli, J. Pergel. /1974/. Statistical problems of the elementary gaussian processes I. MTA SZTAKI Tanulmányok. /1974/. 22, 5-130.

/1975/. Statistical problems of the elementary Gaussian processes II. MTA SZTAKI Tanulmányok. /1975/. 41.

Arató, M., E. Knuth, P. Töke. /1974/. On stochastic control of a multimprogrammed computer based on a probabilistic model. Stochastic Control Symposium. /1974/. 305-311.

Arató, M., A.N. Kolmogorov and Ja. G. Sinai. /1962/. Estimation of the parameters of a complex stationary Gaussian Markov process.

Dokl. Akad. Nauk SSSR 146 /1962/, 747-750 = Soviet Math. Dokl. 3 /1962/, 1368-1371.

Bachelier, L. /1900/. Theorie de la speculation. Ann. Sci. Norm. Sup. Vol. 3, /1900/, 21-86.

Bagchi, A. Consistent estimates of parameters in continuous time systems, in Analysis and Optimisationof Stochastic Systems, Academic Press, 1979.

Balakrishnan, A.V. /1964/. A general theory of nonlinear estimation problems in control systems. J. Math. Anal. App. 8, 4-30.

/1974/. On the approximation of the integrals using band limited processes. SIAM J. Control, 12 /1974/, 237-251.

/1978/. Parameter estimation in stochastic differential systems: Theory and application. Development in Statistics, Academic Press, vol. 1 /1978/, 1-32.

/1981/. On a class of Riccati equations in Hilbert space. Appl. Math. Optim. /1981/, 159-174.

Barndorff-Nielsen, O. and Schou, G. /1973/. On the parametrization of autoregressive models by partial autocorrelations, J. Multivar. Anal., No. 3.

Bartlett, M.S. /1946/. On the theoretical specification of sampling properties of auto-correlated time series. J. Roy. Statist. Soc., Suppl. 8, 27-41.

/1948a/. A note on the statistical estimation of supply and demand relations from time series. Econometrica. 16, 323-329.

/1948b/. Smoothing periodograms from time series with continuous spectra. Nature. 161, 686-687.

/1950/. Periodogram analysis and continuous spectra. Biometrika. 37, 1-16.

/1967/. Some remarks on the analysis of time series. Biometrika. 50, 25-38.

Baxter, G. /1956/. A strong limit theorem for Gaussian processes, Proc. Amer. Math. Soc., 7, 522-525.

Benczúr, A. and L. Szeidl. /1974/. On absolute continuity of measures defined by multidimensional diffusion processes with respect to the Wiener measure, MTA SZTAKI Közlemények, 13, 5-10.

Brillinger, D.R. /1964/. The generalization of the techniques of factor analysis, canonical correlation and principal components to stationary time series. Invited paper at Royal Statistical Society Conference in Cardiff, Wales. Sept. 29-Oct. 1.

/1964/. The asymptotic behavior of Tukey's general method of setting approximate confidence limits /the jackknife/ when applied to maximum likelihood estimates. Rev. Inter. Statist. Inst. 32, 202-206.

Brillinger, D.R. /1968/. Estimation of the cross-spectrum of a stationary bivariate Gaussian process from its zeros. J. Roy. Statist. Soc., B. 30, 145-159.

/1969a/. A search for a relationship between monthly sunspot numbers and certain climatic series. Bull. ISI. 43, 293-306.

/1969b/. The calculation of cumulants via conditioning. An empirical investigation of the Chandler wobble and two proposed excitation processes. Statistical Congress, ISI, Wien, 1973. Invited paper 8.1. Sect. 1-22.

Brown, B.M. /1974a/. A sequential procedure for diffusion processes. In "Studies in Probability and Statistics" /Papers in honor of E.J.G. Pitman/, Academic Press.

/1974b/. A restricted sequential test. J. Roy. Statist. Soc. Ser. B 36, 455-465.

Brown, B.M. and J.I. Hewitt. /1975/. Asymptotic likelihood theory for diffusion processes. J. Appl. Prob. 12, 228-238.

Cameron, R.H. and W.T. Martin. /1945/. Transformation of Wiener integrals under a general class of linear transformations. Trans. Amer. Math. Soc. 58, 184-219.

Daniels, H.E. /1956/. The approximate distribution of serial correlation coefficients, Biometrika, 43 ,169-185.

Doob, J.L. /1942/. The Brownian Movement and Stochastic Equations. Ann. Math. Stat. 43 , 351-369.

/1944/. The elementary Gaussian processes, Ann. Math. Statist. 15, , 229-282.

Duffin, E.J. /1969/. Algorithms for classical stability problems, SIAM Review, 11, 196-213.

Durbin, J. /1954/. Errors in variables. Rev. Inter. Statist. Inst. 22, 23-32.

/1960/. Estimation of parameters in time series regression models. J. Roy. Statist. Soc., B. 22, 139-153.

/1960/. The fitting of time series models, Rev. Int. Stat. Inst., 28, 233-244.

Fedorov, E. P. /1948/. Trudy Poltavsk Gravimetric. Observatoria 2, 3, /in Russian/.

Gaudi, I.H. On the estimation of regression coefficients in case of an autoregressive noise process. Studi Sci. Math., Hungarica 12 /1977/ 471-475.

Girsanov, I.V. On transformation of one class of random processes with the help of absolutely continous substitution of the measure. Teor. Verojan. i Primen. V, 3 /1960/, 314-330.

Grenander, U. /1950/. Stochastic processes and statistical inference, Arkiv for Matematik, 1, 195-277.

Hajek, J. /1962/. On linear statistical problems in stochastic processes, Czechoslovak Math. Journal, 12, 404-444.

Heidelberger, P. and P.A.W. Lewis. /1981/. Quantile estimation in dependent sequences. RC 9087. IBM Research Report, Yorktown Heights, New York.

Heidelberger, P. and P.D. Welch. /1980/. A spectral method for simulation confidence interval generation and run length control. IBM Research Report RC 8264, Yorktown Heights, New York, Comm. ACM. /1981/, 233-245.

/1980/. On the statistical control of simulation run length. IBM Research Report RC 8571, Yorktown Heights, New York.

Holevo, A.S. /1967/. Estimates of the drift parameters for a diffusion process by the method of stochastic approximation. In "Studies in the Theory of Self-adjusting Systems". Vycis. Centr. Akad. Nauk SSSR, Moscow.

Ibragimov, I.A. /1962/. Some limit theorems for stationary processes. Theor. Probability Appl. 7, 349-382.

/1963/. A central limit theorem for a class of dependent random variables. Theor. Probability Appl. 8, 83-89.

Ibragimov, I.A. and R.Z. Khas'minskii. /1972/. Asymptotic behaviour of statistical estimators in the smooth case. I. Study of the likelihood ratio. Theor. Probability Appl. 17, 445-462.

/1973a/. Asymototic behaviour of statistical estimators. II. Limit theorems for the a posterior density and Bayes estimators. Theor. Probability Appl. 18, 76-91.

/1973b/. On moments of generalized Bayes estimators and maximum likelihood estimators. Theor. Probability Appl. 18, 508-520.

/1974/. On sequential estimation. Theor. Probability Appl. 19, 233-244.

Jeffreys, H. The variation of latitude. Monthly Notices Roy. Astronom. Soc. 102 /1942/, 139-155.

/1968/. The variation of the latitude. Monthly Notices Roy. Astronom. Soc. 41 /1968/, 255-268.

Kailath, T. An innovations approach to least-squares estimation, Parts I, II. IEEE Trans. Automatic Control AC-13 /1968/, 646-660.

The innovation approach to detection and estimation theory. Proc. IEEE 58 /1970/, 680-695.

The structure of Radon-Nikodym derivatives with respect to Wiener and related measures. AMS 42 /1971/, 1054-1067.

Kailath, T. and R. Geesey. An innovations approach to least-squares estimation, Part IV. IEEE Trans. Automatic Control AC-16, /1971/, 720-727.

Kailath, T. and M. Zakai. Absolute continuity and Radon-Nykodym derivatives for certain measures relative to Wiener measure. AMS 42, 1 /1971/, 130-140.

Karhunen, K. /1947/, Ueber lineare Methoden der Wahrscheinlichkeitsrechnung. Ann. Acad. Sci, Fenn., 37.

Kallianpur G. and C. Striebel. Estimation of stochastic systems: Arbitrary system process with additive white noise observation errors. AMS 39 /1968/, 785-801.

Stochastic differential equations occuring in the estimation of continuous parameter stochastic processes. Teoria Verojałn. i Primenen. XIV, 4 /1969/, 597-622.

Kalman, R.E. A new approach to linear filtering and prediction problems. J. Basic, Eng. 1 /1960/, 35-45.

Contributions to the theory of optimal control. Bol. Soc. Mat. Mexicana 5 /1960/, 102-119.

Kalman, R.E. and R.S. Bucy. New results in linear filtering and the prediction theory. Russian transl. Tekhnicheskaja mekhanika 83, ser. D, I /1961/, 123.

Kalman, R.E., P. Falb and M. Arbib. Mathematical System Theory: Russian translation, "MIR", Moscow, 1971.

Kolmogorov, A.N. /1939/. Sur l'interpolation et extrapolation des suites stationnaires, C.R. Acad. Sci. Paris 208, 2043-2045 /1939/.

/1941/. Stationary Sequences in Hilbert Space, Bull. Moscow State Univ. 2, No. 6, 1-40 /1941/.

Interpolation and Extrapolation of Stationary Random Sequences, Izv. Akad. Nauk SSSR. Ser. Mat. 5, 3-14, /1941/.

/1948/. E.E. Slutsky /obituary/ Usp. Mat. Nauk III. 4 /26/ /1948/, 143-151.

/1951/. A simple proof of the ergodic theorem of Birkhoff and Khinchin, Uspehi Mat. Nauk 5, 52-59 /1951/.

/1950/. Unbiased estimates, Izv. Akad. Nauk SSSR Ser. Mat. 14 /1950/, 303-326; English transl., Amer. Math. Soc. Transl. /1/ 11 /1962/, 144-170. MR 12, 116.

Kolmogorov, A.N. and Yu. A. Rozanov. /1960/. On strong mixing conditions for stationary Gaussian random Processes, Theory of Probability and its applications 5, 204-208 /1960/.

Krámli, A. /1971/. On homogeneous Gaussian Markovian processes /in Russian/, Studia Sci. Math. 167-168.

Krámli, A. and J. Pergel /1974/. The connection between Gaussian Markov processes and autoregressive-moving average processes, MTA SZTAKI Közlemények, 13, 53-58.

/1974/. On the Radon-Nikodym derivative of elementary Gaussian processes /in Hungarian/ Alk. Mat. Lapok, 1, No. 1, 45-52.

Kulinic, G.L. /1975/. On an estimation of the drift parameter of a stochastic diffusion equation. Theory, Probability Appl. 20, 384-387.

Kunita, H. and S. Watanabe. /1967/. On square integrable martingales. Nagoya Math. J. 30, 209-245.

Kutoyants Yu. A. /1975a/. On a hypotheses testing problem and asymptotic normality of stochastic integrals. Theory-Probability Appl. 20, 376-384.

/1975b/. Local asymptotical normality fo the diffusion type processes. Izvest Akad Nauk Arminski, SSR 10, 103-112.

/1976/. On asymptotic theory of signal detection. Radio-Technika and Electronica 7, 1458-1466.

/1977a/. Estimation of the trend parameter of a diffusion process in the smooth case. Teor. Veroyat iee Primenen 22, 409-415 /Russian/.

/1977b/. On the estimation of trend parameter. Izv. Akad. Nauk. Armeniski, SSR, 12, 245-251 /Russian/.

/1978/. Parameter estimation for processes of diffusion type. Theor. Probability Appl. /Russian/ 665-672.

Lai, T.L. /1973a/. Gaussian processes, moving averages and quick detection problems. Ann. Probability, 1, 825-837.

/1973b/. Optimal stopping and sequential tests which minimize the expected sample size. Ann. Statistics. 1, 659-673.

Le Breton, A. /1974/. Parameter estimation in a vector linear stochastic differential equation. In "Transactions of the Seventh Prague Conference on Information Theory. Statistical Decision functions and Random Processes" Vol. A, 353-366.

/1976/. On continuous and discrete sampling for parameter estimation in diffusion type processes. In "Mathematical Programming Studies". 5, 124-144.

Le Cam, L. /1953/. "On Some Asymptotic Properties of Maximum Likelihood and Related Bayes Estimates". University of California Publications in Statistics 1, 277-330.

Lehmann, E. and H. Scheffe, Completeness, similar regions, and unbiased estimations. I, II, Sankhya 10 /1950/, 305-340; 15 /1955/, 219-236.

Legostayeva, I., A. Shiryaev /1971/. Minimax weights in a trend detection problem of a random process. Theory of Prob. and Applic. 16 /1971/, 344-349.

Linkov, Yu. V. /1975a/. Asymptotic theory of Bayes estimates of parameters of shift diffusion processes. In "Theory of Random Processes". 3, Naukov Dumka, Kiev. /Russian/ 50-54.

/1975b/. Generalised Bayes estimates of parameters of shift diffusion processes. Theor. Probability Math. Statist. 13, Kiev, 92-99. /Russian/.

Linkov,Ju.V./1977/. On statistical estimators for parameters of diffusion processes. In "Abstracts of communications. Second Vilnius Conference on Prob. Theory and Math. Statist". 1, 240-241. /Russian/.

Linnik, Ju. V. /1950/. On a question of the statistics of dependent events. Izv. Akad. Nauk SSSR Ser. Mat. 14, 501-522. /Russian/.

Luvsanceren, S. /1954/. Maximum likelihood estimates and confidence regions for unknown parameters of a stationary Gaussian rocess of Markov type, Candiate's Dissertation, Moscow State Univ., Moscow, /Russian/.

/1954/. Maximum likelihood estimates and confidence regions for unknown parameters of a stationary Gaussian process of Markov type, Dokl. Akad. Nauk SSSR 98 , 723-726. /Russian/.

Liptser, R.S. /1968/. On extrapolation and filtering of some Markov processes II. Kibernetika 6 /1968/, 70-76.

Liptser, R.S. and A.N. Shirya ev. /1968/. Nonlinear filtering of diffusion type Markov processes. Trudy matem, In-ta im. V.A. Steklo a AN SSSR 104 , 135-180.

/1968/. On filtering, interpolation and extrapolation of diffusion type Markov processes with incomplete data. Teoria Verojatn. i Primenen. XIII, 3 , 569-570.

/1968/. Extrapolation of multivariate Markov processes with incomplete data. Teoria Verojatn. i Primenen. XIII, 1 , 17-38.

/1968/. On the cases of effective solving the problems of optimal nonlinear filtering, integrability, and extrapolation. Teoria Verojatn. i Primenen. XIII, 3 , 570-571.

/1968/. Nonlinear interpolation of the components of diffusion type Markov processes /forward equations, effective formulas/. Teoria Verojatn. i Primenen. XIII, 4 , 602-620.

/1969/. Interpolation and filtering of the jump component of a Markov process. Izv. AN SSSR, ser. matem, 33, 4 901-914.

/1969/. On the densities of probability measures of diffusion type processes. Izv. AN SSSR, ser. matem, 33, 5 1120-1131.

/1970/. Statistics of Conditionally Gaussian Random Sequences. Proc. Sixth Berkeley Sympos. Math. Statistics and Probability , Vol II, Univ. of Calif. Press, 1972, 389-422.

/1972/. On absolute continuity of measures corresponding to diffusion type processes with respect to a Wiener measure. Izv. AN SSSR, ser. Matem. 36, 4 , 874-889.

Mann, H.B. and A. Wald. /1943a/. On stochastic limit and order relationships. Ann. Math. Statist. 14, 217-226.

Mann, H.B. and A. Wald. /1943b/. On the statistical treatment of linear stochastic difference equations. Economietrica. 11, 173-220.

Mehra, R.K. /1974/. Optimal input signals for parameter estimation in dynamic systems - Survey and new results, IEEE Trans. Automatic Control, AC-19, /1974/, No. 6, 753-768.

Gy. Németh, T. /1973/. On estimates of parameters of the second order autoregressive process with continuous time, SZTAKI Közlemények, 10 , 33-43 /in Hungarian/.

Neumann, John von. /1941/. Distribution of the ratio of the mean square successive difference to the variance, Ann. Math. Statist. 12, 367-395.

Novikov, A.A. /1971/. Sequential estimation of the parameters of diffusion-type processes. Teoria Verojatn. i Primenen. XVI, 2 394-396.

/1971/. On stopping times of a Wiener process. Teoria Verojatn. i Primenen. XVI, 3 , 548-550.

/1972/. Sequential estimation of the parameters of diffusion processes. Mathematical Notes 12, 812-818.

/1972/. On an identity for stochastic integrals. Teoria Verojatn. i Primenen. XVII, 4 , 761-765.

/1972/. On the estimation of parameters of diffusion processes, /in Russian/, Studia Sci. Math. Hungar., 7 201-209.

/1979/. On estimates and the asymptotic behavior of nonexit probabilities of a Wiener process to a moving boundary. Math. USSR Sbornik 110 /152/ , No. 4.

/1981/. A martingale approach to first passage problems and a new condition, for Wald's identity Stochastic Differential Systems, Visegrád, 1980 /Ed. by M. Arató, D. Vermes, A.V. Balakrishnan/, 146-156.

/1981/. Martingale approach to first passage problems for nonlinear boundaries, Proceedings of the Steklov Inst. of Math., Vol. 158.

Obukhov, A.M. /1938/. Normally correlated vectors. Izv. Akad. Nauk SSR. Section on Mathematics. 3, 339-370.

/1940/. Correlations theory of vectors. Uchen. Zap. Moscow State Univ. Mathematics Section. 45, 73-92.

Orlov, A.Ja. /1958/. Service des latitudes, Izdat. Akad. Nauk SSSR, Moscow

Pancenko, N.I. /1960/. On the question of the decay of free nutation. Proc. Fourteenth Astronom. Conf. USSR, Izdat. Akad. Nauk SSSR, Moscow, pp. 232-243. /Russian/

Pisarenko, V.F. /1961/. On the problem of discovering a random signal in noisy background, /in Russian/, Radiotehnika i elektronika 6, , 515-528.

Pisarenko, V.F. /1963/. On the estimation of parameters of a stationary Gauss process with spectral density $|p(i\lambda)|^{-2}$, /in Russian/, Litovsk. Mat. Sbornik, 2 , 159-167.

/1970/. Statistical estimates of amplitude and phase corrections. Geophys. J. Roy. Astron. Soc. 20, 89-98.

/1972/. On the estimation of spectra by means of nonlinear functions of the covariance matrix. Geophys. J. Roy Astron. Soc. 28, 511-531.

Pisarenko, V.F. and Yu.A. Rozanov. /1963/. On some problems for stationary processes which reduce to integral equations related to the Wiener-Hopf equation. Problemi Peredachi Informatsi 14, 113-135

Pontrjagin, L.S., A.A. Andronov and A.A. Vitt. /1934/. Statistische Aufassung dynamischer Systeme, Phys. Z. Sowjetuinon 6 1-24.

Prakasa Rao, G.L.S. /1972/. Maximum likelihood estimation for Markov processes. Ann. Inst. Statist. Math. 24, 333-345.

/1973/. On the rate of convergence of estimators for Markov processes. Z. Wahrscheinlichkeitstheorie Verw. Gebiete. 26, 141-152.

/1977/. Berry-Esseen type bound for density estimators of stationary Markov processes. Bull. Math. Statist. 17, 15-21.

/1978b/. Density estimation for Markov processes using Delta-sequences. Ann. Inst. Statist. Math. 30, 321-328.

Prokhorov, Yu.V. /1956/. Convergence of random processes and limit theorems of probability theory. Teoria Verojatn. i Primenen. 1, 2 , 177-238.

Ramsey, Fr. /1974/. Characterization of the partial autocorrelation function, Ann. Stat., 2, No. 6, 1296-1301.

Rao, M.M. /1963/. Inference in stochastic processes, I. Teor. Verojatn. i Primenen. 8, 282-298.

/1966/. Inference in stochastic processes, II. Zeit. Wahrschein 5, 317-335.

Ratkó, I. and M. Ruda. /1974/. On an estimate for the parameter of a multidimensional stationary Gaussian process. Közlemények, MTA SZTAKI, 13 /1974/, 21-30.

Rozanov, Yu.A. /1957/. On the linear interpolation of stationary processes with discrete time. Dokl. Akad. Nauk SSSR 116, 923-926.

/1958/. Spectral theory of multi-dimensional stationary random processes discrete time. Uspehi Mat. Nauk 13, 2 /80/, 93-142.

/1959/. The linear extrapolation of multi-dimensional stationary processes of rank 1 with discrete time. Dokl. Akad. Nauk SSSR 125, 277-280.

Rozanov, Yu.A. /1959/. On the extrapolation of generalized stationary random processes. Theory of probability and its applications 4, 426-431.

Ruda, M. /1974/. Parameter estimation in the first order autoregressive process. Biometrika 61 /1974/, 632-633.

Rykhlova, L.V. /1967/. Soviet Astron. Journal, 12 /1967/, 989.

/1969/. Soviet Astron. Journal, 13 /1969/, 544.

/1970/. Astron. J. 47 /1970/, 1426. /with V.V. Nesterov/.

Shaman, P. [1] /1969/. On the inverse of the covariance matrix of a first order moving average. Biometrika 56 /1969/ 595-600.

/1973/. On the inverse of the covariance matrix for an autoregressive-moving average process. Biometrika, 60, No. 1, 93-196.

Shiryaev, A.N. /1960/. Problems of spectral theory of higher moments I. Teoria Verojatn. i Primenen. V, 3 /1960/, 293-313.

/1966/. On stochastic equations in the theory of conditonal Markov processes. Teoria Verojatn. i Primenen. XI, 1 /1966/, 200-206.

/1966/. Stochastic equations of nonlinear filtering of jump Markov processes. Problemy peredachi informatsii. II, 3 /1966/, 3-22.

/1967/. New results in the theory of controlled random processes. Trans. 4th Prague Confer. Inform. Theory /1965/, Prague, 1967, 131-203.

/1968/. Studies in the statistical sequential analysis. Matem. zametki 3, 6 /1968/, 739-754.

/1969/. Statistical Sequential Analysis. "Nauka", Moscow.

/1970/. Sur les Equations Stochastiques aux Derivees Partielles. Actes Congres Intern. Math., 1970.

/1971/. Statistics of diffusion type processes. Proc. Second Japan - USSR Sympos. Probab. Theory, I /1971/, 69-87.

/1972/. Statistics of diffusion type processes. In "Proc. Second Japan - USSR Symp. on Prob. Theory", 397-412, Springer-Verlag, Berlin Lecture Notes in Mathematics, 330.

Slutsky, E. /1929/. Sur l'extension de la theorie de periodogrammes aux suites des quantites dependentes . Comptes Rendues, 189, 722-733.

/1934/. Alcuni applicazioni di coefficienti di Fourier al analizo di sequenze eventuali coherenti stazionarii . Giorn. d. Instituto Italiano degli Atuari. 5, 435-482.

/1935/. On the eleven year periodicity of sunspots. Dokl. A. Na /1935/ IV /IX/ No. 1-2 /70-71/ 37-40.

Smoluchowski, M.V. /1906/. Drei Vortrage uber Diffusion, Brownsche Bewegung and Koagulation von Kolloidteilchen. Physik. Zeit. 17 /1906/, 557-585.

Siddiqui, M.M. /1957/. On the inversion of the sample co variance matrix in a stationary autoregressive process, Ann. Math. Stat., 29, 585-588.

Strasser, H. /1977/. Improved bounds for the equivalence of Bayes and maximum likelihood estimation, Teoria Ve royat. iee primenen. 22, 358-370.

Striebel, Ch. /1959/. Densities for stochastic processes, Ann. Math. Stat., 30, 559-567.

Taraskin, A.F. /1970/. The anymptotic normality of stochastic integrals and estimates of the coefficient of diffusion process transfer. In "Mathematical Physics" No. 8 Naukova Dumka, Kiev, 149-163 /Russian/.

/1971a/. Statistical problems for a class of stochastic differential equations. In "Mathematical Physics", No. 10, Naukova Dumka, Kiev, 91-99 /Russian/.

/1971b/. Parameter estimation by method of maximum likelihood for a stationary process. In "Mathematical Physics", No. 9, Naukova Dumka, Kiev, 123-131 /Russian/.

/1974/. On the asymptotic normality of vector-valued stochastic integrals and estimates of drift parameters of a multidimensional diffusion process. Theor. Probability Math. Statist. 2, 209-224.

/1975/. On confidence regions for parameters of diffusion type Markov processes. "Matematika" No. 1, Kuybüshev. 3-9 /Russian/.

Vilenkin, S.Ja. /1959/. On estimating the mean of stationary processes, Teor. Verojatnost. i Primenen. 4 /1959/, 451-453 - Theor. Probability Appl. 4 /1959/, 415-416.

Volkonskii, V.A. and Ju.A. Rozanov. /1959/. Some limit theorems for random functions. I, Teor. Verojatnost. i Primenen 4 /1959/, 186-207 - Theor. Probability Appl. 4 /1959/, 178-197.

Yamada, T. and Watanabe, S.H. /1971/ On the uniqueness of solution of stochastic differential equations , J. Math. Kyoto Univ. 11/1971/, 155-167

Yershov, M.P. /1969/. Nonlinear filtering of Markov processes. Teoria Verojatn. i Primenen. XIV, 4 /1969/, 757-758.

/1970/. Sequential estimation of diffusion processes. Teoria Verojatn. i Primenen. XV, 4 /1970/, 705-717.

/1972/. On representations of Ito processes. Teoria Verojatn. i Primenen. XVII, 1 /1972/, 167-172.

/1972/. On absolute continuity of measures corresponding to diffusion type processes. Teoria Verojatn. i Primenen. XVII, 1 /1972/, 172-178.

/1972/. Stochastic Equations. Proc. Second Japan-USSR Sympos. Probab. Theory, Kyoto, I /1972/, 101-106.

Yule, G. /1927/. On a method investigating preiodicities. Philos. Trans. Roy. Soc. A. 226 /1927/, 267-298.

Wald, A. /1948/. Asymptotic properties of the maximum likelihood estimate of an unknown parameter of a discrete stochastic process. Ann. Math. Statist. 20, 595-601.

/1949/. Note on the consistency of the maximum likelihood estimate. Ann. Math. Statist. 20, 595-601.

Walker, A.M. /1960/. Some consequences of superimposed error in time series analysis, Biometrika, 47 /1960/, 33-43.

/1962/. Large-sample estimation of parameters for autoregressive process with moving average residuals, Biometrika, 49 /1962/, 117-131.

Walker, A.M. and A. Young. /1955/. The analysis of the observations of the variation of latitude, Monthly Notices Roy. Astronom. Society 115 /1955/, 443-59.

/1957/. Further results on the analysis of the variation of latitude, ibid. 117 /1957/, 119-141.

White, J.S. /1956/. Approximate moments for the serial correlation coefficients, Ann. Math. Statist., 27 /1956/, 798.

/1957/. A t-test for the serial correlation coefficient, Ann. Math. Statist., 28 /1957/, 1046-1048.

/1958/. The limiting distribution of the serial correlation coefficient, Ann. Math. Statist., 29 /1958/, 1188-1197.

Whittle, P. /1952a/. Some results in time series analysis. Skand. Aktuar. 35, 48-60.

/1952b/. The simultaneous estimation of a time series' harmonic and covariance structure. Trab. Estad. 3, 43-57.

/1953/. The analysis of multiple stationary time series. J. Roy. Statist. Soc., B. 15, 125-139.

/1954/. A statistical investigation of sunspot observations with special reference to H. Alven's sunspot model. Astrophys. J. 120, 251-260.

Wonham, W.M. /1965/. Some applications of stochastic differential equations to optimal nonlinear filtering. SIAM J. Control, 2 /1965/, 347-369.

/1968/. On the separation theorem of stochastic control. SIAM J. Control, 6 /1968/, 312-326.

/1968/. On a matrix Riccati equation of stochastic control. SIAM J. Control, 6 /1968/, 681-697.

AUTHORS INDEX

Lecture Notes in Control and Information Sciences

Edited by A. V. Balakrishnan and M. Thoma

Vol. 22: Optimization Techniques
Proceedings of the 9th IFIP Conference on
Optimization Techniques,
Warsaw, September 4–8, 1979
Part 1
Edited by K. Iracki, K. Malanowski, S. Walukiewicz
XVI, 569 pages. 1980

Vol. 23: Optimization Techniques
Proceedings of the 9th IFIP Conference on
Optimization Techniques,
Warsaw, September 4-8, 1979
Part 2
Edited by K. Iracki, K. Malanowski, S. Walukiewicz
XV, 621 pages. 1980

Vol. 24: Methods and Applications
in Adaptive Control
Proceedings of an International Symposium
Bochum, 1980
Edited by H. Unbehauen
VI, 309 pages. 1980

Vol. 25: Stochastic Differential Systems –
Filtering and Control
Proceedings of the IFIP-WG7/1 Working Conference
Vilnius, Lithuania, USSR, Aug. 28 – Sept. 2, 1978
Edited by B. Grigelionis
X, 362 pages. 1980

Vol. 26: D. L. Iglehart, G. S. Shedler
Regenerative Simulation of Response
Times in Networks of Queues
XII, 204 pages. 1980

Vol. 27: D. H. Jacobson, D. H. Martin, M. Pachter, T. Geveci
Extensions of Linear-Quadratic Control Theory
XI, 288 pages. 1980

Vol. 28: Analysis and Optimization of Systems
Proceedings of the Fourth International
Conference on Analysis and Optimization of Systems
Versailles, December 16–19, 1980
Edited by A. Bensoussan and J. L. Lions
XIV, 999 pages. 1980

Vol. 29: M. Vidyasagar,
Input-Output Analysis of Large-Scale
Interconnected Systems –
Decomposition, Well-Posedness and Stability
VI, 221 pages. 1981

Vol. 30: Optimization and Optimal Control
Proceedings of a Conference Held at
Oberwolfach, March 16–22, 1980
Edited by A. Auslender, W. Oettli, and J. Stoer
VIII, 254 pages. 1981

Vol. 31: Berc Rustem
Projection Methods in Constrained
Optimisation and Applications
to Optimal Policy Decisions
XV, 315 pages. 1981

Vol. 32: Tsuyoshi Matsuo,
Realization Theory of
Continuous-Time Dynamical Systems
VI, 329 pages, 1981

Vol. 33: Peter Dransfield
Hydraulic Control Systems –
Design and Analysis of Their Dynamics
VII, 227 pages, 1981

Vol. 34: H.W. Knobloch
Higher Order Necessary Conditions
in Optimal Control Theory
V, 173 pages, 1981

Vol. 35: Global Modelling
Proceedings of the IFIP-WG 7/1 Working
Conference Dubrovnik, Yugoslavia,
Sept. 1–5, 1980
Edited by S. Krčevinac
VIII, 232 pages, 1981

Vol. 36: Stochastic Differential Systems
Proceedings of the 3rd IFIP-WG 7/1
Working Conference
Visegrád, Hungary, Sept. 15–20, 1980
Edited by M. Arató, D. Vermes, A.V. Balakrishnan
VI, 238 pages, 1981

Vol. 37: Rüdiger Schmidt
Advances in Nonlinear
Parameter Optimization
VI, 159 pages, 1982

Vol. 38: System Modeling and Optimization
Proceedings of the 10th IFIP Conference
New York City, USA, Aug. 31 – Sept. 4, 1981
Edited by R.F. Drenick and F. Kozin
XI, 894 pages. 1982

Vol. 39: Feedback Control of
Linear and Nonlinear Systems
Proceedings of the Joint Workshop
on Feedback and Synthesis of
Linear and Nonlinear Systems
Bielefeld/Rom
XIII, 284 pages. 1982

Vol. 40: Y.S. Hung, A.G.J. MacFarlane
Multivariable Feedback:
A Quasi-Classical Approach
X, 182 pages. 1982

Vol. 41: M. Gössel
Nonlinear Time-Discrete Systems –
A General Approach by
Nonlinear Superposition
VIII, 112 pages. 1982

Vol. 42: Advances in Filtering and
Optimal Stochastic Control
Proceedings of the IFIP-WG 7/1
Working Conference
Cocoyoc, Mexico, February 1–6, 1982
VIII, 391 pages. 1982

Lecture Notes in Control and Information Sciences

Edited by A. V. Balakrishnan and M. Thoma

Vol. 43: Stochastic Differential Systems
Proceedings of the 2nd Bad Honnef Conference
of the SFB 72 of the DFG at the University of Bonn
June 28 – July 2, 1982
Edited by M. Kohlmann and N. Christopeit
XII, 377 pages. 1982.

Vol. 44: Analysis and Optimization of Systems
Proceedings of the Fifth International
Conference on Analysis and Optimization of Systems
Versailles. December 14–17, 1982
Edited by A. Bensoussan and J. L. Lions
XV, 987 pages, 1982

Vol. 45: M. Arató
Linear Stochastic Systems
with Constant Coefficients
A Statistical Approach
IX, 309 pages. 1982